# PAUL HALMOS
## Celebrating 50 Years
## of Mathematics

John H. Ewing • F.W. Gehring

Editors

# PAUL HALMOS
# Celebrating 50 Years
# of Mathematics

With 67 Photographs

Springer-Verlag

New York Berlin Heidelberg London Paris
Tokyo Hong Kong Barcelona Budapest

John H. Ewing
Department of Mathematics
Indiana University
Bloomington, IN 47405
USA

F.W. Gehring
Department of Mathematics
University of Michigan
Ann Arbor, MI 48109
USA

Part I of "Paul Halmos by Parts: Interviews by Don Albers" was taken from *Mathematical People: Profiles and Interviews* (D.J. Albers, G.L. Alexanderson, eds.), Birkhäuser-Boston, 1985.

Dust jacket and frontispiece photograph by Glenn Matsumura.

Library of Congress Cataloging-in-Publication Data
Paul Halmos: Celebrating 50 years of mathematics / John H. Ewing, F.W. Gehring, editors.
    p.    cm.
Includes bibliographical references and index.
ISBN 0-387-97509-8
    1. Mathematical analysis.   2. Halmos, Paul R. (Paul Richard),
1916–    .  3. Mathematicians – United States – Biography.   I. Halmos,
Paul R. (Paul Richard), 1916–    .  II. Ewing, John H.  III. Gehring,
Frederick W.   IV. Title: Paul Halmos: Celebrating 50 Years of Mathematics.
QA300.5.H35      1991
515 – dc20                                                                    90-25802

Printed on acid-free paper.

Photocomposed from a LaTex file.
Printed and bound by Braun-Brumfield, Inc., Ann Arbor, MI.
Printed in the United States of America.

9 8 7 6 5 4 3 2

ISBN 0-387-97509-8 Springer-Verlag New York Berlin Heidelberg
ISBN 3-540-97509-8 Springer-Verlag Berlin Heidelberg New York

# Preface

Paul Halmos will celebrate his 75th birthday on the 3rd of March 1991. This volume, from colleagues, is an expression of affection for the man and respect for his contributions as scholar, writer, and teacher. It contains articles about Paul, about the times in which he worked and the places he has been, and about mathematics.

Paul has furthered his profession in many ways and this collection reflects that diversity. Articles about Paul are not biographical, but rather tell about his ideas, his philosophy, and his style. Articles about the times and places in which Paul has worked describe people, events, and ways in which Paul has influenced students and colleagues over the past 50 years. Articles about mathematics are about all kinds of mathematics, including operator theory and Paul's research in the subject. This volume represents a slice of mathematical life and it shows how many parts of mathematics Paul has touched.

It is fitting that this volume has been produced with the support and cooperation of Springer-Verlag. For over 35 years, Paul has contributed to mathematics publishing as founder and editor of many outstanding series. Paul began his editorial career in 1955 as one of the editors for the Ergebnisse series published by Springer-Verlag. In the late 1950's he became an editor with J.L. Kelly of the University Series in Higher Mathematics published by Van Nostrand. Shortly afterwards, he founded and edited with F.W. Gehring the Van Nostrand Mathematical Studies, the first paperback series of mathematical "notes." Later, he founded and edited the Springer series Graduate Texts in Mathematics and Problem Books in Mathematics. He was also editor for the Springer series Undergraduate Texts in Mathematics and Universitexts. As a writer Paul has touched many mathematicians; as an editor he has touched *every* mathematician.

We are grateful for the opportunity to have worked with Paul, to have learned from him, and to have had him as a colleague.

We join Paul's many colleagues, friends, and admirers in sending him congratulations on the occasion of his 75th birthday and in wishing him continued success in his multifaceted activities.

John Ewing
Fred Gehring

# Contents

## Some Mathematics

# Mathematician, Writer, and Editor

*It is true that a mathematician, who is not somewhat of a poet, will never be a perfect mathematician.*
—Weierstrass

*I'm a great believer in luck, and I find the harder I work the more I have of it.*
—Thomas Jefferson

# Paul Halmos by Parts

## Interviews by Donald J. Albers

Paul R. Halmos is Professor of Mathematics at Santa Clara University and Past-Editor of the *American Mathematical Monthly*. He received his Ph.D. from the University of Illinois and has held positions at Indiana, Illinois, Chicago, Michigan, Hawaii, and Santa Barbara. He has authored 10 books and 117 articles. He is a member of the Royal Society of Edinburgh and of the Hungarian Academy of Sciences.

The writings of Halmos have had a large impact on both research in mathematics and the teaching of mathematics. He has won several awards for his mathematical exposition, including the Chauvenet Prize, the Lester R. Ford Award (twice), the Pólya Award, and the Steele Prize.

I first interviewed Halmos in August of 1981 in Pittsburgh at the combined annual summer meetings of the Mathematical Association of America and the American Mathematical Society. During the course of that first interview, Halmos confessed to being a **maverick mathologist**. A **mathologist** is a pure mathematician and is to be distinguished from a **mathophysicist**, who is an applied mathematician. (Both terms were coined by Halmos.) A few of his statements from that interview help to underscore his maverick nature:

"I don't think mathematics needs to be supported."

"If the NSF had never existed, if the government had never funded American mathematics, we would have half as many mathematicians as we now have, and I don't see anything wrong with that."

"The computer is important, but not to mathematics."

In Part I that follows, Halmos in his inimitable style talks about teaching mathematics, writing mathematics, and doing mathematics. After a short time with him, I was convinced that he is a **maverick** and a **mathologist**.

In 1984 Halmos moved from Indiana to Santa Clara, less than 30 minutes from my house. In short order I found myself visiting with him, seeking his advice on publishing matters, and getting to know him better. After a few years and several chats with Paul, it was clear that the first interview had missed many fascinating aspects of Halmos. A few months ago, *before* the idea for this book was born, I asked Paul if he would be willing to do a follow-up interview. He agreed.

The result of that second interview is PART II: IN TOUCH WITH GOD. In PART II, Halmos talks about the best and worst parts of being a mathematician, cats, the law, doing mathematics, the root of *all deep* mathematics, and smelling mathematicians.

PART II makes it clear that Paul Halmos after 50 years of doing mathematics is lively, challenging, playful, and still very much a **maverick mathologist**.

Don Albers
December 7, 1990

# Part I: Maverick Mathologist

## A DOWNWARD-BOUND PHILOSOPHER

**Albers**: *You have described yourself as a downward-bound philosopher. What does that mean?*

**Halmos**: Most mathematicians think of a hierarchy in which mathematics is above physics, and physics is higher than engineering. If they do that, then they are honor-bound to admit that philosophy is higher than mathematics. I started graduate school with the idea of studying philosophy. I had studied enough mathematics and philosophy for a major in either one. My first choice was philosophy, but I kept a parallel course with mathematics until I flunked my master's exams in philosophy. I couldn't answer all the questions on the history of philosophy that they asked, so I said the hell with it — I'm going into mathematics. I made philosophy my minor, but even that didn't help; I flunked the minor exams too.

**Albers**: *So mathematics was not your original calling, if you like?*

**Halmos**: As a philosophy student, I played around with symbolic logic and was fascinated with all the symbols in *Principia Mathematica*. Even as a philosopher, I tended toward math.

**Albers**: *You are the third Hungarian we have interviewed.*

**Halmos**: I reject the appellation.

**Albers**: *We know that you were born in Hungary and that you lived there until the age of 13, but you still reject the appellation.*

**Halmos**: I don't feel Hungarian. I speak Hungarian, but by culture, education, world view, and everything else I can think of — I feel American. When I go to Hungary, I feel like an American tourist, a stranger. I speak English with an accent, but I speak it infinitely better than I speak Hungarian. I can control it, and I cannot do that in Hungarian. In every respect, except accent, I am an American.

**Albers**: *You may not claim Hungary, but I wouldn't be surprised if Hungary claims you. In fact, you are a member of the Hungarian Academy of Science.*

**Halmos**: I was elected a member of the Hungarian Academy of Science only a couple of years ago, in recognition of my work, I hope, but I am sure that my having been born in Hungary helped. In theory, it needn't help, as there are a certain number of foreign members elected each year. But if they are in some sense ex-Hungarians or have Hungarian roots, that doesn't hurt. I am not ashamed of my Hungarian connection, but just as a matter of fact I try to straighten out my friends and tell them that they shouldn't attribute to my country of origin whatever properties they ascribe to me.

**Albers**: *How did you come to leave Hungary?*

**Halmos**: I give full credit to my father. In 1924, when he was in his early forties, he left Hungary, where he had been a practicing physician with a flourishing practice. The country was at peace and in good shape, but he thought it was a sinking ship. He arranged for his practice to be taken over by another physician, who was also foster father to his three boys, of whom I am the youngest. (My mother died when I was six months old, and I never knew her.) He came to this country with the feeble English that he had learned. After working as an intern at an Omaha hospital for a year in order to prepare for and pass the state and national boards, he started a practice in Chicago. Five years later he became a citizen and imported his sons. Coming to America wasn't a decision on *my* part; it was a decision on *his* part. It turned out to be a very smart move.

**Albers**: *Did you have any glimmerings of strong mathematical interests as a child? We know the Hungarians do a remarkably good job of producing superior mathematics students.*

**Halmos**: Yes and no. I cannot give credit to the Hungarian system, which I admire and about which I am somewhat puzzled (as are most Americans), as to how they produce Erdös's, Pólya's, and Szegö's, and dozens more that most of us can rattle off. I know the rumor that they look for them in high school and encourage them and conduct special examinations to find them. Nothing like that had a chance to happen to me. By the age of 13 I was exposed to a lot more mathematics than American students are exposed to nowadays, but not more than American students were exposed to in those days. I was exposed to parentheses and quadratic equations, two linear equations in two unknowns, a few applied ideas, and the basic things in physics. I remember that I enjoyed drawing the design of a water pump and other things like that. I was good at it, the way good students in calculus are good in calculus in our classrooms, but not a genius. I just enjoyed it and fooled around with it. In mathematics classes, I usually was above average. I was bored when class was going on, and I did things like take logarithms of very large numbers for fun.

The American system in those days was eight years of elementary school and four years of high school. In Europe it was the other way around — four years of elementary school followed by eight years of secondary school, adding up to the same thing. I left Hungary when I was in the third year of secondary school, which would have been the equivalent of the seventh grade in this country.

**Albers**: *So the 13-year-old Halmos came to the U.S. and entered a high school in the Chicago area. You spoke Hungarian and German and knew a little Latin, and yet instruction was in English. That must have posed a few problems.*

**Halmos**: For the first six months it was a hell of a problem. On my first day, somebody showed me to a classroom in which, I still remember, a very nice man was talking about physics. I listened dutifully for the first hour and didn't understand a single word of what was being said. At the end of the hour, everyone got up and went to some other room, but I didn't know where to go, so I just sat there. The instructor, Mr. Payne, came over to my seat and asked me something and I shrugged my shoulders helplessly. We tried various languages. I didn't know much English, and he didn't know German. We both knew a few Latin words and a few French words, and he finally succeeded in telling me that I had to go to Room 252. I went to Room 252, and that was my first day in an American high school. Six months later I spoke rapid, incorrect, ungrammatical, colloquial English.

**Albers**: *Were there any special events in high school that stand out in your memory?*

**Halmos**: Well, there was a little chicanery surrounding my admission. I explained this business of eight years followed by four in this country, and four years followed by eight in Europe. There was some confusion about that. I *hinted* to the school authorities that I had completed three years of *secondary* school, and I was *believed.* There was, to be sure, a perfunctory examination of my record, and, after being translated by an official in the Hungarian consulate, it said three years of secondary school. That means in effect that I skipped four grades at once, and I went from what was the equivalent of the seventh grade to the eleventh grade; and a year and a half later, at the age of fifteen, I graduated from high school.

## A COLLEGE FRESHMAN AT AGE FIFTEEN

**Albers**: *So you were a very young high-school graduate.*

**Halmos**: Yes. I entered the University of Illinois at the age of fifteen.

**Albers**: *That's very young to be entering college. Did that produce any difficulty?*

**Halmos**: There were no problems. I was tall for my age and cocky. I pretended to be older and got along fine.

**Albers**: *When did you become interested in mathematics and philosophy?*

**Halmos**: I started out in chemical engineering, and at the end of one year decided that it was for the birds; I got my hands too dirty. That's how mathematics and philosophy came into the act.

**Albers**: *Can you remember what attracted you to mathematics and philosophy? Can you separate them?*

**Halmos**: It is difficult. I remember calculus was not easy for me. I was a routine calculus student — I think I got B's. I didn't understand about limits. I doubt that they taught it. At that time, they probably wouldn't have dared. But I was good at integrating and differentiating things in a mechanical sense. Somehow I like it. I kept fooling around with it. In philosophy, it was symbolic logic that interested me. What attracted me is hard to say, just as it is hard for any of us to say what attracts us to a subject.

There was something about abstraction. I like the cleanness, the security of the ideas. When I learned something about history, I was at the least very suspicious; and strange as it may sound, when I learned something about physics and chemistry, I was most suspicious: I was practically doubtful, and I thought it might not even be true. In mathematics and in that kind of philosophy (logic), I knew exactly what was going on.

## "SUDDENLY I UNDERSTOOD EPSILONS!"

**Albers**: *Was there some point when you decided that you were going to be a mathematician?*

**Halmos**: There was *no* point when I decided that I was going to be an *academic*. That somehow was just taken for granted, not by anybody else, but by me. I just wanted to take courses and see what happened. I was studying for a master's and flunking the master's exam in philosophy, but nothing would stop me. I continued taking courses. I finished my bachelor's quickly, in three years instead of four. As a first-year graduate student, I took a course from Pierce Ketchum in complex function theory. I had absolutely no idea of what was going on. I didn't know what epsilons were, and when he said take the unit circle, and some other guy in class said "open or closed," I thought that silly guy was hair-splitting, and what was he fussing about. What difference did it make? I really didn't understand it.

Then one afternoon something happened. I remember standing at the blackboard in Room 213 of the mathematics building talking with Warren Ambrose and suddenly I understood epsilons. I understood what limits were, and all of the stuff that people had been drilling into me became clear. I sat down that afternoon with the calculus textbook by Granville, Smith, and Longley. All of that stuff that previously had not made any sense became obvious; I could prove the theorems. That afternoon I became a mathematician.

**Albers**: *So there was a critical point. You even remember the room number.*

**Halmos**: I *think* I remember the room number.

**Albers**: *After earning your Ph.D., you became a fellow at the Institute for Advanced Study, where you served as an assistant to Johnny von Neumann. How did you come to be his assistant? What was it like being an assistant to someone with that kind of power?*

**Halmos**: Let me back up a little. I got my Ph.D. in 1938, and preparatory to graduation, I applied for jobs. Xerox was not known in those days, and secretarial service was not available to starving graduate students. I typed 120 letters of application, mailed them out, and got two answers, both *no*. I got no job. The University of Illinois took pity on me and kept me on for one year as an instructor. So in '38–'39 I had a job, but I kept applying. I

did get a job around February or March at a state university. I accepted it
without an interview. It was accomplished with correspondence and some
letters of recommendation. Two months later my very good friend Warren
Ambrose, who was one year behind me, got his degree. He had been an
alternate for a fellowship at the Institute for Advanced Study; and when
the first choice declined, he got the scholarship, and that made me mad. I
wanted to go, too! I resigned my job, making the department head, whom
I had never met, very unhappy, of course. In April, I resigned my job,
and went to my father and asked to borrow a thousand dollars, which in
those days was a lot of money. The average annual salary of a young Ph.D.
was then $1,800. I wrote Veblen and asked if I could be a member of the
Institute for Advanced Study even though I had no fellowship. It took him
three months to answer. He answered during summer vacation, and said
"Dear Halmos, I just found your letter, and I guess you mean for me to
answer. Yes, of course, you are welcome." That's all it took; I moved to
Princeton.

But, of course, Veblen wasn't giving me anything except a seat in the
library. Six months after I got there, the Institute took pity on me and gave
me a fellowship. During the first year, I attended Johnny von Neumann's
lectures, and in my second year I became his assistant. I followed his lectures
and took careful notes. The system of the Institute was that each professor
had an assistant assigned to him. The duties of the assistant depended
upon the professor. Einstein's assistant's duties were to walk him home
every day and talk German to him. Morse's assistant's duties were to do
research with him — eight hours a day sitting with Morse and listening to
him talk and talk. Von Neumann's assistant had very little to do — just go
to the lectures and take notes; and sometimes those notes were typed up
and duplicated. Von Neumann's assistant that year was Hugh Dowker, who
is a mathematician *par excellence*, but not in the least interested in matrices
and operator theory and all those things that von Neumann lectured on.
On the other hand, I was fascinated by them; that was my subject. So, I
took careful notes and Dowker used them and took them to Johnny. There
was no duplicity about it. He told Johnny what he was doing. When his
job was up, I became Johnny's assistant.

How was it? Scary. The most spectacular thing about Johnny was not his
power as a mathematician, which was great, or his insight and his clarity,
but his rapidity; he was very, very fast. And like the modern computer,
which no longer bothers to retrieve the logarithm of 11 from its memory
(but, instead, computes the logarithm of 11 each time it is needed), Johnny
didn't bother to remember things. He computed them. You asked him a
question, and if he didn't know the answer, he thought for three seconds
and would produce the answer.

## INSPIRATIONS

**Albers**: *You have described an inspirational day with Warren Ambrose when you decided to become a mathematician. Are there other individuals who have been inspirations for you?*

**Halmos**: I'm not prepared for this question. Therefore, my answer is bound to be more honest than for any other question. The first two names that occur to me are two obvious ones. The first is my supervisor, Joe Doob, who is only six years older than I. I was 22 when I finished my Ph.D., and he was 28, both young boys from my present point of view. He arrived at the University of Illinois when he himself was about 25. I was already at the stage where I was signed up to do a Ph.D. thesis with another professor. I remember having lunch with Joe one day at a drugstore and hearing him talk about mathematics. My eyes were opened. I was inspired. He showed me a kind of mathematics, a way to talk mathematics, a way to think about mathematics that wasn't visible to me before. With great trepidation, I approached my Ph.D. supervisor and asked to switch to Joe Doob, and I was off and running.

The other was Johnny von Neumann. The first day that I met him he asked if it would be more comfortable for me to speak Hungarian, which was his best language, and I said it would not. So we spoke English all the time. And as I said before, his speed, plus depth, plus insight, plus inspiration turned me on. They — Doob and von Neumann — were my two greatest inspirations.

**Albers**: *In 1942, you produced a monograph called "Finite-Dimensional Vector Spaces." Was it a result in part of notes you had taken?*

**Halmos**: Yes. Von Neumann planned a sequence of courses that was going to take him four years. He began at the beginning with the theory of linear algebra — finite-dimensional vector spaces from the advanced point of view. And just as van der Waerden's book was based on Artin's lectures, my book was based on von Neumann's lectures and inspired completely by him. That's what got me started writing books.

**Albers**: *Most people who read that book remark that it is written in an unusual way; the Halmos style is quite distinctive. I studied from your book, and I still remember that it gave me fits because your problems were not of the classical type. You didn't set* **prove** *or* **show** *exercises; more often than not you gave statements that the student was to prove if true or disprove if false. I am sure that it was deliberate, and it seems to underscore a philosophy of teaching that you have spoken about in a recent article in the Monthly, "The Heart of Mathematics." In that article, you said that it is better to do substantial problems on a lesser number of topics than to do oodles of lesser exercises on a larger number of topics. Had you thought a great deal about that before writing problems for "Finite-Dimensional Vectors Spaces"?*

**Halmos**: No. That wasn't a result of thought; it was just instinctive somehow. I felt it was the right way to go, and 30 years later I summarized in expository articles what I have been doing all along. You said it very well. I strongly believe that the way to learn things is to do things — the easiest way to learn to swim is to swim — you can't learn it from lectures about swimming. I also strongly believe that the secret of mathematical exposition, be it just a single lecture, be it a whole course, be it a book, or be it a paper, is not the beautifully written sentence, or even the well-thought-out paragraph, but the architecture of the whole thing. You must have in mind what the lecture or the whole course is going to be. You should get across *one* thing. Determine that thing and then design the whole approach to get at it. Instinctively in that book, and I must repeat it was inspired by von Neumann, I was driving at one thing — that matrix theory is operator theory in the most important and the most translucent special case. Every single step, and in particular every exercise (they were not different from any other step), was designed to shed light on that end.

**Albers**: *In the article, "The Heart of Mathematics," you discussed courses that went down as low as calculus. What do you think about that approach for precalculus or for high-school algebra? Would you also advocate that approach for such courses?*

**Halmos**: Yes and no. I think, and I repeat, the only way to learn anything is do that thing. The only way to learn to bicycle is ride a bicycle. The only way to teach bicycling is to put challenges in front of the prospective

"I am proud to be a teacher. — Teaching is an ephemeral subject. It is like playing the violin. The piece is over, and it's gone. The student is taught, and the teaching is gone."

bicyclist and make him conquer them. So, yes, I believe in it. I have tried it, not only in calculus, but in as low-level college courses as precalculus and high-school trigonometry with a great deal of joy and enthusiasm many times. To the extent possible, I have tried to follow that kind of system. But let's be honest. The so-called Moore method, which is a way to describe the Socratic question-asking, problem-challenging approach to teaching, doesn't work well when you have 40 people in the class, let alone when you have 140. It is beautiful if you have 2 people sitting at 2 ends of a log, or 10 or 11 sitting in a classroom facing you. Obviously, there are practical problems that you have to solve, but they can be solved. Moore, for instance, did teach first-year calculus that way. So a one-word answer to your question is, *yes*, I do advocate it in all teaching; but *no*, one has to be careful. One has to be wise. One has to face realities and adapt to economic circumstances.

## MATHOPHYSICS

**Albers**: *A few have said that you have been a strong exponent of what is called the New Mathematics.*

**Halmos**: Absolutely not! I was a reactionary all the time. The old mathematics was just fine. I think high-school students should be taught high-school geometry *à la* Euclid. You should teach them step one, reason; step two, reason; and all that stuff. I thought that was wonderful. I got my training that way. Morris Kline and I hardly know each other, but we seem to disagree on everything; and he is (a) strongly against the New Math, and (b) strongly against many things I advocate. It's quite possible that people who agree with him identify me as a champion of the New Math because we disagree on most things.

**Albers**: *You say you think that you and Kline disagree on just about everything. He stood in strong opposition to the New Math, and you just said that you were absolutely not an exponent of the New Math. Now there is some agreement there.*

**Halmos**: I hate to admit it. He may be against the right thing, but he certainly is against it for the wrong reason.

**Albers**: *It seems to me that you and Kline have another strong point of agreement, which surprised me a bit. In your article, "Mathematics as a Creative Art," that appeared in American Scientist back in the late sixties, you surprised me by saying that virtually all of mathematics is rooted in the physical world. Kline, as you know, wrote a book entitled "Mathematics and the Physical World."*

**Halmos**: I think we understand different things by it. I get the feeling that Kline either thinks, or would love to think, that all mathematics is not only rooted in the physical world but must aim toward it, must be applicable to it, and must touch that base periodically.

**Albers**: *So he is what you would call a mathophysicist?*

**Halmos**: And how! But it's another thing to say, almost a shallow, meaningless thing to say, that we are human beings with eyes, and we can see things that we think are outside of us. Our mathematics — our instinctive, unformulated, undefined terms — come from our sense impressions; and in that sense at least, a trivial sense, mathematics has its basis in the physical world. But that is an uninformative, unhelpful, shallow treatment.

## What Is Mathematics?

**Albers**: *This prompts the next question for which I can't expect you to give a complete answer in such a short time, but I will ask it anyhow. What is mathematics to you?*

**Halmos**: It is security. Certainty. Truth. Beauty. Insight. Structure. Architecture. I see mathematics, the part of human knowledge that I call mathematics, as one thing — one great, glorious thing. Whether it is differential topology, or functional analysis, or homological algebra, it is all one thing. They all have to do with each other, and even though a differential topologist may not know any functional analysis, every little bit he hears, every rumor that comes to him about that other subject, sounds like something that he does know. They are intimately interconnected, and they are all facets of the same thing. That interconnection, that architecture, is secure truth and is beauty. That's what mathematics is to me.

## Federal Support of Mathematics?

**Albers**: *In "Mathematics as a Creative Art," you were addressing lay readers when you said: "I don't want to teach you what mathematics is, but that it is." This reflects a concern that you had at the time about mathematics in the mind of the layman. (You said, "A layman is anyone who is not a mathematician.") Is your concern still there — that a great body of intelligent, well-educated people don't know perhaps that your subject **is**? Is that concern stronger or weaker than it was in '68, when you wrote the article?*

**Halmos**: The same, I would say. Let me first of all explain that I am a maverick among mathematicians. I don't think it is vital and important to explain to members of Congress and administrators in the National Science Foundation what mathematics is and how important it is and how much money it must be given. I think we have been given too much money. I don't think mathematics needs to be supported. I think the phrase is most offensive. Mathematics gets along fine, thank you, without money, and I look back with nostalgia to the good old days, three or four hundred years ago, when only those did mathematics who were willing to do it on their own time.

In the fifties and sixties, a lot of people went into mathematics for the
wrong reasons, namely that it was glamorous, socially respected, and well-
paying. The Russians fired off Sputnik, the country became hysterical, and
then NSF came along with professional, national policies. Anything and
everything was tried; nothing was too much. We had to bribe people to
come to mathematics classes to make it appear respectable, glamorous,
and well-paying. So we did. One way we did it, for instance, was to use a
completely dishonest pretense — the mission attitude towards mathematics. The way it worked was that I would propose a certain piece of research,
and then if it was judged to be a good piece of research to do, I would get
some money. That's so dishonest it sickens me. None of it was true! We
got paid for doing research because the country wanted to spend money
training mathematicians to help fight the Russians.

Many young people of that period were brought up with this Golden
Goose attitude and now regard an NSF grant as their perfect right. Consequently, more and more there tends to be control by the government of
mathematical research. There isn't strong control yet, and perhaps I'm just
building a straw man to knock down. But time and effort reporting is a
big, bad symptom, and other symptoms are coming I am sure. Thus, I say
that it was on balance a bad thing. If the NSF had never existed, if the
government had never funded American mathematics, we would have half
as many mathematicians as we now have, and I don't see anything wrong
with that. Mathematics departments would not have as many as 85 and
100 people in some places. They might have 15 or 20 people in them, and I
don't see anything wrong with that. Mathematics got along fine for many
thousands of years without special funding.

**Albers**: *But we certainly have seen a great increase in demand for mathematical skills which means a need for people who are able to teach mathematics. You certainly need someone to deliver the mathematics.*

**Halmos**: That's a different subject. The demand for teaching mathematics that seems to be growing is again because of a perceived threat by the
Russians and Chinese. In other words, we want people in computer science;
we want people in statistics; and we want people in various industrial and
other applications of mathematics. We have to teach them trigonometry
and other subjects so that they can do those things. That's not mathematics; that's a trade. It's doing mathematicians good only insofar as it
enables them to buy an extra color TV set or more diapers for the baby.

**Albers**: *Can we return to your concern about letting layman know that
mathematics is?*

**Halmos**: My interest was more on the intellectual level. I have absolutely
no idea of what paleontology is; and if somebody would spend an hour with
me, or an hour a day for a week, or an hour a day for a year, teaching it
to me, my soul would be richer. In that sense, I was doing the same thing
for my colleague, the paleontologist: I was telling him what mathematics

"Writing is permanent. The book, the paper, the symbols on the sheets of papyrus are always there. — Writing is very hard work for me, but I love it."

is. That, I think, is important. All educable human beings should know what mathematics *is* because their souls would grow by that. They would enjoy life more, they would understand life more, they would have greater insight. They should, in that sense, understand all human activity such as paleontology and mathematics.

## WHY WRITE ABOUT MATHEMATICS?

**Albers**: *So you were performing a service to paleontologists perhaps by explaining to them that your subject is rather than what it is. How do you explain the motivation for your other writing activities? Writing is hard work. In fact, when I reread your "How to Write Mathematics" last night, I was more convinced than ever that you must work very hard when you write. Why do you do it? Now you aren't talking to paleontologists; you are talking to mathematicians.*

**Halmos**: It is the same thing. Why do I do it? It is a many-faceted question with many answers. Yes, writing is very hard work, and so is playing the piano for Rubinstein and Horowitz, but I am sure they love it. So is playing the piano for a first-year student at the age of ten, but many of them love it. Writing is very hard work for me, but I love it. And why do I do it? For the same reason that I explain mathematics to the paleontologist. The answer is the same — it is all communication. That's important to me. I want to make things clear. I enjoy making things clear. I find it very difficult to make things clear, but I enjoy trying, and I enjoy it even more on the rare occasions when I succeed. Whether it is making clear to a medical doctor or to a paleontologist how to solve a problem in the summation of geometric series, or explaining to a graduate student who has had a course in measure theory why $L_2$ is an example of a Hilbert space, I regard them as identical problems. They are problems in communication, explanation, organization, architecture, and structure.

**Albers**: *So you enjoy doing it. It makes you feel good, and it makes you feel perhaps even better if you can sense that the receiver understands. That sounds like a teacher — the classical reasons that people give for teaching — the joy of seeing the look of understanding.*

**Halmos**: That's very good. Yes, I accept the word. I am proud to be a teacher and get paid for being a teacher, as do many of us who make a living out of mathematics. But it is also something else. Mathematics is, as I once maintained, a creative art and so is the exposition of mathematics. Teaching is an ephemeral subject. It is like playing the violin. The piece is over, and it's gone. The student is taught, and the teaching is gone. The student remains for a while, but after a while he too is gone. But writing is permanent. The book, the paper, the symbols on the sheets of papyrus are always there, and that creation is also the creation of the rounded whole.

**Albers**: *You've also written about talking mathematics. Based on what you have just said, my strong suspicion is that you get a greater joy out of writing than talking, although you also seem to have a lot of joy when you talk about mathematics.*

**Halmos**: They're nearly the same thing, but writing is more precise. By more precise, I mean the creator has more control over it. I myself feel that I am a pretty good writer, A− or B+; and a good, but less good speaker, B or B−; and therefore, I enjoy writing more. But they are similar and are part of the art of communication.

**Albers**: *A short time ago, someone talked with me about your book "Naive Set Theory." She said that it has a smooth, conversational style, and is in some ways like a bedtime story. What motivated you to write it?*

**Halmos**: *Naive Set Theory* was the fastest book I ever wrote. *Measure Theory* took 18 months of practically full-time work and *Naive Set Theory* took 6 months. Bedtime story is an apt description, for most of it was written while perched on the edge of a bed in a rented house in Seattle, Washington. It was being written because I had just recently learned about axiomatic set theory, which was a tremendous inspiration to me. It was a novelty to me. I didn't realize that it existed and what it meant, and at once I wanted to go out and tackle it. So I wrote it down. It wrote itself. It seems 100% clear to me that you have to start here and you have to take the next step there, and the third step suggested itself after the first two. I had almost no choice, and, as I keep emphasizing, that's the biggest problem of writing, of communication — the organization of the whole thing. Individual words that you choose you can change around; you can change the sentence around. The structure of the whole thing you cannot change around — that's what was created while I perched on the edge of the bed the first day, and from then on the book wrote itself.

**Albers**: *Was that a unique writing experience for you?*

**Halmos**: In that respect, yes, because it had a much better defined subject than usual. When I wrote on Hilbert space theory, I had, subjectively speaking, an infinite area from which to carve out a small chunk. Here was an absolute, definite thing. There is much more to axiomatic set theory than I exposed in *Naive Set Theory*, but it was a clearly defined part of it that I wanted to expose, and I did.

## Is Applied Mathematics Bad?

**Albers**: *You have recently written an article with another intriguing title, "Applied Mathematics is Bad Mathematics."*

**Halmos**: I have been sitting here for the last 55 minutes dreading when this question was going to be asked.

**Albers**: *What do you mean when you say applied is bad?*

**Halmos**: First, it isn't. Second, it is. I chose the title to be provocative. Many mathematicians, whom everybody else respects and whom I respect, agree with the following attitude: There is something called mathematics — put the adjective "pure" in front of it if you prefer. It all hangs together. Be it topology, or algebra, or functional analysis, or combinatorics, it is the same subject with the same facets of the same diamond; it's beautiful and it's a work of art. In all parts of the subject the language is the same; the attitude is the same; the way the researcher feels when he sits down at his desk is the same; the way he feels when he starts a problem is the same. The subject is closely related to two others. One of them is usually called applied mathematics, and its adherents frequently deny that it exists. They say there is no such thing as applied mathematics and that there isn't any difference between applied mathematics and pure mathematics. But, nevertheless, there is a difference in language and attitude. I am about to say a bad word about applied mathematicians, but, believe me, I mean it in a genuinely humble way. They are sloppy. They are sloppy in perhaps the same way that you and I are sloppy, as ordinary mathematicians are sloppy compared with the requirements of a formal logician. And a formal logician would probably be called sloppy by a computing machine.

There are at least three different kinds of language, which can roughly be arranged in a hierarchy: *formal logic* (that's equated nowadays with computer science) *mathematics*, and *applied mathematics*. They have different objectives; they are different facets of beauty; they have different reasons for existence and have different manners of expression and communication.

Since communication is so important to me, that is the first thing that jumps to my eye. A logician just cannot talk the way a topologist talks. An an algebraist couldn't make like an applied mathematician to save his life. Some geniuses like Abraham Robinson can be both. But they are different people being those different things. So, in that sense, what I wanted to say in that article is that there are at least two subjects. Now I am saying there are three or more, and I wanted to call attention to what I think the differences are.

"When I was forty, I had every disease in the book, I *thought* hypochondriacally. I went to the doctor with a brain tumor, with heart disease, with cancer, and everything else, I *thought*. He examined me and said, 'Halmos, there isn't anything wrong with you. Go take a long walk.'"

There is a sense in which applied mathematics is like topology, or algebra, or analysis, but (and shoot if you must this old grey head) there is also a sense in which applied mathematics is just bad mathematics. It's a good contribution. It serves humanity. It solves problems about waterways, sloping beaches, airplane flights, atomic bombs, and refrigerators. But just the same, much too often it is bad, ugly, badly arranged, sloppy, untrue, undigested, unorganized, and unarchitected mathematics.

**Albers**: *Computers are still relatively new objects within our lifetimes and intimately linked to what many call applied mathematics. What do you think of them? Are they important to you?*

**Halmos**: Who am I? A citizen or mathematician? As a mathematician, no, not in the least.

**Albers**: *Let's take something specific, the work of Appel and Haken and the computer.*

**Halmos**: On the basis of what I read and pick up as hearsay, I am much less likely now, after their work, to go looking for a counterexample to the four color conjecture than I was before. To that extent, what has happened convinced me that the four color theorem is true. I have a religious belief that some day soon, maybe 6 months from now, maybe 60 years from now, somebody will write a proof of the four color theorem that will take up 60 pages in the *Pacific Journal of Mathematics*. Soon after that, perhaps 6 months or 60 years later, somebody will write a four-page proof, based on concepts that in the meantime we will have developed and studied and understood. The result will belong to the grand, glorious, architectural structure of mathematics (assuming, that is, that Haken and Appel and the computer haven't made a mistake).

I admit that for a number of my friends, mostly number theorists and topologists, who fool around with small numbers and low-dimensional spaces, the computer is a tremendous scratch pad. But those same friends, perhaps in other bodies, got along just fine 25 years ago, before the computer became a scratch pad, using a different scratch pad. Maybe they weren't as efficient, but mathematics isn't in a hurry. Efficiency is meaningless. Understanding is what counts. So, is the computer important to mathematics? My answer is *no*. It is important, but not to mathematics.

**Albers**: *Do you sense the same attitude about computers among most of your colleagues?*

**Halmos**: I think the ones who share my attitude are perhaps in the minority.

**Albers**: *There are now mathematicians who seem to have a hybrid nature. Let's take someone like Don Knuth, who earned his Ph.D. in mathematics, and along the way discovered the art of computing.*

**Halmos**: It's not fair arguing by citing examples of great men. How could I possibly disagree? Don Knuth is a great man. Computer science is a great science. What else is there to say? In many respects, that science touches mathematics and uses mathematical ideas. The extent to which the big architecture of mathematics uses the ideas rather than the scratch pad aspect of that science is, however, vanishingly small.

Nevertheless, the connection between computer science and the big body of pure mathematics is sufficiently close that it cannot be ignored, and I advise all of my students to learn computer science for two reasons. First, even though efficiency is not important to mathematics, it may be important to them; if they can't get jobs as pure mathematicians, they need to have something else to do. Second, to the layman, the difference between this part and that part and the third part of something, all of which looks like mathematics to him, looks like hair-splitting. My students and all of us should represent this science in the outside world.

**Albers**: *It is rumored that you're one of the world's great walkers.*

**Halmos:** That's certainly false. I enjoy walking very much. It is the only exercise I take. I do it very hard. I walk four miles every day at a minimum. I just came from a ten-day holiday, most days of which I walked fast for 10, 12, 15 miles, and I got hot and sweaty. I live it because I am alone; I can think and daydream; and because I feel my body is working up to a healthy state. To call me one of the world's great walkers is an exaggeration, I'm sure.

**Albers:** *Have you been a walker, in a strong sense, for many years?*

**Halmos:** Twenty-five years. When I was 40, I had every disease in the book, *I thought* hypochondriacally. I went to the doctor with a brain tumor, with heart disease, with cancer, and everything else, *I thought.* He examined me and said, "Halmos, there isn't anything wrong with you. Go take a long walk." So I took a five-minute walk. And then the next week, I increased it to 6 minutes, and 7, and 8, and 9, until I got to 60, and I stopped. On weekends, I walk greater distances. When I was young, I drank like a fish, smoked heavily, and had every other vice that you can imagine. Then when I started worrying about such things, I really started worrying about such things. How old are you?

**Albers:** *I just turned 40.*

**Halmos:** Then start worrying!

**Albers:** *I fully expect you soon to write another article that would pick up on two previous articles. You have done "How to Write Mathematics" and "How to Talk Mathematics." May I soon expect to see "How to Dream Mathematics"?*

**Halmos:** I'm ahead of you, but I haven't written the article. I have half-planned an entire book. If I live long enough and really have the guts to stand up in public to do it, I might write a book on how to be a mathematician. I have outlined it on paper.

It will include all aspects of the profession, except how to do research. I won't pretend to tell anyone how to do that. What I think I can do is describe the mechanical steps that people go through (and apparently we all have to go through) to do research, to be a referee, to be an author, to write papers, to teach classes, to deal with students — in short, to be a member of the profession — the most glorious profession of all.

# Part II: In Touch with God

## THE BEST PART

**Albers**: *What's the best part of being a mathematician?*

**Halmos**: No answer is going to roll off my tongue — I have never thought about such a question. What comes first to mind is being alone in a room and thinking. When things go wrong, the weather is bad, or my cat is sick, I enjoy being alone in a room, making marks on paper or fiddling with my Macintosh, thinking even about silly little calculus problems. I almost always wake up in the middle of the night, go to the john, and then go back to bed and spend a half hour thinking, not because I decided to think; it just comes. And I might think about a calculus problem or some more genuine research level mathematics. But that's a joy. What's the best part of being a mathematician? I'm not a religious man, but it's almost like being in touch with God when you're thinking about mathematics. God is keeping secrets from us, and it's fun to try to learn some of the secrets. Everything else is fun. It was fun when I was young to try to get to be a big-shot by going to meetings, by taking committee assignments seriously, by working at them, eventually by becoming a sub-junior member of the Council of the AMS, and then it was fun to have been a big-shot on the Council and other things. And it was fun to go to meetings, meet people, hear new ideas, and exchange ideas. I used to say that I enjoyed teaching, and maybe I still do, but I'm somehow less sure of that than I used to be. Sure, when I was in my twenties and tried to inculcate some calculus in those recalcitrant students, I didn't enjoy every step of it. But it seems to me that more often I ran into people who wanted to know things, and more often I ran into people who were properly prepared. Now they are neither curious nor prepared. So teaching has become less the good part of being a mathematician.

Thinking, being in contact with the mathematical world, being a member of an 'in' group are good parts. Learning mathematics is always extraordinarily hard work. I can't easily read mathematics, I can't listen to lectures. The only thing I enjoy is a kind of mathematical gossip, when people sit in easy chairs with their feet up on something and tell me their mathematics; then I can learn.

## THE WORST PART

**Albers**: *What's the worst part?*

**Halmos**: Whew! I wasn't expecting that. The first thing that comes to my mind is self-contradictory. The worst part has to do with the best part — a part of the best part that I did not yet mention: competitiveness. I like competitiveness. I am competitive. I want to beat the other guys. At the same time, I don't like it. Of course, that's not unique to mathematics, it's the same in many human affairs. There are some people who have talent, but who are so competitive, that although I can learn something from them, I don't want to. What else is a bad part? It's a little bad — I wouldn't put it into the worst part — that we are so unrecognized by the world. If you say you're a physicist, even relatively uneducated people have some idea of what you're saying. If you say you're a scientist, everybody has some idea of what you're saying. If you say you're a mathematician, even physicists often don't know what you're talking about. And the oft suggested remedy for that, that we should write a lot more propaganda, tell a lot more people what mathematics is and prove to them that it's useful and challenging and interesting is something I've always deplored. I think in our first long chat I went on record as saying I'm not for PR in mathematics, but it would be nice if the effect that the PR people want to achieve had already been achieved.

**Albers**: *How prominent do you think competition is in the mathematical community?*

**Halmos**: All pervasive. Everybody, all levels, all the time.

Halmos with Pizzicato. Halmos likes cats because "they look nice, they're interesting, they're loving, they're loveable, and somehow one is in touch with another soul."

**Albers**: *Do you think it differs very much from, say, physics?*

**Halmos**: Physics, I'm sure, does not differ. Other disciplines, I can't be sure. One of my friends who wrote a lot was a historian, to be sure a historian more or less of science, and he gave me an idea that in his business the same competitiveness existed. Yet mathematics is different from all other forms of human endeavor in the sense that it is (well, except for some of the arts) the only thing that a person does all by himself. Therefore, this competitive aspect of the profession might be greater in mathematics, but that's a dim feeling.

## HALMOS AND LAW

**Albers**: *What other careers do you think you might have been good at?*

**Halmos**: I wasn't expecting that question, either, but I'm ready with an answer. I was always interested in law. I took some courses in law, not many, two or three or four in the course of my life, separated by years, and I used to daydream about how some of the reasoning in law is like mathematical reasoning and how I could make a serious project out of it, a contribution by making it completely mathematical. That is not original with me. Things like that exist. People have tried it. I think by and large they've failed, though I cannot off-hand think of a reference.

In a complicated trial, lawyers want to establish that someone is guilty, and in order to do that, they call this witness. For that witness it is necessary to establish credibility, and to do that, they have to cite a certain piece of history, and so on. It seems to me that this is a theorem to be proved and a sequence of lemmas that are necessary to prove it. The lawyers, the good ones, know this and you can almost hear them say 'theorem, lemma, proof,...' and so on. To a layman the whole thing might seem to end in a fizzle after a long sequence of steps and corollaries. The very last thing is a piddly little sentence that was just needed to fill the gap, but this little last thing proves the first, great, big thing. That is exactly how mathematicians proceed. Incidentally, it's bad exposition when they proceed that way, but they often do. They state a big theorem and at the end of 30 or 300 pages at a trivial level they finish the proof of the big theorem. And that parallel called my attention to the similarity of the logical attitude of lawyers and mathematicians. That's one field.

The other field is perhaps more related to mathematics, and that's linguistics. And there, too, people have tried and to some extent succeeded and to some extent failed. Languages have always interested me, too.

**Albers**: *When did law first enter your thoughts?*

**Halmos**: Before I went to college, I thought of going to law school.

**Albers**: *Do you remember what motivated your decision to go the other direction because we know you didn't really start in mathematics either?*

**Halmos**: Maybe you know. I no longer remember anything. I just somehow oozed, like a bit of jelly, into mathematics. I was just good at solving algebra problems in high school and moderately good at solving calculus problems in college. I was too unimaginative to stop school, so I went to graduate school. The next thing I knew, I was a mathematician.

**Albers**: *As I recall, you originally started out in chemical engineering.*

**Halmos**: That was just a silly, temporary, mistake. I liked math, and as a kid that meant engineering.

**Albers:** *For what it's worth, in the now fairly large collection of mathematicians that we've interviewed, the field that many of them started in was chemistry.*

**Halmos**: I offer one explanation that wasn't in my head 50 years ago. Chemical formulas, balancing them, are lovely things. The only things I remember from my freshman years in college and year or so in high school was the beauty of chemical formulas and Euclidean geometry with statements in the left-hand column and reasons in the right-hand column. Those were good, hard things that you could trust.

## PRACTICE, PRACTICE

**Albers**: *The image that you project in writing and giving public addresses is thorough professionalism.*

**Halmos**: I regard that as one of the nicest compliments. When I want to pat myself on the back, I don't say I'm a great mathematician, but I say I'm a pro.

**Albers**: *I mentioned to Jerry Alexanderson that you had told me that you had rehearsed your talk for MAA's 75th Birthday several times, and Jerry said, "Well, it really shows, doesn't it?" Do you practice in the privacy of your home or office — in front of a mirror? I don't think many mathematicians would do that, and it may show that they don't.*

**Halmos**: It's an old gag that it's very hard work for an actor or any other public performer to be spontaneous. I try very hard in my writing and in my public speaking to be spontaneous, by which I mean that I prepare everything to within an inch of its life. It's not always fun to give the same performance over and over again, but I have done it a few times, and I noticed how it improves each time. I know how to time the laughs. I know when to shrug my shoulders and throw up my hands and grin and look sad and so on. It's all ham acting, and it's important, not because people will pat you on the back and applaud you, but because it contributes to communication. I say things better if people understand them better. My recent talk, "Has Progress in Mathematics Slowed Down?" involved condensing the 2 hours and 20 minutes (I timed it) that it would have taken to present my entire Monthly article to 29 minutes. I practiced that talk at least 20 times, sometimes just by sitting here at my desk and reading it out loud and other times by doing the same thing but recording it on a cassette recorder and then listening to it and making notes as I listened as to what required change, and what I had to say more clearly. Preparation is vital and important and an indispensable part of professional life. You say that not many people do it. I wonder. One of the greatest ham actors was Emil Artin. Von Neumann was less of a ham and more sure of himself. He thought he could get away with things. He thought he could think them up on the spur of the moment. And much of the time he could, but I saw him give bad lectures and get badly confused — rarely, but it did happen. Preparation, including rehearsal, is vital, and I think your surprise over my practicing may be unjustified.

**Albers**: *Does practicing extend to your writing as well?*

**Halmos**: Yes, even if somehow less so because the pressure is less and you have more time to write. When you're preparing a talk, you might have six months' notice. And also in the talk, you're publicly exposed. But I feel that the kind of changes and revisions I make, going round and round as I do are the same when I prepare a talk or an article.

**Albers**: *How about in writing? How many drafts do you usually prepare before you feel that you're there?*

**Halmos**: That's an unanswerable question because there isn't something called Draft 1 and then Draft 2 and then Draft 3. There is something called Draft 1 all right except I prefer to call it Draft Zero. And then I change a sentence, and then I change a paragraph, then I change a page, then I have to change two pages, and it's unclear when it becomes a different draft. Every single word that I publish I write at least six times.

## DOING MATHEMATICS

**Albers**: *I want to talk about how you do mathematics and how you did it. Has it changed over the years? Did you do it differently at 40 than you did at 30 than at 50 than at 60 than at 70?*

**Halmos**: Most of the questions that you ask or imply I don't know the answers to. I don't know how I did it when I was 20, which was when it mattered. That's when I started. I remember improving. I remember at 28 or 30 thinking, 'Gee, now I know how to do this better than I did back then.' What do I do? The hardest part of doing mathematics for me is finding a question. When writing my Ph.D. thesis with Doob, he had to tell me the problem. He was sore at me for having to tell me the problem. He knew that a good mathematician thinks of his own problems. I don't know when that came along, but I think it came along quite early in my life, and it certainly has been increasing if anything. I feel joyous, I want to run up a flag and sing, when I think of a question. Never mind the answer;

"I'm not a religious man, but it's almost like being in touch with God when you're thinking about mathematics."

the answer will come or something will come. That's something I've been saying for many decades.

Then what do I do? Well, from here on I'll have to be either vague or cliche-ridden. I look for examples. Since a lot of my work has been in operator theory and infinite dimensional Hilbert spaces, and since the most easily accessible part of that is matrix theory and finite dimensional vector spaces, I start by looking at a 2 by 2 matrix. Sometimes I look at a 4 by 4 matrix. That's when things get out of control and too hard. Usually 2 by 2 or 3 by 3 is enough, and I look at them, and I compute with them, and I try to guess the facts. First, think of a question. Second, I look at examples, and then third, guess the facts. I felt better about the other questions you asked me, whether I had thought about them before or not, because I heard myself speak. On this one I just hear an ocean of cliches, everybody has said these things, and I can't add much to them.

**Albers**: *It is hard for most mathematicians to explain what their subject is to nonspecialists for some very obvious reasons, not the least of which is language, if you're outside the field. How would you describe, let's say, to a freshman or sophomore high-school student, how a professional mathematician really does his subject?*

**Halmos**: High-school students are too easy. Those guys I can talk to. The people who are hard are medical doctors, grocery clerks, automobile mechanics, and lawyers. People, white collar or blue, who have no idea what mathematics is like, are tough. All they know is that it's an obscene word that other people didn't talk about. To seniors in high school or college freshmen, I can explain how mathematics is done, but not the way mathematicians sometimes try, by talking about a given complex structure. That's a bunch of nonsense. You can't tell people outside the business what the actual theorems in the business are. The best you can do is to communicate the spirit of mathematics which is: find the question, look for examples, guess the answer, and go on from there. I give talks on problems. In the course of the centuries, I've accumulated a few hundred problems. Those problems range from very fancy stuff (the fanciest things I ever knew about Hilbert spaces) down to the stuff that your mother-in-law, the grocery clerk, and the medical doctor could understand. Just puzzles. Puzzles that please people. And sometimes they say, "How do people think of such things?" Well, that's question one. Where did the question come from? And then, "Gee, I don't know how to think about that. How would you ever find out?" And so you look for examples. And in these puzzle talks, problem talks that I give, in effect I reach or try to reach audiences just such as you and I described, high-school freshmen or grocery clerks, and it can be done. But it has to be done in spirit, not in detail, and done at a level that has a chance of reaching them.

**Albers**: *Where do you think mathematics is going, and then closely allied to that, where do you think it should go?*

**Halmos**: I have an instinctive, emotional reaction to both parts. I don't think it's going anywhere, and that's exactly where it should go. In other words, I'm giving the completely reactionary, classical, pure mathematicians's answer. Mathematics just is, we nibble away at it, I don't think we direct it worth a damn, and it seems to me as silly to ask where is it going as to ask where is the dawn going. You might say it's going to the morning and to noon, but it isn't going anywhere, it just is. Of course, I know some people would say, "Well, it's going to more and more applications, going towards more and more abstraction, and it should go that way or the other way." My emotional reaction to all of those things is that it's baloney.

## THE ROOT OF ALL DEEP MATHEMATICS

**Albers**: *In the conclusion of "Fifty Years of Linear Algebra," you wrote: "I am inclined to believe that at the root of all deep mathematics there is a combinatorial insight ... I think that in this subject (in every subject?) the really original, really deep insights are always combinatorial, and I think for the new discoveries that we need the pendulum needs to swing back, and will swing back in the combinatorial direction." I always thought of you as an analyst.*

**Halmos**: People call me an analyst, but I think I'm a born algebraist, and I mean the same thing, analytic versus combinatorial-algebraic. I think the finite case illustrates and guides and simplifies the infinite.

Some people called me full of baloney when I asserted that the deep problems of operator theory could all be solved if we knew the answer to every finite dimensional matrix question. I still have this religion that if you knew the answer to every matrix question, somehow you could answer every operator question. But the 'somehow' would require genius. The problem is not, given an operator question, to ask the same question in finite dimensions — that's silly. The problem is — the genius is — given an infinite question to think of the right finite questions to ask. Once you thought of the finite answer, then you would know the right answer to the infinite question.

Combinatorics, the finite case, is where the genuine, deep insight is. Generalizing, making it infinite is sometimes intricate and sometimes difficult, and I might even be willing to say that it's sometimes deep, but it is nowhere near as fundamental as seeing the finite structure.

**Albers**: *Seeing the finite structure brings me back to your work style. When you're thinking of problems, do you often see the problems in visual terms? What kinds of images are you holding in your head or playing with on paper? What's bouncing around in your head?*

**Halmos**: I haven't the faintest idea. I remember quoting John Thompson when he was once asked what does he see when he thinks about a group. His answer was a huge German, capital G. Whether he was joking I do not know. I'm a very bad geometer. For the calculus problem that I keep mentioning, it would have helped to see some pictures. I tried to draw them, but I'm bad at drawing them. I don't mean that I'm not artistic. I mean that I didn't see the mathematics. I didn't see whether the curve was convex or concave or whatever. And if somebody else draws pictures for me, I can't absorb them, I can't see them. Nevertheless, some kind of geometric picture seems to be necessary even for geometric idiots like myself. So to some extent, I see a picture in the plane. They're actions, they're movements, and sometimes I try to see them. I see points moving and usually I visualize a rigid translation which is not even a linear transformation. But much more than that, I think I have some kind of symbolic sense, a sense of symmetry. If there is a capital 'A' there and at the same time a little 'a' there, then the next time I see a capital 'B', I look for the little 'b.' Letters of the alphabet and mathematical symbols, and their symmetry are what I visualize. What are the marks that I put on paper? Well, I'm stuck. To a large extent they are words. It helps to slow down my mind which jumps around in a strange fashion by writing down "the question is ..." etc. It helps to scribble a letter and to say small a is less than capital A, and I look at that formula, and follow it with a question mark.

**Albers**: *Over the years, some moments must have been brighter and more exciting than others. Which of those moments stand out for you?*

**Halmos**: Once or twice in my life when I did prove a theorem that I was struggling with for many weeks or months preceding, I remember. There I remember a particular minute, or well, more like an hour that

I felt good and rushed to tell Ginger about something or just plain felt unusually good. Okay, so that's one kind of accomplishment, a piece of mathematical accomplishment. Then, of course, there is recognition, the payoff to competitiveness. When you've competed and won, that can be good. I received the Steele Prize for mathematical exposition, and not long before, but some non-trivial time before, maybe a few months or even a couple of years I somehow felt that it was high time I got it. I scribbled out my acceptance of it. When I did receive it, I was pleased, and I would count that as one of my joyous moments. My secretary at the time practically fainted when upon being notified of it, I showed her the scribbles and proved to her that I had expected it.

**Albers**: *During the first half of this interview, Pizzicato, one of your two cats, rested on your lap. Is there anything about cats that especially appeals to you?*

**Halmos**: Do animals have souls? People debate the subject, and I stand firmly on the affirmative. But what is it that appeals? Well, they look nice, they're interesting, they're loving, they're lovable, and somehow one is in touch with another soul. They enlarge one's life a little bit.

## SEVENTY-FIVE AND WORRYING

**Albers**: *Seventy-five is approaching. How old do you feel?*

**Halmos**: For some reason something like that popped into my head the other day, and I made up an epigram. Not a funny one, not a long one, but I thought it was good at the time. 'I'm older than I look, I look older than I feel, I feel older than I act, and I act about 30.' How old do I feel? People near 75 have creaks and minor aches and pains, and I have them. Therefore, sometimes I feel 85. But most of the time, although my mathematics is not that of a 25-year-old, I can think, I can walk, I can with moderation eat and drink and watch movies and enjoy people and cats and enjoy life, and I don't feel any different from how I felt (please notice that I said 'from' rather than 'than') between the ages of 25 and 65.

**Albers**: *Do you ever worry about getting old?*

**Halmos**: I worry all the time. I worry about dying. Somebody tells me that I'm going to have to die sometime or other, and I resent it. And I especially resent that it might be very soon, like within five years, which is considered more than normal. It could be 10 years, or maybe 20 years, but I still resent it. And yes, I worry about various things — my brain, my eyesight. I think those are the two main things. I don't want to go ga-ga, and if I don't, I don't want to go blind either, so I can keep reading and writing. If those two prayers are answered, I'd like to be healthy enough so that I can pick up a pen and write or type or take a little walk. So I worry about those things. My father lived to be 80, and he was by no means in

"I don't have a computer; I have the world's most intelligent typewriter...
I am very dependent on my Macintosh."

perfect health the last 10 years of his life, but he was not bedridden. He
took tiny little walks just around his 20-foot garden, but even so, he moved,
and his brain was clear. He died relatively suddenly of a major stroke, and
that's the way to go.

## FATHER

**Albers**: *We haven't talked about your father very much. Since he lived to
be 80, he got to see you develop and enjoy real prominence. What did he
think of your activity?*

**Halmos**: We didn't pay much attention to each other. When I was eight
years old, my father left his three sons in Hungary and emigrated to America as a widower, and for the next five years didn't see them. He had five
years to become a citizen so that he could import his family. So, in those
crucial years from 8 to 13. I didn't have a father. I had various substitutes,
of course. Even before that, I didn't feel very close to him; he was a very
busy medical man, and I remember somehow being ordered into his presence. I think he was being dutiful. And maybe he loved me. And then I
lived in Chicago in his house for a year and a half. Then I moved away and
went to college, so that takes me up to the age of about 16 or 17. He did
not play an important role in my life, not that I know of, and I don't think
he was all that impressed by whatever I accomplished. For one thing, it
was not clear to him that I accomplished anything. He was not a rich man,
but he was a medical doctor, an accepted and good one, and so he was

much richer in relation to society than you and I are ever likely to be. He knew that I wasn't and probably never would be; he didn't count things by money necessarily, but that was a symptom, an indication.

**Albers**: *What's on Halmos's agenda for the future?*

**Halmos**: I've been boring my friends about a book called "LINEAR ALGEBRA PROBLEM BOOK" ("LAPB"), which will have 50 or 75 sections and maybe 10 or 15 of them are written, and those 10 or 15 took a couple of years. So at that rate, it'll be another six or eight years. Maybe I can speed up the rate. There is an actual piece of mathematical research I'm thinking about, but that's very near the beginning. And I keep doing piddling little things. Academic Press is going to publish a long scientific dictionary having thousands of pages. You know about it? They asked me to write an article of 300 words. Well, that's 300 words. This sheet of paper on which I type probably has 200 words on it. It took me 3 days to write 300 words. What do I see in the future? Well, I think I see that kind of thing continuing to come in. Extensive correspondence with, for instance, one of my very good friends, Max Zorn, who is almost exactly ten years older than I. He seems to enjoy receiving my letters.

It's hard for me to get used to the absence of pressure. I always put myself under pressure, and of course, I blamed the world. The world is putting on the pressure. Well, now I'm beginning to realize that the world is not putting on the pressure. If I never published anything, not even an elementary textbook, if I never again answered a letter, if I never did anything any more except drink my beer and watch the telly, nobody would, I think, think any the worse of me. But I keep putting myself under a little pressure and keep doing these small piddling jobs.

**Albers**: *What should I have asked you about that I didn't?*

**Halmos**: You might have asked about the contradiction of my having a computer. I say I don't have a computer. I have the world's most intelligent typewriter, but other people call it a computer. Although I agitated against them and denied that they have anything to do with mathematics, I am very dependent on my Macintosh.

**Albers**: *To what extent to you depend on your Macintosh?*

**Halmos**: I spend, I sometimes exaggerate and say ten hours every day in front of the machine, but a minimum of four, and more like six every day. I live here.

**Albers**: *How many hours do you work per day?*

**Halmos**: I don't know what the word means. I get up early, usually Ginger and I get out of bed at 5:30, and by 7:00 or 7:30 I'm here at my desk. And some days I have classes; other days I go to seminars. Every day I try to get my walk. But I keep doing those things, including time out for an hour for lunch. The actual eating takes 20 minutes and my nap takes 20 minutes and piddling around takes an hour in the middle of the day. Well, some of that is leisure time that I spend with the gang. So discounting the lunch and the walk, I do something that people might call work from 7:30 a.m. to 5:30 p.m. which is 10 hours minus the things I've said to minus. But of course, it isn't all work. A lot of it is, as I said, correspondence. That's as close as I can come to an answer.

## SMELLING MATHEMATICIANS

**Halmos:** One thing you didn't ask about I want to speak about very briefly. The thing you didn't ask me about, but you almost asked me and somehow told me is that I'm a mathematician or a good mathematician or a great mathematician or something, and I say that one of my best professional qualities, though people sometimes resent it when they learn that I think that, is evaluation. I say I can tell a mathematician, and I can smell one. I'm on record as having said, "Give me an hour alone with a student, and I can tell how good a mathematician the student will be," but better than that, give me a few minutes or an hour with a so-called mathematician, and I can tell you if he's really a mathematician. And in particular, I claim I know me; I know exactly how good a mathematician I am. It embarrasses me when sycophantic admirers compare me with Milnor or Gauss — that's just plain silly.

I haven't been as prolific as some mathematicians, and I have not been as deep. I'm a mathematician and I know just how un-great I am.

*Department of Mathematics*
*Menlo College*
*Atherton, CA 94027*

# Paul Halmos's Expository Writing

## Leonard Gillman

Paul Halmos is a redoubtable expositor. This essay presents mini-reviews of some of his works. It is a natural sequel to my article in *Selecta—Expository writing,* Springer, 1983.

I stated in *Selecta* that Halmos's renown as an expositor ranged over several decades. It now encompasses several-plus-one decades.

To be a good expository writer you have to be enthusiastically interested in truly helping your readers, recognizing who they are and adapting to them. Organize your material effectively; know what to say or leave out and when and how to say it or leave it out; maintain a flow of clear prose; and adhere to correct grammar. (The last rule should not have to be stated, except that so many writers violates it.) Revise and revise; hold up your "final" draft for a day for another look before sending it in. Halmos practices these precepts and others, and adds his personal *je ne sais quoi.*

Halmos has received expository awards from the Mathematical Association of America and the American Mathematical Society—from MAA, a George Pólya award, two Lester R. Ford awards, and a Chauvenet Prize; and from AMS, a Leroy P. Steele Prize.

The Pólya award is given for articles published in the *College Mathematics Journal,* and the Ford award for articles in the *American Mathematical Monthly.* (There is also a Carl B. Allendoerfer award for articles in *Mathematics Magazine,* MAA's remaining journal; one of Halmos's outstanding failings is that he has never won this one.) The Chauvenet Prize is awarded for articles from a wider range of books and journals, and is the MAA's highest prize for exposition.

The Steele Prize is the only AMS award for exposition. The award to Halmos was made at the 1983 summer meeting, in Albany. Here is the citation:

> The award for a book or substantial survey or research-expository paper is made to PAUL R. HALMOS for his many graduate texts in mathematics, dealing with finite dimensional vector spaces, measure theory, ergodic theory and Hilbert space. Many of these books were the first systematic presentations of their subjects in English. Their felicitous style and content has had a vast influence on the teaching of mathematics in North America. His articles on how to write, talk and publish mathematics have helped all mathematicians to communicate their ideas and results more effectively.

Leonard Gillman, 1989

Half or more of Halmos's 180 works are expository—35 of his articles; at least half of his dozen books; and all 50 reviews (counting the article that consists of two dozen reviews as two dozen reviews). I have selected about 30 of them for discussion. I exclude his mathematical books, most of which have been well known for decades.

In listing works I give the title and date as well as a reference to the complete bibliography at the end of this section of the volume. The prefix B in a reference signifies a book, and R indicates a review.

# Expository papers

The four sublists that follow are intended to include all of Halmos's expository articles to date. Articles marked [s] were reprinted and discussed in *Selecta*, and most of them are listed in small type and are not reported on here; those marked [×] should have been in *Selecta* but were inadvertently omitted.

### Somewhat technical articles

[25]  Measurable transformations, 1949 [s]

[50]  Recent progress in ergodic theory, 1961[S]

[53]  What does the spectral theorem say?, 1963[S]

[54]  A glimpse into Hilbert space, 1963[S]

[69]  Finite dimensional Hilbert spaces, 1970[s]

[111]  Fifty years of linear algebra: a personal reminiscence, 1988

*Finite dimensional Hilbert spaces* [69] won a Ford award.

*Fifty years of linear algebra: a personal reminiscence* [111] is a warm account of various problems in operator theory, classified as "naturally finite" (the solution in the finite case contains the core of the idea), "superfinite" (the answer and the methods in the infinite case are different from the finite ones), and "infinite" (visible in the finite situation in a degenerate form only).

**Articles typical of** *The American Mathematical Monthly*

[13]  The foundations of probability, 1944[s]

[41]  The basic concepts of algebraic logic, 1956[×]

[85]  American mathematics from 1940 to the day before yesterday (with J. H. Ewing, W. H. Gustafson, S. H. Moolgavkar, W. H. Wheeler, and W. P. Ziemer), 1976[s]

[87]  Logic from A to G, 1977[S]

[88]  Bernoulli shifts, 1977[S]

[89]  Fourier series, 1978[S]

[90]  Arithmetic progessions (with C. Ryavec), 1978[S]

[91]  Invariant subspaces, 1978[S]

[92]  Schauder bases, 1978[S]

[93]  The Serre conjecture (with W. H. Gustafson and J. M. Zelmanowitz), 1978[S]

[105]  The work of F. Riesz, 1983[S]

[114]  Has progress in mathematics slowed down?, 1990

Ten of these twelve articles appeared in the *American Mathematical Monthly*, long the world's foremost journal for expository articles addressed to the college mathematics teacher.

*The foundations of probability* [13] won a Chauvenet Prize.

*The basic concepts of algebraic logic* [41] is an extraordinarily readable, 25-page introduction to the relatively unknown subject of algebraic logic (and in fact the term was apparently coined by Halmos for the article). It starts from zero and patiently prepares the reader for each new concept, culminating in a discussion of the Gödel completeness theorem and the Gödel incompleteness theorem. As a model of exposition it is equal to *The foundations of probability* (supra). I conjecture it failed to win a prize because its subject matter was somewhat off the beaten track.

*American mathematics from 1940 to the day before yesterday* [85] was presented (in abbreviated form) as a lecture by Halmos at the 1976 annual meeting of MAA, in San Antonio, whose theme was the U.S. Bicentennial. The paper won a Lester R. Ford award for its six authors.

*Has progress in mathematics slowed down?* [114] was presented (in greatly abbreviated form) to the 75th anniversary meeting of the Mathematical Association of America, at Columbus, August 1990. It is similar to *American mathematics from 1940 to the day before yesterday* (supra) but covers twice as many subjects (including five of the earlier ten), deals with a period of twice as many years, and is concerned with mathematics the world over rather than just in America. There are 22 subjects, divided among 9 "concepts," 2 "explosions," and 11 "developments." $K$-theory is a concept, and the independence of the continuum hypothesis is a development. (I would have called it an explosion.) The two explosions are the Appel–Haken proof of the four-color theorem, and Gerd Faltings's proof of the Mordell conjecture. (It was already known that Mordell's conjecture implies that for every $n > 2$, Fermat's Last Theorem "for that $n$" has only a finite number of independent solutions.) The author's answer to the question of the title is a resounding no.

The style is generally similar to that of [85]—in fact, four of the articles are taken from there almost verbatim—but some of it is tighter (index theorem) and some is looser ($K$-theory). There are also some trenchant comments about the character and mathematical importance of catastrophe theory, chaos, and the proof of the four-color theorem. Each essay concludes with a single reference to which the reader can turn for more information—a very welcome feature; 17 of the 22 are to the *Encyclopedic Dictionary of Mathematics* [R22] and 5 are to the *Mathematical Intelligencer*.

In *Selecta,* I described the earlier paper as "thrilling." I now say the same about this one.

## Advice and comments

[68] How to write mathematics, 1970[s]

[81] How to talk mathematics, 1974[s]

[82] What to publish, 1975[s]

[83]  The teaching of problem solving, 1975 [S]

[98]  The heart of mathematics, 1980 [S]

[100]  Does mathematics have elements?, 1981 [S]

[101] Think it gooder, 1982[×]

[103] The thrills of abstraction, 1982[s]

[115] The calculus turmoil, 1990

*How to write mathematics* [68]: I reiterate my admonition in *Selecta* that almost all mathematical writers should look it over every so often. To cut down your excuse about nonavailability, I point out that it was reprinted in a 1973 AMS booklet with the same title (authors Steenrod, Halmos, Schiffer, and Dieudonné) as well as in *Selecta*.

*Think it gooder* [101] was written as a sort of rebuttal to an adjoining article by George Piranian called *Write it better*. George said good English is important. Paul said, what do you mean, good English is important?—good mathematics is more important. They are both right.

There is some new information about *The thrills of abstraction* [103]. Halmos's "automathography" [B12, 402] discloses that the article was first submitted to *The Mathematics Teacher*, the journal for secondary-school teachers published by the National Council of Teachers of Mathematics. According to the referees:

> The author apparently feels she/he is illustrating the thrill and power of abstraction. While his/her examples contain the potential for doing that, I don't feel that the presentation creates that effect... Much of the problem with the paper is that of a meandering style. The progression of ideas is not clear... The mathematical topics focused on are of only moderate interest...

—and the paper was firmly rejected. He thereupon sent the same article, word for word, to the *Two-Year College Mathematics Journal* (now the *College Mathematics Journal*), which accepted it—and it won a Pólya prize.

*The calculus turmoil* [115]. The author believes that the calculus turmoil is both misplaced and exaggerated: what is wrong is not that calculus students flunk and textbooks are dull but that students have no respect for intellectual effort. They come to the university not just badly trained but almost negatively trained. He endorses teaching the classical basic techniques of calculus, as well as "rigged integration techniques and tedious calculations—they are as rigged and as tedious as finger exercises for the pianist, and just as indispensable." There is a disease, but calculus is neither its cause nor its main symptom. We mathematicians can help cure it, not by rewriting calculus books but by encouraging and properly training prospective school teachers and insisting on higher standards in the schools, and by enforcing the prerequisites in our own courses.

I myself am already on public record as espousing such views, so will confirm my comments to the following personal item: As a pianist well beyond the level corresponding to calculus students, I thoroughly enjoy practicing finger exercises daily—and I even keep rigging up new ones.

## Popular articles

[44]  Nicolas Bourbaki, 1957 [S]

[45]  Innovation in mathematics, 1958 [×]

[65]  Mathematics as a creative art, 1968 [S]

[78]  The legend of John von Neumann, 1973 [S]

[99]  Applied mathematics is bad mathematics, 1981 [S]

(∗)  Paul Halmos: Maverick Mathologist (interview by Donald J. Albers), 1982 [s]

[109]  Why is a congress?, 1987

[110]  How to remember Walter Kaufmann-Bühler, 1987

Halmos's automathography [B12, 238] reports that *Innovation in mathematics* [45] "was ruthlessly edited by a junior editor; the published version is watery, badly organized, badly expounded. ... Much of the article is not my diction, not my arrangement, not my style." The article gives examples of new mathematical facts and new proofs of old facts, and of sources of mathematical innovation, and a lot of pure Halmos comes through despite the editorial vandalism.

An updated version of the Albers interview (∗) is included in this volume.

*Why is a congress?* [109] presents personal impressions of the 1986 International Congress in Berkeley, including general comments and specific assessments. Examples:

> The ideal number of lines on one transparency is seven or eight; with the twenty-odd that a typewriter can put on them, they get beyond the comprehension threshold.

> The Skorohod address was a catastrophe

—mostly because of difficulty in coordinating the translation from the Russian.

The best talk I heard at the Congress (and quite possibly the best one in my life) was Hendrik Lenstra's *Efficient algorithms in number theory*. The subject was deep mathematics (elliptic curves and their relation to primality testing and factorization), the attitude was "practical" in the sense that it had to do with the computer approach to the subject and its efficiency, and the delivery was beautifully organized, clear, and witty.

Judy Grabiner's *The centrality of mathematics in the history of Western thought* deserves special mention. [She is] an expert historian of mathematics with an unusually broad understanding of mathematics. The work she told us about is an impressive piece of scholarship, and her telling was, like Lenstra's, clear and witty. She talked about history as organized thought rather than as a list of facts. She read her paper ... but she did it well; the presentation was alive.

*How to remember Walter Kaufmann-Bühler* [110] is a memorial to a close friend and publishing colleague. It is refreshingly written, in the present tense, and totally devoid of maudlin hyperbole. Here is the closing paragraph:

Walter is a great mathematical editor. I am glad I know him as publisher, editor, author, and friend. He makes my soul richer. I am sorry he isn't with us today. I miss him.

Had Walter lived, he would surely have been the Springer editor for the present volume.

# Popular books

[B12] I want to be a mathematician, 1985

[B13] I have a photographic memory, 1987

*I want to be a mathematician* [B12] carries the subtitle "an automathography," which the author defines as "a mathematical biography written by its subject." It is a candid account of a mathematician's life and inner thoughts. Mathematicians will be fascinated by it, and nonmathematicians will enjoy it almost as much.

*I have a photographic memory* [B13] is a collection of over 600 photos of mathematicians, often more than one to a photo, taken by the author (with two obvious exceptions). Every mathematician will recognize a good number of the people and even more of the names; but these won't mean much to people outside the profession.

# Book reviews

(The date shown is the date of the review)

[R2]  H. Tietze, *Gelöste und ungelöste mathematische Probleme aus alter und neuer Zeit,* Vols. I, II, 1951

[R4]  N. Bourbaki, *Intégration,* 1953

[R19]  R.D. Traylor, *Creative teaching: Heritage of R.L. Moore,* 1974

[R22]  S. Iyanaga and Y. Kawada, *Encyclopedic Dictionary of Mathematics,* 1981

[112]  P.R. Halmos, *Some books of auld lang syne,* 1988

Book reviews are excellent examples of expository writing. Halmos has written two dozen, listed in a section of their own in the bibliography, plus another two dozen in *Some books of auld lang syne* [112]. I have picked out about 20 from the total that I found so enjoyable and instructive that I will just quote from them with only occasional comments.

First let us look at what Halmos thinks a book review should say.

> The review should tell its readers where the subject of the book comes from, where it is now, and where it is going. What are its connections with other, perhaps better known, subjects? What makes it interesting? ... As for judgements, ... on balance, the review should be the reviewer's exposition rather than a critique of the author's.

—*What a book review is, and is not,* Notices AMS **22** (1975) 283, written when he was book reviews editor of the *Bulletin AMS.*

> [The idea of a review is] to provide, for the busy specialist in something else, some compromise between knowing nothing at all about the subject and spending a hard month reading the whole book.

—*Statement of policy,* Amer. Math. Monthly **89** (1982) 3–4, his opening statement as *Monthly* editor.

> A book review is sometimes like an auction catalogue: the item for sale is described in legalistically accurate detail. That helps you decide whether you want to buy the product, but it's rather boring to read. The best kind of review is like a preview at the movies: you are shown a few samples, and, in addition, several tempting new angles that are not actually in the production. If it's well done, the result is fun to see in its

own right, and you don't begrudge the time it takes even if you never go back to get the whole picture. An encyclopedia, unlike a book about topology or a movie about Timbuktu, cannot be compressed by a factor of several hundred; that's why the sequel is very little like a preview and much more like a catalogue.

—Introduction to his review of the *Encyclopedic Dictionary* [R22].

And now for the reviews themselves.

*Tietze* [R2]

> This is one of the deepest and at the same time the most charming of the popular books on mathematics that I have ever seen. ... Tietze presents fourteen lectures, on topics as diverse as digital representations of integers, distribution of primes, properties of geodesics on surfaces, dimension theory, the regular 17-gon, solution of equations by radicals, and the concept of infinity. ... Even the trite and altogether too famous problems (such as angle trisection, ...) receive fresh treatment.

*Bourbaki* [R4]

> Putting ourselves in the place of a student, we must ask: "Is the subject important, is the book clearly written, and is the material well organized?" Putting ourselves in the place of the supervisor of a Ph.D. thesis in one of the applications of integration ..., we must ask: "Is this the point of view that will help a student to understand and to extend his field of interest?" I say that the answer to the student's question is yes and the answer to the professor's question is no ...

and the reviewer continues with four-and-a-half pages of detailed reasons, including one of my favorite nuggets of exposition:

> Owing, no doubt, to the authors' predilection for using as definiens what for most mathematicians is the definiendum, there are many spots at which the treatment appears artificial.

*Traylor* [R19]

> The book consists of over 450 pages (typewritten), of which the first 200 are badly organized, repetitious, and mostly boring prose, and the rest are lists: ... Moore's students (one page), Moore's publications (five pages), Moore's students' students, onto the sixth generation, and their publications (242 pages), index of names (15 pages), and footnotes (nine pages). ...

I want to know more about Moore ... his views on the
Russell paradox, the empty set, nonseparable compact Haus-
dorff spaces, the Brouwer fixed-point theorem, category theory,
Gödel's incompleteness theorem, the continuum hypothesis. ...
I want to know as much as possible about the Moore method.
What, exactly, does he say on the first day of classes? How does
he handle the shrinking violet, the buffoon, and the loudmouth?
... Is it possible to use the Moore method to produce a broadly
educated mathematician? ... I should love to read a good book
entitled "*Creative Teaching: Heritage of R. L. Moore.*" This is
not it.

### *Iyanaga and Kawada* [R22]

[This encyclopedia] is more like a dictionary. It consists of
436 alphabetically arranged, anonymous, short, brisk, current
summaries, with an average length of 3.0849 pages, focused on
definitions rather than explanations. ... The second volume
ends with a huge collection of tables, lists, and indices (over
350 pages of them). ... If you want to learn something from
this encyclopedia, you have to thread your own way through
an unstructured maze. You must turn to one or more of the
lists at the back of the book, make a note of probably dozens of
reference numbers, and then juggle volumes and ruffle pages in
an attempt to organize the material. There are, for instance, 25
articles listed under Algebraic topology. ... The first of them is
Topology; it is a survey of the whole subject in less than two
pages. ... The survey uses over 40 undefined terms. ... If you
need to read the survey, you are not likely to know what most
of those terms mean, and to follow the dagger that refers you
to the index won't help you find out in what order you could
learn them. ... I doubt that this encyclopedia will be of use to
anyone who is not a professional mathematician already. ... I
think that it has some strengths, but on balance it is useless.

Halmos is never afraid to climb out on a limb and then saw it off.
Recall that *Has progress in mathematics slowed down?* [114], written a
few years after this review, lists 22 references, of which no less than 17 are
to this "useless" encyclopedia.

### *Halmos* [112]

Halmos wrote this work for an AMS 100th anniversary volume. He picked
26 famous and influential books ranging over time from 1902 to 1955 and
reviewed them. Dozens of illuminating comments, coupled with incisive
conclusions, make reading them entertaining and informative, and partic-

ularly interesting to those of us who remember the books. Here are some excerpts.

## Calculus

*Granville, Smith, and Longley,* Elements of the differential and integral calculus, 1904

> "A sequence is a succession of terms formed according to some fixed rule or law." If you know what's going on, you can tell what's going on, but, harassed students or jaded mathematicians, most of us agree that if you don't know what's going on, this isn't the place to find out. ... This book was useful once, and I no longer hold it against it that it is not a "rigorous" book.

*Courant,* Differential and integral calculus, 1938

> "We now define the expression $dy = y'dx = h\varphi(x)$ as the differential of the function $y$; $dy$ is therefore a number which has nothing to do with infinitely small quantities." I am not happy. What is $dy$? Is it really a number? ... I didn't like [the book] much. I found it verbose, pedantic, and heavy handed.

Everything is relative. In *The calculus turmoil* [115], Courant is one of the good guys.

*Hardy,* A course of pure mathematics, 1908

> This book ... shows that a conscientious and honest writer who really *really* understands what he is writing about can produce a work of rich exposition that is readable and enjoyable. Hardy's book is the toughest, most challenging, most rewarding, and most mathematical calculus book you could possibly imagine.

## Algebra and number theory

*Bôcher,* Introduction to higher algebra, 1907

> Determinants are assumed known. ... "By the side of these determinants it is often desirable to consider the system of the $n^2$ elements arranged in the order in which they stand in the determinant, but not combined into a polynomial. Such a square array of $n^2$ elements we speak of as a matrix. ... A matrix is not a quantity at all, but a system of quantities. ... Every determinant determines a square matrix, the matrix of the determinant." ... Something about all this I found more bewildering

than helpful, and, in fact, I am prepared to argue that a part of it is outright misleading: what could the definition of determinants be that makes it true that every determinant determines a square matrix? ... Very few people still remember the book, and their memories of it are not always affectionate.

*Dickson,* Modern algebraic theories, 1926

At the universities that I knew about in the 1930s the two main competitors for the official text of a graduate course in algebra were van der Waerden and Dickson, and Dickson wasn't in German. ... I feared and respected the book. The exposition is brutal—correct but compressed, unambiguously decipherable but far from easy to read. ...

With minor exceptions, the contents of the book are still alive. ... They could be polished and modernized, but anyone who could do that could write a better book of his own.

*van der Waerden,* Moderne Algebra, 1931

It was the standard source of quality algebra. ... It has a lot of good material, it is arranged well, it is written clearly—it is just plain good. ... Sprinkled throughout the book there are problems. Their number is not large, but their level is high.

Why is van der Waerden's book no longer as fashionable as it once was? Is the material in it superseded by now, and is the treatment old-fashioned? Are there many books as good or better?

*Halmos,* Finite-dimensional vector spaces, 1942

FDVS was a good book when it appeared, and then it became perhaps the most influential book on linear algebra for a quarter of a century or more. ... It's still a good book, and it's still in print.

## Set theory and foundations

*Hausdorff,* Grundzüge der Mengenlehre, 1914

This book is beautiful, exciting, and inspiring—that's what I thought in 1936 when I read it, avidly, like a detective story, trying to guess how it would end, and I still think so. ... But by now I couldn't use it as a text. ... While the statements and proofs are clean and elegant, the emphasis and the point of view are dated. I am glad Hausdorff wrote the book, and I am glad I read it when I did.

I would have mentioned the world's first presentation of the maximal principle (every partially ordered set has a maximal chain, p. 140) and of the axioms for a topological space (Hausdorff of course, p. 213).

### Real and functional analysis

*Banach,* Théorie des opérations linéaires, 1932

> The influence of the book is difficult to overestimate; it started an avalanche. Many of the problems that Banach spacers have been working on during the last 50 years are raised in the book, and not all the ones raised are solved yet. ... It's a classic, and I think that students living at the end of the twentieth century could still profit from looking at it.

*Kolmogoroff,* Grundbegriffe der Wahrscheinlichkeitsrechnung, 1933

> This is (was?) without doubt one of the most important mathematics books of the century. ... Its expository style is mercilessly concise. ... The dictionary connecting probability and measure theory is easy to learn. Events are sets, ... Those are the secrets that Kolmogoroff's book reveals. By now they are well-known secrets, and Kolmogoroff's exposition can be improved, but the book served an almost unsurpassingly useful purpose, and we should all be grateful for its existence.

*Stone,* Linear transformations in Hilbert space and their applications to analysis, 1932

> The style is formal and ponderous. The definitions are long, and the theorems are longer, ... frequently filling a half page or more. ... Here is how Stone describes his own writing. "In order to compress the material into the compass of six hundred odd pages, it has been necessary to employ as concise a style as is consistent with completeness and clarity."

### Complex analysis

*Whittaker and Watson,* A course of modern analysis, 1902

> That book precipitates no surprise and shows no beauty; the impression it makes is that of calculus made messy. I didn't like it when I studied it, and have never learned to like it since. It isn't and it never was a good place to learn complex function theory from.

*Knopp,* Funktionentheorie, 1930

I like everything about this book. I liked everything about it a
little more than fifty years ago when I learned complex function
theory from it, and I still think it is one of the best teaching
books that I have ever seen. There is nothing fancy about it. ...
What it contains is the standard material in every first course
about complex variable theory, plus a few luxuries that not
everybody must know, minus a few other luxuries that other
books do put in. ... The book even looks good. ... The Knopp
volumes fit easily, both of them together, into the side pocket of
any jacket—they're ideal to read while ... squeezed in one of the
middle seats of a crowded airplane. ... Knopp is one of the few
"dead" books that shouldn't be dead. Unlike Dickson's algebra
book, it would be easy to modernize, and, by the addition of
a third tiny volume, it could make available many of the same
luxuries that Ahlfors and Conway offer. Would anybody care
to try?

## Topology

*Lefschetz,* Algebraic topology, 1942

The book had a strong influence on the subject; it was regarded as the gospel. It was, moreover, not only the new testament, but the old one at the same time. ... It taught us the
facts, the words, and even the right attitudes that we should
adopt. ... There are not very many books like this, and, for
sure, there are not very many authors like Lefschetz who can
write them.

*John L. Kelley,* General topology, 1955

This, the last book of auld lang syne that I am reporting on, is
the youngest of the lot, but more than thirty years of exerting a
strong influence on the terminological and topological behavior
of hordes of students make it auld enough. It's a good book, and
it was a best seller and a trend setter in its day. ... Students
find the book difficult sometimes, but they almost always find
it rewarding. ... It's a good book indeed—a useful book—a
teaching book, a learning book, and a reference book—long
may it wave.

# Halmos's views on exposition

## What expository writing is

Here is Halmos's response on receiving the Steele Prize. The third paragraph gives his thoughts on exposition.

> Not long ago I ran across a reference to a publication titled *A method of taking votes on more than two issues.* Do you know, or could you guess, who the author is? What about an article titled *On automorphisms of compact groups?* Who wrote that one? The answer to the first question is C. L. Dodgson, better known as Lewis Carroll, and the answer to the second question is Paul Halmos.
>
> Lewis Carroll and I have in common that we both called ourselves mathematicians, that we both strove to do research, and that we both took very seriously our attempts to enlarge the known body of mathematical truths. To earn his living, Lewis Carroll was a teacher, and, just for fun, because he loved to tell stories, he wrote *Alice's adventures in wonderland.* To earn my living, I've been a teacher for almost fifty years, and, just for fun, because I love to organize and clarify, I wrote *Finite dimensional vector spaces.* And what's the outcome? I doubt if as many as a dozen readers of these words have ever looked at either *A method of taking votes ...* or *On automorphisms ...*, but Lewis Carroll is immortal for the Alice stories, and I got the Steele Prize for exposition. I don't know what the Reverend Mr. C. L. Dodgson thought about his fame, but, as for me, I was brought up with the Puritan ethic: if something is fun, then you shouldn't get recognized and rewarded for doing it. As a result, while, to be sure, I am proud and happy, at the same time I can't help feeling just a little worried and guilty.
>
> I enjoy studying, learning, coming to understand, and then explaining, but it doesn't follow that communicating what I know is always easy; it can be devilishly hard. To explain something you must know not only what to put in, but also what to leave out; you must know when to tell the whole truth and when to get the right idea across by telling a little white fib. The difficulty in exposition is not the style, the choice of words— it is the structure, the organization. The words are important, yes, but the arrangement of the material, the indication of the connections of its parts with each other and with other parts of mathematics, the proper emphasis that shows what's easy and what deserves to be treated with caution—these things are much more important.

But there I go again being expository—or should I say meta-expository? Enough of this foolishness; I think I had better stop now and go prove a theorem about automorphisms of compact groups.

## What expository writing is not

*Can research be expounded?*, *Notices AMS* **30** (1983) 600–601, is a letter from Halmos to the *Notices* criticizing the *Bulletin's* newly introduced "Research-Expository Articles." Most of the pieces are not expository, he explains, but survey articles, which are very different things:

> A survey tells the history of a subject, contains a detailed, scholarly bibliography, and, in between, it defines, it states, it proves, and it is mercilessly complete. A survey is, in effect, a mini-encyclopedia. A good thing—yes, sometimes—but not for exposition, not for learning.

Every would-be expositor should read the letter in full.

## What hard work is

When you listen to Halmos lecture or read his writing, doesn't it all seem spontaneous, as though he just tosses it off? Here is how his automathography [B12, 401] describes the "spontaneity":

> As for work, I got my first hint of what that means when Carmichael told me how long it took him to prepare a fifty-minute invited address. Fifty hours, he said; an hour of work for each minute of the final presentation. When, many years later, six of us wrote our "history" paper ("American Mathematics from 1940 ..."), I calculated that my share of the work took about 150 hours; I shudder to think how many man-hours the whole group put in. A few of my hours went toward preparing the lecture (as opposed to the paper). I talked it, the whole thing, out loud, and then, I talked it again, the whole thing, into a dictaphone. Then I listened to it, from beginning to end, six times—three times for spots that needed polishing (and which I polished before the next time), and three more times to get the timing right (and, in particular, to get a feel for the timing of each part). Once all that was behind me, and I had prepared the transparencies, I talked the whole thing through one final rehearsal time (by myself—no audience). That's work.

As I remarked in *Selecta*, this is surely a lesson for all the rest of us.

*Department of Mathematics*
*University of Texas*
*Austin, TX 78712*

# Paul Halmos: Words More than Numbers

## W.M. Priestley

Let us begin in the Halmos fashion by considering a problem. What word best fits in the three blank spaces below?

> For a true _____, each work should be a new beginning, where he tries for something that is beyond attainment. He should always try for something that has never been done or that others have tried and failed. It is because we have had such great _____s in the past that a _____ is driven far out past where he can go, out to where no one can help him.

Do you think that "mathematician" belongs here? And do you think that the resulting passage fairly describes Paul Halmos as a true mathematician? If so, we have agreed upon a solution to our problem.

However, the solution is not unique. The quotation is paraphrased from Ernest Hemingway's 1954 Nobel Prize acceptance speech and "writer" is the word that Hemingway used. Now we have a new problem. Does the resulting passage still fairly describe Paul Halmos?

This paper is not about a mathematician, but a writer; not about a writer of mathematics books, but of popular works about mathematics. Halmos began regularly publishing such works roughly 20 years ago, and they now make up a sizeable part of his writings. Most of those to be discussed here may be found in Chapters III and IV of his *Selecta: Expository Writing* [8].

## Mathematics and Writing

How is mathematics connected with writing? In *On Writing Well* William Zinsser [13, p. 53] says:

> All writing is ultimately a question of solving a problem. It may be a problem of where to obtain the facts, or how to organize the material. It may be a problem of approach or attitude, tone or style. Whatever it is, it has to be confronted and solved.

Solving problems is something mathematicians do; they also value brevity and clarity, as writers do. Wasn't it Pascal's training in mathematics that made him, at the end of a long letter, offer his apology for not having had time to write a short one? Gauss remarked that writing concisely takes far more time than writing at length, but he was never satisfied until he had

W.M. Priestley, 1972

said as much as possible in a few words. Likewise, Bertrand Russell stated that he wished to say everything in the smallest number of words in which it could be said clearly.

Russell, who won the Nobel Prize for literature in 1950, made a striking observation [12, p. 101] about the attitude fostered by the discipline of mathematics:

> It is one of the rarest gifts to be able to hold a view with conviction and detachment at the same time. Philosophers and scientists more than other men strive to train themselves to achieve it, though in the end they are usually no more successful than the layman. Mathematics is admirably suited to foster this kind of attitude. It is by no means accidental that many great philosophers were also mathematicians.

"Holding a view with conviction and detachment at the same time" is a carefully worded phrase that probably derives from Russell's study of mathematical logic that made him abandon Platonism. "Telling the truth" is what one would really like to say, but, as von Neumann remarked (because of his own studies in logic?), truth is much too complicated to allow anything but approximations.

Halmos studied logic, yet is an "unreconstructed die-hard Platonist" [7, p. 251] regarding truth in mathematics. Regarding truth in nonmathematical writings, one supposes he would agree that good approximations must often suffice. In any case, Halmos has given us many good ones — usually with conviction and often with detachment. Some are sharp.

# Writing About Writing

It takes courage to meet Hemingway's requirement of a true writer: to try something that has never been done or that others have tried and failed. Halmos's undertaking of a comprehensive paper on the devilish topic of writing mathematics required considerable courage. No one, so far as I know, had dared try it before. Yet Halmos, in "How to Write Mathematics" [4], led us unpretentiously through this morass, casting light in every direction, until we emerged more conscious and more thoughtful about what we said and what we wrote.

Incidentally, Halmos here [4, p. 125] identified the major problem in calculus reform:

> Calculus books are bad because there is no such subject as calculus; it is not a subject because it is many subjects.... Any one of these is hard to write a good book on; the mixture is impossible.

Producing a good calculus book may be the ultimate challenge for a true writer.

Quoting an excerpt out of context cannot illustrate Halmos's style very well. Flashes of wit and wisdom gently radiate throughout his writings, but they are always subordinate to a natural grace that eases the reader's way from start to finish. "Let the scintillations of your wit be as the coruscations of summer lightning," goes the jovial old maxim, "lambent but innocuous." It is the work as a whole that should be offered — not a series of epigrams or witticisms to be read in fits and starts. The holistic style of exposition may have come naturally to Halmos, but it is also consciously cultivated, as indicated in an interview [1] done in 1981:

> I strongly believe that the secret of mathematical exposition, be it just a single lecture, be it a whole course, be it a book, or be it a paper, is not the beautifully written sentence, or even the well-thought-out paragraph, but the architecture of the whole thing. You must have in mind what the lecture or the whole course is going to be. You should get across *one* thing. Determine that thing and then design the whole approach to get at it.

I admire none of Halmos's essays more than "How to Write Mathematics". As pleasing today as when it appeared in 1970, it is a precursor of the current high interest in writing throughout the curriculum. "Choose someone ... whose writing can touch you and teach you," advised Halmos [4, p. 151], "and adapt and modify his style to fit your personality and your subject..." Good reading goes hand in hand with good writing.

No one can talk about writing in mathematics without taking this paper into account and risking his contribution's being labelled as nothing more than a footnote to Halmos. This paper, in other words, has become a classic.

## Writing About Talking

While spending the fall semester of 1973 at Indiana University I made a trivial contribution to Halmos's writing of a short piece entitled "How to Talk Mathematics". I learned a nontrivial lesson in return.

Along with other newcomers I was sent an early draft of this article by Paul, who asked us to criticize it unmercifully. If I was surprised that an eminent mathematician and gifted writer would ask for comments from small fry, I was even more surprised to find that the draft was disorganized and, in one place, unclear.

The only thing I could contribute — aside from pointing out the fuzzy passage — was a remark about how embarrassing it was when a speaker does not to know how to start. (Halmos, characteristically, was preoccupied with condemning those who do not know how to stop.) I had just been to a mathematics meeting where, after a flowery introduction, the principal speaker arrived at the microphone and began by saying, "Uhhhh, Uhhh,..., I guess I'll begin..."

I wrote back to Paul to suggest that embarrassment might be avoided if a speaker memorized his opening line. Paul's draft had inveighed against memorization because it stifles spontaneity; nevertheless he went out of his way to thank me for my response.

I supposed that would be the last I would hear of it. I left Indiana after spending a semester there on sabbatical leave, and, except for one occasion that I will mention later, my only other personal contact with Paul has been through letters and through brief, chance meetings at conventions.

The following spring when the finished paper appeared in the *Notices,* it bore little resemblance to what I had seen. The published work was a deft reorganization and condensation of the loose ends of the draft. My suggestion, for example, was encapsulated in 14 words [5, p. 157]:

> One good trick to overcome initial stage-fright is to memorize one sentence, the opener.

Like Halmos's other publications, this one seemed so unlabored that it appeared to have written itself. Having had an insider's look at its inception, I knew how much labor must have gone into giving that appearance. It is a good lesson to learn that a gifted writer's first draft may be no more distinguished than ours. The difference between the gifted and us is probably due less to their superior imagination than to their unwillingness (inability?) to give up on the problems they encounter. Halmos's final bit of advice in *I Want To Be a Mathematician* [9] is "you must never give up."

Eventually, a reprint of "How to Talk Mathematics" arrived in my mail. Beside the 14 words above Paul had written my initials, followed by an exclamation mark.

## Mathology: A Creative Art

"Artist" may have come to mind as a solution to the problem posed at the beginning of this article. The painter Paul Klee said that art is "exactitude winged by intuition", a phrase that applies equally well to good mathematics and, according to Zinsser [14, p. 55], to good writing.

In 1968 Halmos published an article entitled "Mathematics as a Creative Art" [3] that was based upon a speech addressed to nonmathematicians ("laymen"). In it he coined the terms "mathophysics" and "mathology" to help explain the difference between the layman's conception of mathematics and the pure mathematician's view of his calling. The article culminates with analogies between mathematics and music, mathematics and literature, and — the closest analogy of this type, in Halmos's view — mathematics and painting. His thesis: Mathology is, in the purest sense, an art.

In 1981 Halmos reiterated this thesis in a paper [6] with the provocative title "Applied Mathematics is Bad Mathematics." Here, he flirts dangerously with — yet never embraces — argumentativeness. The continual flirtation is infuriating, but thereby engrossing. Here is how the article begins:

> It isn't really (applied mathematics, that is, isn't really bad mathematics), but it's different.
>
> Does that sound as if I had set out to capture your attention, and, having succeeded, decided forthwith to back down and become conciliatory? Nothing of the sort! The "conciliatory" sentence is controversial...

Who can stop reading here?

Halmos [1] sees pure mathematics as "one great glorious thing" — a seamless web of interconnections to which mathematicians are drawn.

> Mathematics is ... a creative art and so is the exposition of mathematics. [I am proud to be a teacher, but] teaching is an ephemeral subject. It is like playing the violin. The piece is

over, and it's gone. The student is taught, and the teaching is
gone. The student remains for a while, but after a while he too
is gone. But writing is permanent. The book, the paper, the
symbols on the sheets of papyrus are always there, and that
creation is also the creation of the rounded whole.

The suggestion of an analogy between mathematics and poetry did not
entirely please Halmos, however. In 1989 I was working on a paper making
such a suggestion and sent him a draft, hoping for a reaction. I got one, and
rewrote a bit as a result. Despite all his interest in serious writing, Paul
had never gotten into serious poetry: "I find its non-explicitness arcane,
obscure, AARRRGGGGHHHHH." In the same letter, however, he made
the following remark, which I take the liberty of quoting here because it is
a thoughtful footnote to [4]:

> If the use of [connotation] is poetry, then I insist on being a poet
> when I write and I appreciate poetry when I read. A cleverly
> chosen word that means other things than the one it explicitly
> says — that suggests a whole aura, an ambience, an atmosphere
> — that puts the reader in the right frame of mind to appreciate
> and understand the denotation — that's a good thing. That's
> style, that's poetry if you like, that's efficient communication.

# Memorable(?) Word Battles

Lost causes are strange attractors. Why take up a cause whose defeat is
certain? Sometimes you can't help yourself. Sometimes it is a noble thing
to do. Sometimes both.

Those who love language must defend it, even if they run the risk of being
labelled as pedants, ignored, and forgotten. I remember (does anyone else?)
some years ago when H.S.M. Coxeter gently explained why *dilatation* was
a proper word, whereas *dilation* was not legitimately conceived, springing
from false analogy with *calculation.* Today practically everyone, even Halmos, speaks of dilations. The nickname has become legitimate. Coxeter's
battle has been lost and will soon be forgotten.

Halmos has been involved in such losing battles. The word *eigenvalue* is
an abomination made by lumping together an adjective from one language
with a noun from another. Halmos fought against the word, using instead
"proper value", but in 1967 he finally acceded [2, p. x]:

> For many years I have battled for proper values, and against
> the one and a half times translated German–English hybrid that
> is often used to refer to them. I have now become convinced that
> the war is over, and eigenvalues have won it...

Of course, all of us had long ago acceded to the abomination of *television,* not to mention *automobiles.* (Does anybody remember a battle over these Greek–Latin hybrids? Was there a battle?)

Halmos says somewhere that every writer should try to prove at least one thing. Here is my contribution to the mathematical theory of linguistics.

**THEOREM.** Halmos will lose the battle about *hopefully.*

**PROOF.** Let us first recall what the battle is about by considering the following sentence:

> Hopefully, Halmos is writing a calculus book.

Does this mean that we are hopeful that Halmos is writing a calculus book? Most young readers would think so. Most older readers would interpret the sentence differently, viz., that Halmos is full of hope as he writes a calculus book. Along with a waning number of others, Halmos still insists that *hopefully* may properly have only the latter interpretation, the newer usage being "illogical and ugly" [11, p. 107].

However, we shall soon reach the point where it will be distracting for a reader to meet the word *hopefully* used in its waning sense. Hopefully, we will survive. (Did the preceding sentence distract you?) No good writer willingly distracts his reader, as Halmos says [11, p. 106]. Soon, therefore, good writers will either never use *hopefully* at all, or use it only in its newer sense.  ■

On a more serious note, Halmos has been sympathetic to the feelings behind the movement toward gender-free language, but he worries about the deleterious effect of a quick fix. One cannot put up with old textbooks that speak of the problem of "the milkmaid and the cow", but "the milkperson and the milkanimal" is not better.

It is the "he or she" and "his/her" controversy that is most urgent nowadays, and Halmos [9, p. 372] holds his view with conviction.

> [T]here is nothing sexist about the correct use of the classical English neuter pronoun. It's a mere historical accident that it has the same form as the masculine pronoun, and it has no implications whatever about sex. The use of the ugly "he or she" would call attention to sexism where none was intended and none would have been perceived (except by pathologically sensitive and grammatically uneducated readers)...

Halmos adopts an extraordinary tactic in this battle. In the preface to his *magnum opus* [9, p. 1], displayed where no one can miss it in the center of the page, he quotes a note on pronouns to much the same effect, but written with the authority of a female grammarian who is also a feminist.

Who will join this battle? Will the battle be remembered?

# Automathography

In *I Want To Be a Mathematician: An Automathography* we encounter a unique work, a 400-page book about a life in mathematics that intentionally plays down other aspects of that life:

> Sure, I had parents (two), and wives (two, one at a time, the present one for forty years), and cats (eight, two at a time, the present two for three years).... I like Haydn, long walks, Nero Wolfe, and dark beer, and for a few years I tried TM. All that is true, but it's none of your business — that's not what this book is about.

The lead sentence is "I like words more than numbers, and I always did." Halmos intends this to imply that he likes the conceptual more than the computational in mathematics, and he likes to understand and clarify mathematics more than to discover it [9, pp. 3–4].

To write this book Halmos had to render images about people, places, and feelings that his previous essays and articles rarely dealt with. Here is his evocation of the atmosphere of a special place [9, p. 84]:

> The center of the life of all the mathematicians in Princeton, both University and Institute, was Fine Hall (the old Fine Hall), which is still my Platonic ideal of a mathematics building. Dark corridors, leaded windows, heavy furniture, worn carpets; the common room always open, always in use; the library up to date, complete, run with an iron hand by Bunny Shields, tiny, white haired, probably born looking as if she were in her late 50's, severe, but always helpful.

The range of Halmos's prose has grown, and the book is not only good, but needed and welcome. It is a new kind of autobiography that took courage to write. To help you find out what is in it, Halmos has devised a radical index. If index writing has not bloomed into an art form in *I Want To Be a Mathematician,* it has at least taken a quantum leap forward. From "academic titles, call me mister" to "Zygmund, A., at faculty meetings" this one is actually worth reading.

Under the entry "words" are the listings "Artin yes, Courant no, Lefschetz no, ..., Zariski yes" — each followed by an appropriate page number. Under "book reviewers" are the listings "with a conscience" and "without a conscience". Under the entry "Uruguay" are 29 listings. (Halmos kept his readers in Uruguay too long, I thought.) When you finish the book a good test of comprehension would be to see how many of the items in the index you can recall without referring to the text. It is clear that an imaginative index can be a tool of learning.

*I Want To Be a Mathematician* is uneven, but that could hardly be prevented, given the diversity of subjects that Halmos wishes to treat in a book whose ambitious goal is to tell everything there is to know about a life in mathematics as student, scholar, and senior. He finally [9, p. 400] says this:

> What does it take to be [a mathematician]? I think I know the answer: you have to be born right, you must continually strive to become perfect, you must love mathematics more than anything else, you must work at it hard and without stop, and you must never give up.

While it is evident from the book's title that Halmos is not claiming to have met these criteria, he does not fall far short. On a walk in September, 1976, Paul had the misfortune of being knocked down and hurt by a California motorcyclist. Nevertheless, he kept the commitment he had made to me to speak at Sewanee soon thereafter. When he arrived at the Nashville airport after the long trip, his face was ashen and he was in obvious pain, leaning heavily on a cane. However, he rested enough to put up a brave front, gave fine talks, and charmed everybody. Only Paul would take a snapshot of his audience before beginning his talk.

His leg was continually hurting him, though, and he couldn't keep up the front forever with me, his ever-present escort. At one point it became obvious that I was trying to avoid asking why he had put himself through such torture to journey from California into the backwoods of Tennessee to talk about mathematics. In answer to my unspoken question Paul said softly, "I have never missed an engagement." As it turned out, had he missed this particular engagement, two textbooks would probably never have been published.

The world's only mathography was followed in 1987 by a supplement called *I Have A Photographic Memory* [10], which contains some 600 vignettes Halmos wrote to accompany his snapshots of mathematicians taken over a period of some 40 years. In his book mirth is never far away. "If your favorite mathematician is not here, please forgive me," Halmos writes in the preface, "and if your picture is not flattering, blame yourself and me in equal proportion."

# Summary

Forthright and resolute in his convictions, Paul Halmos is nevertheless disarming, engaging, concise, and — above all — clear in his informal, sometimes conversational style of writing. This directness and lightness of touch, coupled with his sparing but effective use of a scintillating wit, produces the characteristic tone (eigentone?) that endears him to many.

Halmos is unique among authors of popular works on mathematics in his central focus upon mathematics as a profession. To the layman he writes about what a mathematician really does; to the student he writes from the heart about what it means to be a complete mathematician. For him — as is evident in innovative works ranging from essay to autobiography — love of mathematics involves love of language.

Is Paul Halmos a true writer? Let the true reader decide.

## REFERENCES

1. D.J. Albers and P. Halmos, Maverick mathologist, *Two-Year College Math. J.* **13** (1982), 226–242.

2. Paul R. Halmos, *A Hilbert Space Problem Book,* Van Nostrand, Princeton, 1967.

3. Paul R. Halmos, Mathematics as a creative art, *Am. Sci.* **56** (1968), 283–293.

4. Paul R. Halmos, How to write mathematics, *Enseign. Math.* **16** (1970), 123–152.

5. Paul R. Halmos, How to talk mathematics, *Notices Am. Math. Soc.* **21** (1974), 155–158.

6. Paul R. Halmos, Applied mathematics is bad mathematics, *Mathematics Tomorrow,* Springer-Verlag, New York, 1981, pp. 9–20.

7. Paul R. Halmos, The thrills of abstraction, *Two-Year College Math. J.* **13** (1982), 243–251.

8. Paul R. Halmos, *Selecta: Expository Articles,* Springer-Verlag, New York, 1983.

9. Paul R. Halmos, *I Want To Be A Mathematician: An Automathography,* Springer-Verlag, New York, 1985.

10. Paul R. Halmos, *I Have a Photographic Memory,* American Mathematical Society, Providence, 1987.

11. Donald E. Knuth, Tracy Larrabee, and Paul M. Roberts, *Mathematical Writing,* MAA Notes #14, Mathematical Association of America, Washington, D.C., 1989.

12. Bertrand Russell, *Wisdom of the West,* Rathbone Books Limited, London, 1959.

13. William Zinsser, *On Writing Well,* Second edition, Harper and Row, New York, 1980.

14. William Zinsser, *Writing to Learn,* Harper and Row, New York, 1988.

*Department of Mathematics*
*The University of the South*
*SPO Box 1217*
*Sewanee, TN 37375*

# Bibliography of Paul Halmos

## Articles

1. Note on almost-universal forms, *Bull. Am. Math. Soc.* **44** (1938), 141–144.

2. Invariants of certain stochastic transformations: the mathematical theory of gambling systems, *Duke Math. J.* **5** (1939), 461–478.

3. On a necessary condition for the strong law of large numbers, *Ann. Math.* **40** (1939), 800–804.

4. Statistics, set functions, and spectra, *Mat. Sbornik* **9** (1941), 241–248.

5. The decomposition of measures, *Duke Math. J.* **8** (1941), 386–392.

6. The decomposition of measures, II [with Warren Ambrose and Shizuo Kakutani], *Duke Math. J.* **9** (1942), 43–47.

7. Square roots of measure preserving transformations, *Am. J. Math.* **64** (1942), 153–166.

8. On monothetic groups [with H. Samelson], *Proc. Natl. Acad. Sci.* **28** (1942), 254–257.

9. Operator methods in classical mechanics, II [with John von Neumann], *Ann. Math.* **43** (1942), 332–350.

10. On automorphisms of compact groups, *Bull. Am. Math. Soc.* **49** (1943), 619–624.

11. Approximation theories for measure preserving transformations, *Trans. Am. Math. Soc.* **55** (1944), 1–18.

12. Random alms, *Ann. Math. Stat.* **15** (1944), 182–189.

13. The foundations of probability, *Am. Math. Monthly* **51** (1944), 493–510.

14. In general a measure preserving transformation is mixing, *Ann. Math.* **45** (1944), 786–792.

15. Comment on the real line, *Bull. Am. Math. Soc.* **50** (1944), 877–878.

16. The theory of unbiased estimation, *Ann. Math. Stat.* **17** (1946), 34–43.

17. An ergodic theorem, *Proc. Natl. Acad. Sci.* **32** (1946), 156–161.

18. Functions of integrable functions, *J. Indian Math. Soc.* **11** (1947), 81–84.

19. On the set of values of a finite measure, *Bull. Am. Math. Soc.* **53** (1947), 138–141.

20. Invariant measures, *Ann. Math.* **48** (1947), 735–754.

21. The range of a vector measure, *Bull. Am. Math. Soc.* **54** (1948), 416–421.

22. On a theorem of Dieudonné, *Proc. Natl. Acad. Sci.* **35** (1949), 38–42.

23. Application of the Radon–Nikodym theorem to the theory of sufficient statistics [with L.J. Savage], *Ann. Math. Stat.* **20** (1949), 225–241.

24. A nonhomogeneous ergodic theorem, *Trans. Am. Math. Soc.* **66** (1949), 284–288.

25. Measurable transformations, *Bull. Am. Math. Soc.* **55** (1949), 1015–1034.

26. Normal dilations and extensions of operators, *Summa Brasiliensis Math.* **2** (1950), 125–134.

27. Commutativity and spectral properties of normal operators, *Acta Sci. Math. (Szeged)* **12** (1950), 153–156.

28. The marriage problem [with Herbert E. Vaughan], *Am. J. Math.* **72** (1950), 214–215.

29. Measure theory, *Proc. Int. Congress Math.*, Vol. 2, 1950, p. 114.

30. Algunos problems actuales sobre operadores en espacios de Hilbert, *UNESCO Symposium*, Punta del Este, Montevideo, 1951, pp. 9–14.

31. Spectra and spectral manifolds, *Ann. Soc. Polonaise Math.* **25** (1952), 43–49.

32. Commutators of operators, *Am. J. Math.* **74** (1952), 237–240.

33. Square roots of operators [with Günter Lumer and Juan J. Schäffer], *Proc. Am. Math. Soc.* **4** (1953), 142–149.

34. Commutators of operators, II, *Am. J. Math.* **76** (1954), 191–198.

35. Polyadic Boolean algebras, *Proc. Natl. Acad. Sci.* **40** (1954), 296–301.

36. Polyadic Boolean algebras, *Proc. Int. Math. Congress* (Amsterdam), Vol. 2, 1954, pp. 402–403.

37. Square roots of operators, II [with Günter Lumer], *Proc. Am. Math. Soc.* **5** (1954), 589–595.

38. Algebraic logic, I. Monadic Boolean algebras, *Compos. Math.* **12** (1955), 217–249.

39. Predicates, terms, operations, and equality in polyadic Boolean algebras, *Proc. Natl. Acad. Sci.* **42** (1956), 130–136.

40. Algebraic logic (II). Homogeneous locally finite polyadic Boolean algebras of infinite degree, *Fund. Math.* **43** (1956), 255–325.

41. The basic concepts of algebraic logic, *Am. Math. Monthly* **63** (1956), 363–387.

42. Algebraic logic, III. Predicates, terms, and operations in polyadic Boolean algebras, *Trans. Am. Math. Soc.* **83** (1956), 430–470.

43. Algebraic logic, IV. Equality in polyadic algebras, *Trans. Am. Math. Soc.* **86** (1957), 1–27.

44. Nicolas Bourbaki, *Sci. Am.* **196** (1957), 88–99.

45. Innovation in mathematics, *Sci. Am.* **199** (1958), 66–73.

46. Products of symmetries [with Shizuo Kakutani], *Bull. Am. Math. Soc.* **64** (1958), 77–78.

47. Von Neumann on measure and ergodic theory, *Bull. Am. Math. Soc.* **64** (1958), 86–94.

48. Free monadic algebras, *Proc. Am. Math. Soc.* **10** (1959), 219–227.

49. The representation of monadic Boolean algebras, *Duke Math. J.* **26** (1959), 447–454.

50. Recent progress in ergodic theory, *Bull. Am. Math. Soc.* **67** (1961), 70–80.

51. Injective and projective Boolean algebras, *Proceedings of Symposia in Pure Mathematics,* Vol. 2, *Lattice Theory* (1961), pp. 114–122.

52. Shifts on Hilbert spaces, *J. Reine Angew. Math.* **208** (1961), 102–112.

53. What does the spectral theorem say? *Am. Math. Monthly* **70** (1963), 241–247.

54. A glimpse into Hilbert space, *Lectures on Modern Mathematics,* Wiley, New York, Vol. 1, 1963, pp. 1–22.

55. Partial isometries [with J.E. McLaughlin], *Pacific J. Math.* **13** (1963), 585–596.

56. Algebraic properties of Toeplitz operators [with Arlen Brown], *J. Reine Angew. Math.* **213** (1963), 89–102.

57. Numerical ranges and normal dilations, *Acta Sci. Math.* (*Szeged*) **25** (1964), 1–5.

58. On Foguel's answer to Nagy's question, *Proc. Am. Math. Soc.* **15** (1964), 791–793.

59. Cesàro operators [with Arlen Brown and A.L. Shields], *Acta Sci. Math.* (*Szeged*) **26** (1965), 125–137.

60. Commutators of operators on Hilbert space [with Arlen Brown and Carl Pearcy], *Can. J. Math.* **17** (1965), 695–708.

61. Invariant subspaces of polynomially compact operators, *Pacific J. Math.* **16** (1966), 433–437.

62. Invariant subspaces, *Abstract Spaces and Approximation,* Birkhäuser, Basel, 1968, pp. 26–30.

63. Irreducible operators, *Michigan Math. J.* **15** (1968), 215–223.

64. Quasitriangular operators, *Acta Sci. Math.* (*Szeged*) **29** (1968), 283–293.

65. Mathematics as a creative art, *Am. Sci.* **56** (1968), 375–389.

66. Permutations of sequences and the Schröder–Bernstein theorem, *Proc. Am. Math. Soc.* **19** (1968), 509–510.

67. Two subspaces, *Trans. Am. Math. Soc.* **144** (1969), 381–389.

68. How to write mathematics, *L'Enseign. Math.* **16** (1970), 123–152.

69. Finite dimensional Hilbert spaces, *Am. Math. Monthly* **77** (1970), 457–464.

70. Powers of partial isometries [with L.J. Wallen], *J. Math. Mech.* **19** (1970), 657–663.

71. Ten problems in Hilbert space, *Bull. Am. Math. Soc.* **76** (1970), 887–933.

72. Capacity in Banach algebras, *Indiana Univ. Math. J.* **20** (1971), 855–863.

73. Reflexive lattices of subspaces, *J. London Math. Soc.* **4** (1971), 257–263.

74. Eigenvectors and adjoints, *Linear Algebra Applic.* **4** (1971), 11–15.

75. Continuous functions of Hermitian operators, *Proc. Am. Math. Soc.* **31** (1972), 130–132.

76. Positive approximants of operators, *Indiana Univ. Math. J.* **21** (1972), 951–960.

77. Products of shifts, *Duke Math. J.* **39** (1972), 779–787.

78. The legend of John von Neumann, *Am. Math. Monthly* **80** (1973), 382–394.

79. Limits of shifts, *Acta Sci. Math. (Szeged)* **34** (1973), 131–139.

80. Spectral approximants of normal operators, *Proc. Edinburgh Math. Soc.* **19** (1974), 51–58.

81. How to talk mathematics, *Notices Am. Math. Soc.* **21** (1974), 155–158.

82. What to publish, *Am. Math. Monthly* **82** (1975), 14–17.

83. The teaching of problem solving, *Am. Math. Monthly* **82** (1975), 466–470.

84. Products of involutions [with W.H. Gustafson and H. Radjavi], *Linear Algebra Applic.* **13** (1976), 157–162.

85. American mathematics from 1940 to the day before yesterday [with J.H. Ewing, W.H. Gustafson, S.H. Moolgavkar, W.H. Wheeler, and W.P. Ziemer], *Am. Math. Monthly* **83** (1976), 503–516.

86. Some unsolved problems of unknown depth about operators on Hilbert space, *Proc. R. Soc. Edinburgh* **76A** (1976), 67–76.

87. Logic from A to G, *Math. Mag.* **50** (1977), 5–11.

88. Bernoulli shifts, *Am. Math. Monthly* **84** (1977), 715–716.

89. Fourier series, *Am. Math. Monthly* **85** (1978), 33–34.

90. Arithmetic progressions [with C. Ryavec], *Am. Math. Monthly* **85** (1978), 95–96.

91. Invariant subspaces, *Am. Math. Monthly* **85** (1978), 182–183.

92. Schauder bases, *Am. Math. Monthly* **85** (1978), 256–257.

93. The Serre conjecture [with W.H. Gustafson and J.M. Zelmanowitz], *Am. Math. Monthly* **85** (1978), 357–359.

94. Integral operators. Hilbert space operators, *Proceedings Long Beach California (1977), Lectures Notes in Mathematics,* Springer-Verlag, Berlin, 1978, Vol. 693, pp. 1–15.

95. Ten years in Hilbert space, *Integral Eq. Operator Theory* **2** (1979), 529–564.

96. Limsups of Lats, *Indiana Univ. Math. J.* **29** (1980), 293–311.

97. Finite-dimensional points of continuity of Lat [with J.B. Conway], *Linear Algebra Applic.* **31** (1980), 93–102.

98. The heart of mathematics, *Am. Math. Monthly* **87** (1980), 519–524.

99. Applied mathematics is bad mathematics, *Mathematics Tomorrow,* Springer-Verlag, New York, 1981, pp. 9–20.

100. Does mathematics have elements? *Math. Intell.* **3** (1981), 147–153; *Bull. Aust. Math. Soc.* **25** (1982), 161–175.

101. Think it gooder, *Math. Intell.* **4** (1982), 20–21.

102. Quadratic interpolation, *J. Operator Theory* **7** (1982), 303–305.

103. The thrills of abstraction, *Two-Year College Math. J.* **13** (1982), 243–251.

104. Asymptotic Toeplitz operators [with José Barría], *Trans. Am. Math. Soc.* **273** (1982), 621–630.

105. The work of F. Riesz, *Colloq. Math. Soc. János Bolyai* **35** (1980), Functions, Series, Operators, Budapest (Hungary), 37–48.

106. BDF or the infinite principal axis theorem, *Notices Am. Math. Soc.* **30** (1983), 387–391.

107. Weakly transitive matrices [with José Barría], *Illinois J. Math.* **28** (1984), 370–378.

108. Subnormal operators and the subdiscrete topology, *International Series of Numerical Mathematics,* Birkhäuser, Basel, 1984, Vol. 65, pp. 49–65.

109. Why is a congress? *Math. Intell.* **9** (1987), 20–27.

110. How to remember Walter Kaufmann-Bühler, *Math. Intell.* **9** (1987), 4–10.

111. Fifty years of linear algebra: a personal reminiscence, *Texas Tech Univ. Math. Ser.* Visiting Scholars Lectures, 1986–1987, **15** (1988), 71–89.

112. Some books of auld lang syne. A century of mathematics in America, Am. Math. Soc., Providence, Part I (1988), pp. 131–174.

113. Vector bases for two commuting matrices [with José Barría], *Linear Multilinear Algebra* **27** (1990), 147–157.

114. Has progress in mathematics slowed down? *Am. Math. Monthly* **97** (1990), 561–588.

115. The calculus turmoil, *FOCUS*, October 1990.

116. Bad products of good matrices, *Linear Multilinear Algebra,* to appear in 1991.

117. Large intersections of large sets, *Am. Math. Monthly,* to appear in 1992.

# Reviews

1. F.J. Murray, An introduction to linear transformations in Hilbert space, *Bull. Am. Math. Soc.* **48** (1942), 204–205.

2. H. Tietze, Gelöste und ungelöste mathematische Probleme aus alter und neuer Zeit, *Bull. Am. Math. Soc.* **57** (1951), 502–503.

3. S. Ríos, Introducción a los métodos de la estadística, *J. Am. Stat. Assoc.* **48** (1953), 154–155.

4. N. Bourbaki, Intégration, *Bull. Am. Math. Soc.* **59** (1953), 249–255.

5. A.A. Fraenkel, Abstract set theory, *Bull. Am. Math. Soc.* **59** (1953), 584–585.

6. M. Picone and T. Viola, Lezioni sulla teoria moderna dell'integrazione, *Bull. Am. Math. Soc.* **59** (1953), 94.

7. H. Wang and R. McNaughton, Les systèmes axiomatiques de la théorie des ensembles, *Bull. Am. Math. Soc.* **60** (1954), 93–94.

8. A.C. Zaanen, Linear analysis, *Bull. Am. Math. Soc.* **60** (1954), 487–488.

9. G. Pólya, Mathematics and plausible reasoning, *Bull. Am. Math. Soc.* **61** (1955), 243–245.

10. H. Levi, Elements of algebra, *Bull. Am. Math. Soc.* **61** (1955), 245–247.

11. W.H. Gottschalk and G.A. Hedlund, Topological dynamics, *Bull. Am. Math. Soc.* **61** (1955), 584–588.

12. N.I. Achieser and I.M. Glasmann, Theorie der linearen Operatoren im Hilbert–Raum, *Bull. Am. Math. Soc.* **61** (1955), 588–589.

13. S. Ríos, Introducción a los métodos de la estadística (Segunda parte), *J. Am. Stat. Assoc.* **50** (1955), 1002.

14. H. Hermes, Einführung in die Verbandstheorie, *Bull. Am. Math. Soc.* **62** (1956), 189–190.

15. A. Tarski, Logic, semantics, metamathematics, *Bull. Am. Math. Soc.* **63** (1957), 155–156.

16. N. Dunford and J.T. Schwartz, Linear operators, Part I: General theory, *Bull. Am. Math. Soc.* **65** (1959), 154–156.

17. K. Jacobs, Neuere Methoden und Ergebnisse der Ergodentheorie, *Bull. Am. Math. Soc.* **68** (1962), 59–60.

18. H. Helson, Lectures on invariant subspaces, *Bull. Am. Math. Soc.* **71** (1965), 490–494.

19. R.D. Taylor, Creative teaching: heritage of R.L. Moore, *Historia Math.* **1** (1974), 188–192.

20. J. Dieudonné, Panorama des mathématiques pures. Le choix Bourbachique, *Bull. Am. Math. Soc.* **1** (1979), 678–681.

21. A.M. Gleason, R.E. Greenwood, and L.M. Kelly, The William Lowell Putnam mathematical competition, problems and solutions: 1938–1964, *Am. Math. Monthly* **88** (1981), 450–451.

22. *Encyclopedic Dictionary of Mathematics,* edited by Shôkichi Iyanaga and Yukiyosi Kawada; translation reviewed by K.O. May. *Math. Intell.* **3** (1981), 138–140.

23. H. Furstenberg, Recurrence in ergodic theory and combinatorial number theory, *Math. Intell.* **4** (1982), 52–54.

24. Serge Lang, The beauty of doing mathematics, *Am. Math. Monthly* **94** (1987), 88–92.

25. Introduction, *Reviews in operator theory 1980–86,* American Mathematical Society, Providence, RI, 1989, pp. vii–viii.

# Books and Notes

1. *Finite Dimensional Vector Spaces,* Princeton University Press, Princeton, 1942. Second edition: Van Nostrand, Princeton, 1958. Reprint: Springer-Verlag, New York, 1974.

2. *Measure Theory,* Van Nostrand, New York, 1950. Reprint: Springer-Verlag, New York, 1974.

3. *Introduction to Hilbert Space and Theory of Spectral Multiplicity,* Chelsea, New York, 1951.

4. *Lectures on Ergodic Theory,* Mathematical Society of Japan, Tokyo, 1956. Reprint: Chelsea, New York.

5. *Entropy in Ergodic Theory* (Mimeographed notes), University of Chicago, 1959.

6. *Naive Set Theory,* Van Nostrand, Princeton, 1960. Reprint: Springer-Verlag, New York, 1974.

7. *Algebraic Logic,* Chelsea, New York, 1962.

8. *Lectures on Boolean Algebras,* Van Nostrand, Princeton, 1963. Reprint: Springer-Verlag, New York, 1974.

9. *A Hilbert Space Problem Book,* Van Nostrand, Princeton, 1967. Reprint: Springer-Verlag, New York, 1974. Second edition: Springer-Verlag, New York, 1982.

10. *Invariant Subspaces 1969* (notes), Seventh Brazilian Mathematical Colloquium, Poços de Caldas, 1969, pp. 1–54.

11. *Bounded Integral Operators on $L^2$ Spaces* [with V.S. Sunder], Springer-Verlag, Berlin, 1978.

12. *I Want To Be a Mathematician,* Springer-Verlag, New York, 1985. Reprint: Mathematical Association of America, 1988.

13. *I Have a Photographic Memory,* American Mathematical Society, Providence, RI, 1987.

# The Way It Was

*A good storyteller is a person who has a good
memory and hopes other people haven't.*
—Irvin S. Cobb

*It's a poor sort of memory that only works backwards.*
—Lewis Carroll

# Paul Halmos:
# Through the Years

1929, Basel

1938

1941

Ginger and Paul, 1950,
Cambridge

1952, Gainesville

1958, Princeton

Ginger and Paul, 1960

1969, Eugene

Ginger and Paul, 1973

1970, Fort Worth

1980, St. Andrews

1988, Auburn

# Paul Halmos as Graduate Student

## Warren Ambrose

I first encountered Paul Halmos when he and I were both undergraduate students at the University of Illinois, both majoring in philosophy. We were together in several undergraduate philosophy classes in the years 1933–34 and 1934–35. He certainly impressed me and other students in these classes by his remarks in class and out, and by what I considered to be his "elegant European manner." I was born and raised in Illinois, and he was the first European I had ever seen. This "European behavior" was for me excessively polite. He once, at a later date, characterized the difference in our codes of politeness by summarizing our ways of asking directions from a stranger as follows. He would formulate the request as: "Pardon me, Sir, but I wonder if you would be so kind as to inform me where I might find...," where I would say: "Hey, where's..."? He also dressed more elegantly than we ordinary "Illini." It was not at all unusual for him to appear in class in pressed white flannel pants and a jacket of similar elegance. But even more striking than his clothes was his intelligence, which he did not show off but which stood out very naturally and appropriately in philosophy classes. At that time both he and I had the intention of attending graduate school in philosophy. He graduated in 1934 and attended graduate school in philosophy at Illinois in the year 1934–35. For some reason that I never understood he was not well accepted in the philosophy department that year. It is inexplicable to me because he had the knowledge, the talent, and the will to do well. My own tendency is to blame the philosophy department for his failure and I think that the subsequent history of Halmos and of that department bear me out.

During the year 1934–35 I changed my allegiance from philosophy to mathematics and I had decided by the Fall of 1935 to enter graduate school in math in 1936, which I did. Paul also changed to mathematics in the Fall of 1936. By that time our conversations had become even more frequent and were now concerned largely with which parts of mathematics were more interesting for us. I knew I wanted to be an analyst and Paul had first thought he wanted to be an algebraist. He was searching for "clean" and "not messy" mathematics. He was always a demon for clarity. He found analysis with its $\varepsilon, \delta$ arguments to be "messy." We were both learning the Lebesgue integral at that time and we had many arguments about its clarity. Eventually Paul decided that the Lebesgue integral was clean and neat and had all the other qualifications that he demanded of his mathematics. So life for me proceeded in a more serene fashion, and Paul

Ambrose, Halmos, and Sherman, 1941

was converted to analysis. The arguments that I had tried to use to convert him to analysis became much clearer in his version than they had ever been in mine. It was at this time that I began to appreciate Paul's gift for mathematical exposition.

In Paul Halmos's autobiography there are a few negative remarks about the political involvement of mathematicians, with which I am not in agreement, so I would like to mention that here. We and a group of friends often met Friday nights for political and other discussions. The political opinions expressed there covered a large spectrum. I don't understand Paul's present distaste for that aspect of our lives.

Another topic that I wish to mention is the appreciation that Paul and I both feel to Joe Doob for what he has taught us, mainly in mathematics but also outside mathematics. Of all the people who have influenced our mathematical and other intellectual development, Joe stands out. Before Doob, neither of us had ever encountered such a forceful thinker. All our previous opinions about the world had to be modified by the serious criticism he made of them.

Paul Halmos received his Ph.D. degree in 1938 and I received mine in 1939, both as students of Joe Doob, in the field of measure-theoretic probability. Paul Halmos did not always receive the appreciation he receives today. When he was a graduate student at Illinois almost all graduate students there taught one or two elementary courses per semester, for which they received tuition and a stipend. This was standard treatment and the ranks of these so-called teaching assistants were filled with people who had little talent or love for mathematics. But, for some reason nobody understood, Halmos, who had both love and talent for mathematics, was not given such classes to teach, with the exception of one course for one semester. His knowledge of mathematics and his obvious teaching abilities were far above those of most of the teaching assistants. I don't understand why the rest of us did not protest this injustice, but, as far as I can remember, nobody said a word or raised a voice.

Joe Doob, 1955

Paul spent the academic year 1938–39 as an instructor at Illinois while I spent that academic year teaching at the University of Alabama and finishing my Ph.D. thesis, receiving my degree in 1939. In the spring of 1939 Paul had received a teaching position for the year 1939–40 at Reed College in Oregon and I had accepted to teach a second year at the University of Alabama. But Paul and I had also applied for fellowships at the Institute of Advanced Study in Princeton for the year 1939–40 without any serious hope of receiving one. To both our surprise I received, in June 1939, an offer of a fellowship at that Institute for 1939–40. This was for me an offer of a fellowship to heaven. When Paul learned of my offer, after a few days of what must have been the world's most serious thought (and an equal amount of pain), he wrote to Reed College to resign from his position there and wrote to the Institute for Advanced Study asking permission to be in residence there without stipend for the year 1939–40 (an offer that was of course accepted).

Being at the Institute was really the closest thing to heaven for us. There were six permanent professors at the Institute, among the best mathematicians in the world (Albert Einstein, who could be considered either a mathematician or physicist, was one of them) who gave courses when they felt like it on whatever subjects they wished, and a few "permanent members," (Kurt Godel was one of them) who also gave such lectures. One could see, daily, Godel and Einstein walking together on the sidewalks of Princeton between their homes and the Institute. Then there were about ten of us on fellowships, each working on whatever he wished. Besides these Institute members there were the professors and students of mathematics at Princeton University—also a distinguished group. And with all these people there was great interaction, not in any official sense but very freely. So whatever mathematics one was interested in, there were plenty of mathematicians to consult with, or argue with. I have never since encountered such an actively interacting group of mathematicians. Wherever one went when he left Princeton, he would never find such a stimulating group of mathematical associates. I would like to clarify here something about my invitation to the Institute, as told to me by Oswald Veblen at that time, in the Fall of 1939. He called me into his office one day to tell me that the Institute's choice of me, rather than Paul, for the invitation was a random choice; they had no information on which to choose one of us over the other and that I just happened to be the lucky choice. But this affair had a happy ending when they found money to give Halmos a stipend for the Spring semester of that year, and to make him von Neumann's assistant. This was wonderful for Paul because he and I both idolized von Neumann and the opportunity to be so near to him and to work with him on the notes for his course was about as nice a gift as one could hope for. It was also very good for von Neumann because his previous assistant neither cared for nor understood the material in that course. This seemed to have been the first time in Paul's career when he received what he deserved and I think it must have been one of the happiest times in his life.

A final remark about Paul. In all the places he has ever been, he and his wife Ginger have made unusual efforts to be kind and helpful to the students. Also he probably has many more friends in the mathematical world than most. Mathematics has become something of a religion for him, a benevolent religion which has generated many books and articles that are interesting to nonmathematicians as well as mathematicians.

*17 rue Emile Dubois*
*75014 Paris, FRANCE*

# Early Days at Mathematical Reviews

## R.P. Boas

*Inflation*

The growth of mathematics is producing discontent:
The inflation rate per annum is pushing 10 per cent.
Faced with so much information, it's not easy to succeed
In locating any theorem that you appear to need.
Our plethora of indices can leave you in the lurch,
For it takes less time to prove it than to go and make a search.
I can offer a solution, but it's totally upsetting:
We need to introduce a way of constantly forgetting
The results that won't be needed for another 20 years,
At the end of which they'll surface with appropriate loud cheers,
While the ones that won't be needed till forever and a day,
Once their authors get their tenure will be firmly thrown away.

Mathematical Reviews (MR) came into being, and developed, through a sequence of accidents. It might never have been started, and certainly would not have been started as soon, if the German government had not attempted, for ideological reasons, to interfere with the editorial policy of the Zentralblatt für Mathematik (Zbl). However, mathematicians, especially in the United States and Great Britain, saw the war coming, and were afraid of losing the Zbl, which they had been accustomed to rely on. I was too young at the time to have known much about the negotiations that went on in 1938–39, but the upshot was that it was decided to start a new reviewing journal in the United States, that the American Mathematical Society would sponsor it, and that Otto Neugebauer, who had founded and edited the Zbl, would edit the new journal from Brown University. I have been told that some prominent American mathematicians worried that a reviewing journal edited in the United States would distract too many American mathematicians from their own research. However, as things turned out, the effect was more positive than negative, at least for me and presumably for many others.

The whole operation had to be created from scratch, since Neugebauer had destroyed all the records of the Zbl except for a card file containing the cumulative author index (which, I have been told, was almost confiscated by the American customs service). The Society had to recruit reviewers, and procure the journals for them to review. Getting hold of the journals

Ralph Boas, 1955

was sometimes difficult, and after the war had started, impossible in many cases. Curiously enough, there were a few journals that could not be obtained at all, not even if MR offered to pay for them.

Even the name of MR was controversial, because a significant number of mathematicians understood "review" in the sense of a critical review, as in "book review," whereas MR's policy has always been *not* to be critical, except by implication. There have, however, been exceptions; see, for example, MR 9, 329. The card that potential reviewers were asked to fill out had an entry, "I am able to review papers in the following languages." Some respondents thought that they were being asked what languages they could write in (originally, reviewers were allowed to write in English, French, German, or Italian, the languages of the International Congress). Nobody foresaw that English was going to become the international language of mathematics. Some reviewers even had the idea that a review ought to be written in the language of the paper under review, although it seemed obvious to most thoughtful people that the most useful review might well be written in a language different from that of the paper. In any case, although the MR office prided itself on being able to read correspondence in many languages, it wrote letters only in English. One chauvinistic mathematician who complained that Neugebauer had written in English instead of Neugebauer's own "mother tongue," was told that it was not a question of Neugebauer's mother's language, but of his secretary's. This was more of a rebuke than it superficially appears, because the mathematician who wanted Neugebauer to write in German was in fact Dutch.

A fallacy that probably still has some support was that the length of a review ought to be directly proportional to the importance of the paper. Neugebauer's principle (never stated in print) was that the most important papers needed only very short reviews (saying, in effect, "read this paper"), whereas a bad paper needed to be reviewed in enough detail so that nobody but the reviewer would have to read it. Although MR tried to review all serious mathematics, whether or not it appeared in mathematical journals, it tried (and I hope still does try) to review articles only for their mathematical content. MR got out of reviewing a paper on "The Rainbow series and the logarithm of aleph-null" on the grounds that it was a privately printed pamphlet, not a real publication.

MR did try quite hard to keep reviews short; in the early days the editors had more time to edit them than they had later. I was still learning the MR principles when Wald's review of the von Neumann–Morgenstern book on game theory came in; it was 12 pages long, which was far longer than any previous review had been. Feller and I spent half a day cutting it down by eliminating redundant material. Wald had written a paper on game theory, inspired by the book, months before the review arrived, and when Feller teased him by asking why he had written the paper before the review, Wald replied that he only had to read one chapter in order to write the paper, but he had to read the whole book in order to write the review.

Feller, 1958

I suppose that the original expectation was that a volume of MR would be about the size of a year's issues of the Zbl, and for the first six years, this was what happened. Nobody anticipated the exponential increase that began about 1948 and is still going on. Naturally MR expected a bulge after the end of the war, but nobody expected that the bulge would never flatten out again. At the very beginning, the editorial side of MR was handled by Neugebauer (who became less active as time went on), a half-time editor (Feller), and a secretary. Neugebauer, whose professional interest was in the history of mathematics, had an assistant, Abraham Sachs, who was originally an Assyriologist. His wife, Janet Sachs, was the first, and for many years the only, secretary for MR; it was she who provided continuity as the editorial staff fluctuated, and I firmly believe that it was only she who kept MR from an untimely collapse.

I was a reviewer from the beginning; since I was young, enthusiastic, and had a light teaching schedule, I wrote reviews very quickly. It should have been obvious that this made me get more and more papers to review, but that aspect never occurred to me. In 1942–43, I was working for the Navy (teaching aviation cadets), and discovered that our commanding officer didn't want any of his staff to publish anything (I never found out why). I was reluctant to give up reviewing, whether from a sense of obligation to Mathematics or because I found many papers that I reviewed to be useful (I cannot remember which), so I told MR that although I would be unable to review for the time being, my friend Dr. Pondiczery had similar interests, and would be happy to fill in for me. That is why, if you look, you will find a number of reviews by Pondiczery (who had been created at Princeton during 1937–38; not all of Pondiczery's output belongs to me, however).

It was presumably because of my reviewing that when, in 1945, Feller was leaving MR and Neugebauer was becoming less involved, the American Mathematical Society reluctantly decided that MR needed a full-time editor, and I got the job. I was rather taken aback to find that I was expected to do the copyediting, which had previously been farmed out, but it was not very time-consuming, and the skill has been useful at times. It was quite essential to have a copyeditor prepare reviews for the printer, because reviews arrived in any of four languages, typed on all kinds of typewriters, and even handwritten. Unlike an ordinary mathematical journal, MR obviously cannot insist on any particular format; the difficulty was greater in the early days when there were only about 300 reviewers. If you looked for one who could handle a particular subject in a particular language, you were quite likely not to find anyone, and had to try to cajole somebody into taking on the job.

In 1945, when I began work at MR, the whole editorial operation was handled by one editor and one secretary, who between them did everything except setting the type and mailing the issues; the financial side was handled by the New York office of the American Mathematical Society. I inadvertently became responsible for having the Society's headquarters located in Providence, because the Society wanted MR moved to New York; when I refused to move, the Society came to Providence. (Eventually something similar happened, when a later executive editor insisted on moving MR to Ann Arbor.)

One of the executive editor's jobs was to translate the titles of articles that were not in one of the four official languages. We needed titles for our records, before we sent the papers to reviewers. In really difficult cases, we guessed, and hoped that the reviewer would correct our mistakes. This did not always work. On one occasion I was rebuked by a reviewer for rendering the Dutch word "Wiskunde" as "Mathematics"; he thought it meant "Science," but at least that time I was right to begin with. (The Dutch mathematical vocabulary was created by Simon Stevin around 1600.) On another occasion I had a Chinese journal that was entirely in Chinese. At that time, C.C. Lin was at Brown, and I took the journal to him, in order to find out who the authors of the articles were. His office was in the applied mathematics building; since classified research was being done there, visitors had to sign in at the front desk, and be escorted to the desired office. This procedure struck me as being silly, so this time I signed in as V.M. Molotov (whose name had been appearing in the news), in cyrillic characters. Nobody noticed.

It sometimes was (and I suppose still is) difficult to decide what part of an author's name is to be taken as "the" name for use in an index. It does seem to be desirable for all publications by the same author to be listed together, and conversely articles by different authors should not be conflated. Generally speaking, MR took the last component of a name to be the "official" one, but this is not always the right thing to do. Even if you know that Hungarians, writing in Hungarian, put the family name first, the difficulty is like the problem that arises if you are a European writing to an American and want to specify the next floor above street level. You think "first floor," but then recall that the American says "second floor;" but then you reflect that this American may be aware of the difference between idioms, so maybe you had better say second floor; then you realize that your correspondent may interpret this as *your* second floor, and so on and on. However, there are all sorts of additional complications. One Indian mathematician, who always publishes under his personal name, says, "Why should I have to use the same name as all my cousins?" I know of at least one case where MR inadvertently indexed an author under his grandfather's name. On the other hand, some Indian names are arranged just like standard American names; the only (almost) reliable procedure is to ask authors how they prefer to be listed, and use cross-references to other plausible versions. Generally speaking, it is safe to disregard prefixes like "Dr.," but it is not always clear what is a title and what is not. In the early days, even before I came to MR, there was a paper by an author whose name began with "Sahib Ram"; somebody thought that "Sahib" was a title (like "Mr.") and listed the author under "Ram." It turned out that Sahib Ram was something like "Lord Rama"; a western parallel would be somebody named "Lord–Jesus."

The original routine at MR, as it was in 1945, was that each Monday Janet Sachs gave me a week's accumulation of journals. I assigned papers to reviewers, assisted by card files of reviewers available for specific fields and languages. Somehow or other, everything eventually got reviewed, although sometimes we had to list a paper by title, or use the author's summary as a review. Once I ended up by translating a (short) paper from Russian in order to get a review from the most satisfactory reviewer, who couldn't (or wouldn't) read Russian. The other side of the coin was that I could assign papers to myself if they looked particularly interesting.

Of course, there was a lot to do after the papers were assigned to reviewers. In those days every paper was microfilmed, and every paper in journals that couldn't be torn up (for instance, the ones borrowed from the library) had to be photocopied. Then the reviews, when they arrived, had to be edited and references had to be checked. MR was very serious about verifying references, but many reviewers were less painstaking. For example, it was not uncommon for a reviewer to refer to the authors of a joint paper as "he." Somehow I had acquired a belief that French and German authors write more carefully than American authors do, but I was to find out that this belief was fallacious. I once calculated that, first and last, I had read every review at least nine times by the time the annual index went to press.

Aside from our attempts at accuracy, the MR office was, in the beginning, pretty informal. Officially, there were just the two of us, but Neugebauer came in frequently, as did Feller at the beginning, and, together with Abe Sachs, we all used to go to lunch together. I relied on Neugebauer to dictate letters to his European friends on behalf of MR. He was given to closing a letter "Unsincerely yours," or beginning one with "Dear - - - -, You are the greatest ass I have ever known," or peppering his letters with swear words. Of course, Janet knew to delete inappropriate remarks. Later, when we had acquired a more conventional secretary, we had, at first, to see that she hadn't put the extraneous material in. Neugebauer once received a letter from Hopf, complaining that Neugebauer could swear in three languages whereas he (Hopf) could not even swear in one. I believe Neugebauer and Hopf had served together in the Austrian army, where Neugebauer, at least, had acquired a pungent vocabulary.

Although we were, under Neugebauer's influence, careful about mathematical detail, we tended to be rather casual about things that seemed less important. Our bills were paid through the American Mathematical Society's office, which got quite upset once when we tried to pay the same bill twice. We at MR, however, felt that the supplier was honest enough not to accept a second payment. Relations between MR and the fiscal arm of the Society were never very cordial, very likely because MR ran at a loss except for one early year. At one point MR asked for a supply of red pencils, and we received a gross of pencils that were red on the outside, but had black leads.

Because I was living in Cambridge and coming to work by train, I used to get to the office about ten o'clock. The chairman of the Mathematics Department at Brown saw me crossing the campus at that hour, and complained. Neugebauer couldn't understand why, because he didn't care when I came in as long as the work got done. I didn't understand either, because I usually used the hour-long train ride to read proofs for MR. The chairman was, in fact, known to be a fussbudget. He once reproved an assistant professor in his department for not having his shoes shined; it was amusing to recall this, a quarter-century later, when I was a chairman myself and (on a hot summer day) one of the teaching assistants was going to his class barefoot.

A few years earlier, when I was a student, I had needed to consult an article in a journal that Harvard didn't have. Since Brown had a very extensive mathematics collection, I had gone to Providence, introduced myself to the chairman, and asked permission to use the library; he showed me where it was. After a while I noticed the chairman and Professor Tamarkin at the door; the chairman was whispering, "He says he's from Harvard; do you know him?" Fortunately Tamarkin, who frequently visited Harvard, did know me. I suppose that actually this wasn't as strange as it seemed then; in the 1940's most people were more trusting than we can afford to be today.

Heinz Hopf (second from left), 1968

At the time that I was working for MR, Feller was writing his book on probability. I used to pick out papers that contained interesting probability problems (frequently from papers that MR was not going to review because they weren't sufficiently mathematical, but were in journals that I had to scan anyway) and save them for Feller. A good many of these found places in the book.

MR kept up with the German literature rather well during the war, because Hopf had sent the principal German journals from Switzerland. However, we had not received anything from Italy, Romania, Japan, and some other less productive countries. One Monday morning there were only three papers for me to assign; but then the mailman came in with a stack of Romanian journals from the war years, and after that we had great quantities of the journals that we had missed. Right after the end of the war, John Wehausen came back from Japan with a suitcase of Japanese journals, which he did not want to have torn up. It was going to cost $300 to photocopy them. It indicates the original level of expenses that I didn't dare to have the copying done until I had permission from the Society, which agonized over the problem for some time before letting me spend such an enormous sum.

It was rather surprising that MR had kept up with the principal Russian journals throughout the war. We even heard that at one point the mathematicians in Kharkov could find out what was being done in Moscow only through MR. However, some Soviet journals were never received at all. What was much more serious was that the Russians decided to stop publishing foreign-language editions of journals, without warning MR. Eventually we were able to fill the gaps by borrowing material from one of the few American libraries that had been subscribing to the Russian versions all along. The more serious problem for MR was to find enough people who could review papers in Russian.

At the beginning, it was fascinating to have the world's mathematical literature flowing across my desk, but by the time I left, there was already much too much. It has even been suggested, almost seriously, that we need to have another reviewing journal to review MR.

*3540 NE 147th Street*
*Seattle, WA 98155*

# Reminiscences

## Irving Kaplansky

John Ewing and Fred Gehring stirred me up with their invitation to write this document, and I am grateful to them. Otherwise I might never have put in writing my debts to Paul Halmos, what I learned from him, and the fun I had with him.

This article is undoubtedly riddled with inaccuracies and the first one may be coming right up. It is natural to start at the beginning: when and where did I first meet Paul? I found no help in *I Want to Be a Mathematician*. My memory is that it was in the summer of 1943 and that he was one of a number of mathematicians who taught at Harvard during that summer. He came over to 17 Sumner Road where I lived (it served as the home for several academic generations of young Harvard mathematicians) and I was introduced to him. I don't remember anything in particular beyond the impression that I had just met a smart guy who was going places.

The scene changes to the University of Chicago in the fall of 1946. He came with the first wave of mathematicians who inaugurated the "Stone Age" (I had already been there for a year). Very soon I was taking full advantage of his presence. At about that time I decided that I had to learn about locally compact groups. I was struggling slowly, but he knew every locally compact group — and the Haar measure on it — by its first name.

Irving Kaplansky, 1941

While at Syracuse Paul published a paper entitled "Comment on the real line" (1944 Bulletin of the AMS). The paper fascinated me. He showed (and at first blush it's surprising) that the real line can be retopologized so as to become a compact abelian group. He then pointed out that very little was known about which abelian groups are capable of supporting a compact topology. This became an obsession and I hacked away at it until I almost got it. As I think about it, this was the first of many instances where I followed in his footsteps. He went to Michigan, Hawaii, Indiana; pretty soon I was visiting and enjoying the reunion. At Santa Barbara we missed each other by a hair. He visited for a month; then I visited for a month, arriving right after he left and moving into the same apartment.

We both loved swimming in Lake Michigan during Chicago's sultry summers and spent many an afternoon at Indiana Dunes State Park. I can date this as before 1951, because Ginger was there but no spouse of mine. There he taught me the game of "Stinky Pinky." A stinky pinky is a rhyming pair of words; one gives a paraphrase and the challenge is to recover the stinky pinky. I believe Paul is responsible for the following excellent double stinky pinky. Give a stinky pinky for an inebriated scoundrel. Answers: a drunk skunk or a plastered bastard. I also learned about plongitudes from him and I believe he coined the word. (A plongitude is the inverse of a platitude. Sample: we had so little to say and so much time in which to say it.) I shall mince no words: Paul Halmos is the wittiest person I know.

He was generous with his time. At least he was to me. In 1950 I decided that the time had come for me to acquire a car (and a driver's license). I was inexperienced and apprehensive. He offered me the use of his car to practice driving and he went with me to boost my morale during the negotiations with car dealers. (He did get a fringe benefit in return: shortly after I bought a dandy little convertible, he got virtually the same car.)

The time came for me to teach Lebesgue integration at Chicago. There wasn't much of a problem deciding which textbook to adopt. I had browsed through *Measure Theory* more than once. But I had to teach from it to get the full measure of the man. Everything was there: the right things in the right order, with the right motivation and side remarks, and lots of stimulating exercises. Years later I tried to imitate him — so have others. But only he has produced the real thing.

We share a love of linear algebra. I think it is our conviction that we'll never understand infinite-dimensional operators properly until we have a decent mastery of finite matrices. And we share a philosophy about linear algebra: we think basis-free, we write basis-free, but when the chips are down we close the office door and compute with matrices like fury.

*I Have a Photographic Memory* has amused, delighted, and instructed a small army of readers. But it is more than a coffee table book. It is going to serve as a documentary reference for the generations of mathematicians yet to come.

Hans Samelson and Marshall Stone, 1950

I have to record one bitch somewhere, so here goes. Along with my photograph, the book says that I know all the songs of the 1930's and 1940's. Not true. Admittedly, I know almost all of them. And at this point I am going to record something that might otherwise be lost forever. Paul's favorite is "The Night is Young." When I want to get a benign smile from him all I have to do is strike up the opening measures. (That brings up a piece of research that the two of us have left unfinished for a long, long time. The lyrics of "The Night is Young" contain the rhyme amorous — glamorous. So does the Gershwin song "Swonderful." Which came first?)

We were together at Chicago for 15 years (1946 to 1961). As I write this, nearly 30 more years have gone by. I look back nostalgically at the first 15 of these 45 years. They were awfully good years. You can't go home again, but we have set up new homes close to each other in a pretty nice spot: the Bay Area. And in the fall of 1989 we at last got together again, unfortunately for only a month plus epsilon. In that short time he was a ball of fire at MSRI. He gave three lectures to standing room only. He talked to everybody. Well, nearly everybody (I think that in these phrases I am imitating Paul again). The lobby, the lunch room, the lecture hall, anywhere: if he was there, there was electricity in the air.

Happy 75th birthday, Paul, and let there be lots more birthdays to come.

*Mathematical Sciences Research Institute*
*1000 Centennial Drive*
*Berkeley, CA 94720*

# Writing Mathematics

## J.L. Kelley

I've only known Paul Halmos for 50 years, but of course I knew about him before we met. In those days, before the war (not the Spanish American War) there weren't many graduate students in mathematics, the American Mathematical Society meetings were much smaller affairs than nowadays, and younger mathematicians almost always stayed in the dormitories of the University where the meeting was held. We all gawked at the famous mathematicians attending the meeting (have you ever wondered what Sierpinski looked like, or Caratheodory, or Fubini?), and we gossiped and ate and drank beer together. So I knew that Joe Doob's student Paul Halmos was one of the 90 or so people in the U.S. to receive Ph.D. degrees in mathematics in 1939.

I *think* I first met Paul in a dormitory room at one of the New England women's colleges. There were three or four doubledecker bunks, Halmos was sitting in the top story of one of them, Seymour Sherman and the long, lovely wife he had just married were sitting on the lower deck of another, Henry Scheffe was there, and I don't remember who the six or eight other people in the room were. Of course that was a long time ago and my memory is chancy, but I do remember for sure another Society meeting 15 or so years later. Halmos and I were leaning against the wall in a corridor outside a meeting room, gossiping. He said, "Kelley, do you remember when we used to walk around the halls gawking at the old guys? Well, we're them."

One part of my relationship with Paul was established early, perhaps the first time I met him. He was the authority on how to write mathematics. He told me that he'd seen an abstract of mine in the PNAS labeled "A decomposition of compact continua and related results on fixed sets under continuous mappings," and he averred that it wasn't a good idea to put the whole paper into the title. This seemed pretty obvious to me after he pointed it out, so I finally published the paper as simply "Simple links and fixed sets." This was the first but not the last thing I learned from Paul about writing.

Paul and I became friends, not just acquaintances, in Chicago. We both became assistant professors at the University of Chicago in 1946. We both found apartments south of the Midway, we walked home together most days, we talked about mathematics and mathematicians, and his wife Ginger and my wife Nancy became good friends. Ginger taught my son to sing

Kelley, 1950

*Ginger Halmos is no good,*
*Chop her up for kindling wood.*
*If the kindling will not burn,*
*Put her in the butter churn...*

I remember a series of conversations about giving a lecture. There was a steady stream of distinguished visitors at Chicago, and we were exposed to some of the best and some of the worst stylists in the art of mathematical lecturing. Here are some of the things I remember being told, probably by Paul.

Minimize or completely avoid writing on the blackboard before the lecture begins. Reading with comprehension usually takes place at roughly the same speed as well organized writing, and one simply cannot say "Let's have a few minutes of silence while you all read and absorb what's on the blackboard." There is a corollary to this commandment: Avoid, as you would the plague, transparencies and magic lanterns. A transparency has all of the disadvantages and none of the advantages of the printed page.

Here are some words of wisdom about the level at which to pitch a general lecture at a university at which you wish to get a job. Spend the first 20 minutes in a careful, lucid exposition that everyone can understand. This will show that you are a great teacher. Spend the second 20 minutes at a level that only experts in the field will follow. This will make them (the experts) feel good and get them brownie points if they discuss the lecture with their colleagues. And make the last 20 minutes completely incomprehensible. This will show that you are a bloody genius and so a suitable colleague.

There is a wide variety of styles of communicating mathematics, quite distinct from the mathematical level of the communication. Here is a brief personal review of a few of them.

The most elegant, perhaps the most beautiful, and surely the most difficult style can be called the "See mom, no hands" mode. The lecturer walks back and forth, making an occasional note on the blackboard, talking with limpid ease about mildly unfamiliar mathematical processes. Then

Emil Artin, 1960

he writes a longer note on the board and underlines it, and you realize it's a theorem. Something like horror overcomes you as you then realize that the notes spread around the blackboard comprise a complete topical outline of the lecture. The moment of truth has arrived and you didn't even know you were at a bullfight.

Emil Artin was perhaps the finest practitioner of this school of lecturing that I have seen, and my only possible criticism of the style is that I can't do it.

It is easy to give examples of how *not* to lecture on mathematics. I mean real examples, making use of the possibilities offered by the subject matter, not just things like mumbling, or writing on the blackboard with one hand and erasing the writing simultaneously with the other. I remember a lecture at a meeting of the Society that began "Let $A$ be c.a.d.d. and...," and another lecture that opened with "Let the notation of $[xx]$ be in force." This sort of opening is guaranteed to quench any possible interest that the announced title of a lecture may have aroused.

The style of lecturing that is easiest for me to follow (and therefore most desirable) could be called semi-formal. H. Hopf and Paul Halmos were outstanding practitioners of the genre, in my time. The semi-formal mode of communication differs from the "See mom, no hands" method in that the listener is informed in advance just where the lecture is going, with broad hints as to how it's going to get there. If the method of proof is of especial interest, the lecturer may employ the famous three step approach, announcing what he is going to do, then doing it, and finally explaining what he has done. Definitions, lemmas, theorems, and perhaps critical bits of arguments are sketched out on the blackboard, and remain there after the lecture is over like the skeleton of a devoured fish. (A digression: no television screen can display a skeleton of information like that remaining on the blackboard after a good lecture.)

The formal style of mathematical lecturing or writing is sometimes called the "Satz–Beweis mode" or the "definition, theorem, proof format." This style can be very refreshing, particularly after struggling through a muddy bit of mathematical writing full of standing hypotheses, unstated assumptions, and imprecise terminology. But a too formal presentation may leave the impression of an endless chain of definitions, theorems, proofs, and Q.E.D.'s, goose stepping toward the horizon like telegraph poles. And the formal style seems to encourage authors to indulge in such oddities as following every theorem with the label "Proof," even if the proof so labeled consists of the single word "Clear" or "Obvious."

Paul Halmos was writing *Measure Theory* in 1946 and I was preparing course notes that eventually became a book, and we frequently talked about the problems of writing mathematics. I remember one conversation about "dangling theorems." Here is an example of a dangling

THEOREM. *The sum of two and three is five.*

This construction reads better than "...of a dangling theorem. THEOREM *The sum of...*" as well as legitimizing the period after "THEOREM," and I argued for its use. Halmos won the argument with a single question: "Some people object strongly to using dangling theorems, but do you know anyone who objects strongly to not using them?"

This sort of admonition served to remind me that we write books to have them read, and so the reader's views should always be decisive. Or maybe almost always. I do remember a very successful author of an elementary series saying, "You try to write a book bad enough that the textbook committee will adopt it, and good enough that it doesn't make you sick to read it."

At the end of the 1946–47 year I moved to Berkeley and so dropped out of the informal Halmos seminar on writing. But Halmos came to Berkeley on sabbatical leave one year and our conversations continued. I remember that he drove me across California and Nevada to an AMS meeting in Colorado. I was grateful for the company and also for the free ride, since I was suffering from financial anemia at the time (it was a recurrent ailment).

T. Crane, 1962

Halmos finished his book on measure theory and published it in Van Nostrand's new University Series in Higher Mathematics. Van Nostrand was an old firm, a family enterprise headed by T. Crane who was, I believe, related to the Van Nostrands. Crane liked to do some of the book agent work of the company himself, and although Van Nostrand had an agent who lived in Berkeley, Crane visited Berkeley faculty members many times. I read several manuscripts for him, and eventually contracted to write a book on topology. The course notes written in Chicago had expanded at Berkeley, I rewrote them into the draft of a book while I was visiting Tulane, and after a complete rewrite following Van Nostrand's reading of the manuscript (by Halmos), the book appeared.

The University Series in Higher Mathematics was very successful and Crane decided to start an Undergraduate Series. He asked me to be the editor of the Series, I wanted to get Halmos to join in, Crane agreed, and so Halmos and I became joint editors. It worked very well and we published some very good books in the Undergraduate Series. Crane was succeeded as president of Van Nostrand by his nephew Ted Crane, and after a few years the company was sold, if memory serves, to Litton Industries. The company offices were moved from Princeton to New York, the administrative arrangements became more and more complicated, and it eventually became impossible. Neither Paul nor I felt we could recommend to our friends that they publish with us because of company red tape and a long slow chain of command.

But all of this is getting away from our subject (the inefficiencies of scale are not peculiar to mathematics publishing), and we've not yet gotten to the important part of mathematical writing. Quite aside from format and style, mathematical writing is supposed to say something. Put another way: the number of ideas divided by the number of pages is supposed to be positive. Good mathematical writing offers a new insight, a new viewpoint, or a new synthesis. And we appreciate good mathematical writing in just the same way that we appreciate good painting or good music; it's just plain beautiful. The final criterion for good mathematical writing, just as that for good mathematics, is beauty.

*Happy Birthday, Paul*

*Department of Mathematics*
*University of California*
*Berkeley, CA 94720*

# Some Early Memories Around Paul Halmos

## Hans Samelson

Paul Halmos 75! To me that brings up another figure: I have known him now for 50 years — a pretty good fraction of my life. He was in fact the first mathematician I met when I arrived in this country, in 1941. I had had the great luck of getting out of Germany in 1936, to study in Switzerland, under Heinz Hopf; and by 1941 it was high time to get out (the Fremdenpolizei called me in frequently to ask me why I was still there). Another piece of luck was that some money was found for a year's stay at the Institute for Advanced Study in Princeton.

So there I was (after getting off the old P.J. and B.; did we walk from the station?), marching up to the Institute one fine day in June 1941, rather uncertain how it would all turn out and quite awed by what I knew about the Institute and by the almost paradisical and also let's say affluent impression the building made on me. Fuld Hall stood then in imposing splendor, with only the well kept grounds around it, and no other buildings far and wide. (It was also unaccustomedly hot.) The way I remember it, I was met by a kind secretary who said to me: "I'll find somebody to show you around — here is Paul Halmos." There appeared a dapper figure, a young man with wavy dark hair and maybe a moustache and glittering glasses, who talked fast, in a simultaneously friendly and sardonic way. He seemed very worldly-wise. He also turned out to be very helpful.

He, and others, helped us find a place to live (to start with by subletting the apartment of Ted Martin, who was just leaving) and to get started on daily life. Paul had a small car, a convertible with a rumble seat, and he was quite generous about giving people rides, in town and in the country around Princeton. Many is the trip I spent in that rumble seat. We soon became good friends. A project that Paul undertook (inadvertently or not) was to introduce me to the U.S. culture in its large and small manifestations, how to behave, and what to expect in many situations (including the ice cream parlor on Nassau Street, the Balt; is it still there?). I remember the occasion when he and somebody else, maybe Jimmy Savage, spent half an hour or so correcting every mistake in my English that I made; it was a sobering and helpful experience.

The Institute was somewhat different then, mainly smaller and maybe more innocent — "Aydelotte gets paid a lot" seems a rather mild and good natured ditty about the then director. And it had no provisions for housing; one just found something in town. (One of the quaint habits of those days had the spouses of the temporary members print up visiting cards, and then go visit the wives of the permanent members, preferably at a time when they weren't home, and leave the visiting cards.)

It wasn't difficult to meet and get to know everybody who was there; in fact it seems to me that a good deal of the time all of us were together. Jimmy Savage, Ellis Kolchin, David Blackwell, Warren Ambrose, George Mackey, Erdoes, Kakutani, Arthur Stone, Dorothy Maharam, Irving Segal, Bob Thrall, Abe Taub, Alfred Brauer, and Deane Montgomery (with whom I soon started collaborating). I am not quite sure I have everybody. The permanent members — those famous names that I had never thought of as people: Hermann Weyl, Oswald Veblen, Marston Morse, J.W. Alexander, von Neumann, also (invisible, to me at any rate) Goedel, and, somehow floating above this group all by himself and out of reach for me, the serene figure of Einstein.

Actually I didn't see much of the permanent members. I wish I had had more contact with Alexander. He was quite aristocratic, with intensely blue eyes, and rather elusive; but the few times I talked to him he gave me excellent answers to my questions. Judging by those answers I think he would have had a great time with surgery, handles, and bordism. In my memory he was the one permanent member who gave a course that year or at any rate started out to give one, on his version of cohomology (it appeared eventually in the Bulletin, I think); three times he started

Hans Samelson, 1960

over again, because he had found a better way of doing it, and then the course lapsed. A real course was that given by Carl Ludwig Siegel, on symplectic geometry (the story was that Siegel heard that some people — the trustees maybe? — were unhappy that he wasn't publishing anything, and said, in effect: "Oh, they want papers? Then they shall have papers" and wrote Symplectic Geometry). It was a difficult course, with the audience sometimes going down to zero, but Siegel keeping on talking. There were courses given at Princeton University of course; our group, or almost all of it, used to troop over to listen to Chevalley. A mathematical tempest in the teapot was provided by the quite heated Alaoglu–Bourbaki priority dispute about the Alaoglu Theorem (compactness of the unit sphere in the dual of a normed linear space with the weak-∗ topology). I never found out why Lefschetz and Bochner weren't talking to each other.

I learned a lot of mathematics from Paul. The triumvirat Hopf, Plancherel, Polya at Zurich was not much interested in things like measure theory, Banach spaces, and spectral theorem. So I was quite deficient there, and picked up what I could from Paul in many discussions. His well ordered understanding of all these subjects was a tremendous help for me (this reminds me that I also heard of Zorn's lemma from him). Some of these discussions took place in his apartment, with hot chocolate in the afternoon, a welcome treat during the cold months. One outcome of these talks was a joint paper on monothetic groups (topological groups with a dense cyclic subgroup). Unbeknownst to us Beno Eckmann in Zurich was writing a paper on the same subject. It wasn't quite the same paper, but it came close.

Much happened at the international front during that year. First Hitler's invasion of the Soviet Union in the summer (June 22) and then Pearl Harbor. The main effect on the Institute was that people like Marston Morse disappeared to Washington. I remember a quite moving small ceremony, when Kakutani had to go back home to Japan. (Some time before that happened the oft-told incident, when Erdoes, Kakutani, and Arthur Stone managed to stumble into an Army installation that was quite off limits for civilians. The officers couldn't make sense of those three characters, who were then detained as potential spies; it needed some higher level intercession to get them out again.) — A somewhat bemusing memory: There were in that period two people in our group who were not allowed into Fine Hall at the University, myself (as a German citizen I was an "enemy alien" and might endanger the war work being done there) and David Blackwell (as a black American).

The year came to an end, and the time came to find a job. Paul worked hard at helping me to write letters and telegrams and suggesting possible places. Some of these were pretty fancy institutions, and I was — as it turned out, rightly — dubious whether anything would come of it. But I appreciated the (difficult, for me) lesson that one should try anyway. In the end my job turned out to be at the University of Wyoming in Laramie, WY.

Jimmie Savage, 1955

(I remember Lefschetz's advice: "You must go there. They have wonderful sunsets." On the day of arrival, Sept. 18, they had a heavy snowstorm.)

Laramie was a hospitable place, interesting and quite different from anything I had known; it consisted in those days of the University people, the railroad people ("Union Pacific"), and the people on the other side of the tracks. The teaching load then was 18 h/w. The most interesting mathematical event was a visit and talk by G.D. Birkhoff. The year went by fast, and the question of what to do next year came up again. It could have been Laramie, or Smith College where Deane Montgomery was on leave now, but it turned out to be Syracuse, NY, where Ted Martin was starting to build up the Mathematics Department. His first appointment was Paul Halmos. Spurred on by Paul he made me an offer, and Paul persuaded me to accept it. The third member that year was Abe Gelbart.

It was the time of the Army's Specialized Training Program (ASTP). Groups of soldiers were coming through who as part of their assignment had to learn calculus in six weeks, and they kept us busy. (There were surprises, like the man in the back row who was always sleeping, but always got A's in the tests, and turned out to have a Ph.D. in physics. Or the day when morale suddenly went to rock bottom, because the Army in its wisdom had decided not to send these specially trained soldiers to jobs where they could use their training, but to send the whole group back to an infantry battalion.)

Paul and I usually had parallel sections, and we cooperated as much as possible, talking about what to present and what not, preparing tests and grading them, and so on. I had been teaching a year in Laramie, but it was all on my own. I didn't really know what other people were doing,

Kakutani, 1965

and of course I hadn't grown up in the system. So it was very helpful for me to be able to observe Paul at work and to get his opinion and advice, always cogent, explicit, and helpful; and he was not afraid to be critical. In between we had seminars. I remember learning about Wiener measure; this was Ted Martin's specialty, but Paul's expertise in measure theory was crucial for me. [We had quite some trouble with the notation $f(\cdot)$ for a function; I guess the idea of a function as an object and the notation were just coming in; until then I had known only the very incomplete description of a function by its image set — "the function $f(X) \subset Y$."]

Life at Syracuse was quite hectic. The idea of building up a department was quite exciting, even though the physical aspect — the somewhat dark and dank basement of the old Hall of Languages — was not so great. (We had a long battle with the Library. They were the proverbial old kind of librarians who don't want any book to get out of their hands, and in particular were dead set against a departmental library.) Many people were coming to the department, for a shorter or longer time. Among them were Loewner, Bers, Ralph Fox, Henry Scheffé, Arthur Milgram (Jim Milgram's father), and many more later. It almost seemed as if anybody who was anything must have spent some time at Syracuse. We had many discussions whom to approach next and how. Paul was of course quite involved in this, since he was very up to date and in touch with the mathematical community, and Ted Martin listened carefully to his suggestions. It was quite an education for me.

After the war was over, Paul felt it was time for a change. He got himself a new car (a Henry J, I think) and started to look for a new job. He found it after a while, at the University of Chicago. He tried to get me a position at the College in Chicago, that then new venture in undergraduate education; but it didn't work out, and I landed in Ann Arbor (where Steenrod was — although only for one more year, before he went to Princeton). Since then we haven't been together for any length of time at any institution; but our paths keep crossing frequently.

As I ponder these old memories, I see how large Paul's contribution to my life on the mathematical and the plain human level has been and it occurs to me that this is a good place to thank him out loud. And I hope that it hasn't been a one-way street, and that I have contributed something to his life.

*Department of Mathematics*
*Stanford University*
*Stanford, CA 94305*

# Recollections of P.R. Halmos at Chicago

## E.H. Spanier

During the period 1944–1947 I was a graduate student at the University of Michigan. It was there that I first heard of P.R. Halmos. He became known to me as an author through his excellent book *Finite Dimensional Vector Spaces* (recommended to me as the best source from which to learn linear algebra). He became a living mathematician to me when I learned in 1946 that he was accepting a position at the University of Chicago as Assistant Professor with a salary of $5,000 (reputed to be the highest salary of any Assistant Professor of Mathematics in the country at the time).

In the spring of 1947 my thesis advisor, Norman Steenrod, was an invited speaker at the AMS meeting at the University of Chicago. I was completing my doctorate and Steenrod urged me to attend the meeting to introduce myself into the mathematical scene.

Since Halmos was at Chicago at that time, I may have met him at the meeting but I have no recollection of having done so. In fact I remember very little of the meeting (the first meeting of the AMS I attended). One event I do remember is a party at the home of J.L. Kelley. It took place at his apartment not far from the campus, and I was impressed by the huge (liter?) size glasses available for beer. That made an impression on me. I was then (and am now) partial to beer drinking.

After getting my doctorate in 1947 I spent a year at the Institute for Advanced Study in Princeton on a Jewett Fellowship of the Bell Laboratories. Halmos was also at the Institute at that time on a Guggenheim Fellowship working on his book on measure theory. We had a good deal of interaction during that period. Paul was on leave from Chicago, and I was considering an offer from Chicago so we talked frequently about life in Chicago (both mathematical and otherwise).

Besides our common interest in Chicago we shared something else. Both of us owned Crosley automobiles. These were subcompact cars manufactured in the late forties by a company with no previous experience in the automotive field and based on an engine developed during the second world war for submarine use.

During the second world war there were no passenger cars manufactured by the automotive industry (which was totally involved in production for the war effort). As a consequence there was a huge demand for automobiles when production was resumed in the late forties. Waiting periods of two years or longer for new cars were common and those few used cars available were quite old dating to prewar production. To fill the need Crosley began to manufacture cars right after the war, and Halmos and I both had them at Princeton.

These cars were much smaller than the other cars and not as sturdily constructed. A Crosley parked between two normal sized cars could be damaged in the following way. When a standard car in front backed up to a Crosley the bumpers wouldn't line up and the bumper of the larger car would pass over that of the smaller possibly causing minor denting to the front of the Crosley. Over a period of time sufficient damage was done so that the hood latch at the top of the front of the Crosley didn't quite close completely. This meant that the hood would be held down only by the secondary latch and this latter could fail if the car went over a bump or encountered a wind (as, for example, when moving at a rate of over 40 mph). Thus, one might be driving along serenely one moment and the next moment find the hood flying open in front of the windshield. This completely blocked visibility and the driver had to stop to put the hood back down. After starting again the same thing might recur obscuring visibility once more. This happened to me and it happened to Halmos.

Ed Spanier, 1955

His solution to this hood problem was ingenious. He procured two trunk hasps which he had mounted on each side of the hood toward the front to secure it down. He told me he wanted to have bright yellow (for lemon) arrows painted on each side pointing to the hasps, but I don't believe he ever did.

After getting to know Paul I learned that he had interesting opinions about all sorts of things — opinions that he could back up rationally. For example, his view about academic positions in mathematics was that it was preferable to have a high salary in a position of low rank. The higher the salary the better for obvious reasons, and the lower the rank the easier it would be to receive outside offers and so to advance rapidly.

We were together at the University of Chicago from 1948 to 1959 and I got to know him fairly well. He was interested in and fascinated by mathematics in all its forms and would ask questions, pose problems, or give solutions at any time or place. He was intrigued by puzzles and gadgets. I remember he had (or was considering buying) a pair of glasses with prisms made so that one could read while lying on one's back. It was always a treat to visit him because of the new toys to be found.

Because he is older than I both chronologically and mathematically I often consulted him about all manner of questions. He was more than willing to respond openly and casually to these questions, and his accessibility and ready wit added much to the attractiveness of Chicago at that time.

One year I decided to buy a typewriter with special symbols for typing mathematical manuscripts. Naturally I went to Paul for his advice about keys to get, arrangement of the keyboard, etc. He had already had two such typewriters made and he kept improving his design. Mine ended up with keys for union, intersection, inclusion, square brackets, curly braces, some sort of an epsilon for set membership, but no special Greek letters. He explained to me that in one paper certain Greek letters are to be used while in another paper other Greek letters. It's better not to have any rather than to have a few which might not fit standard mathematical usage in a particular instance. However, because there is a small limit to the number of special symbols that can be added, it is important to be able to combine symbols (two opposite slant lines make a times sign, a long centered underline and greater than sign make an arrow). All of these facts he had learned by experience and related to me for use in my own design. Later Paul had a third typewriter designed with a special dead key that printed a large square while not advancing the carriage. By enclosing a letter in the large square he had a whole new alphabet, Greek or whatever. Typing a mathematical manuscript was no easy task in those days before word processing and desktop publishing.

Chicago in the fifties was an exciting place to be. The Mathematics Department instituted a series of courses and corresponding comprehensive examinations for the master's degree. These exams were both written and oral and were used as qualifying examinations for doctoral candidates. They insured that our students had a certain breadth of knowledge that we considered essential. There were many excellent students as well as a constant stream of leading mathematicians as visitors.

Paul Halmos, 1950                      Wilfrid Cockcroft, 1950

By and large the members of the Department shared a common view
of mathematics. The faculty was small (12 to 14 members for much of
the time) and cohesive. We had occasional department meetings to award
fellowships or assign courses to faculty for the following year. There was
open discussion and some disagreement, but we functioned harmoniously
and shared the administrative tasks as well as the teaching. I once heard a
young foreign visitor (who had come to Chicago for a quarter after spending
a term in Princeton) express positive views of the Department's size and
morale. He felt it was large enough to represent a significant amount of
mathematics yet small enough so that each one could have some idea of
what everyone else was doing.

Halmos was known as one of the better teachers in the department. He
was certainly one of the most efficient and well-organized, compulsively so.
I remember that in one course he numbered the theorems he proved in
class consecutively and managed to arrange it so that on the last day of
the term a main theorem of the subject was the last theorem proved and
its number was 100 in the list.

At Chicago, at least in the early years, tea was only available in the
Common Room in Eckhart Hall on Thursday, the day of the colloquium.
On other days a few of us, usually including Halmos and myself, would go
to the nearby coffee shop in Hutchinson Common in midafternoon. One
year a British visitor, Wilfrid Cockcroft (now Sir Wilfrid in recognition of

his contributions to education in England), frequently accompanied us for coffee. He was a student of J.H.C. Whitehead at Oxford, sported a reddish moustache, and was spending a postdoctoral year at Chicago.

One day, as we were sitting around the table having our coffee and chatting, I noticed that Cockcroft's moustache was gone and remarked about it. He answered that it had taken me a long time to observe this inasmuch as he had shaved off his moustache a week before. At this Halmos said, "You ain't seen nothing yet," and picked up a paper napkin from in front of him on the table. Holding it to conceal his lower face he turned toward me and asked, "Ed, do I or don't I have a beard?" This took me completely by surprise. I knew he had a moustache, in fact, I didn't remember him without one. But when I first met him he didn't have a beard, then later he had one, then again he didn't,. . . The upshot was that at that moment I didn't know the answer to his question. I tried to remember what he looked like on the way to coffee or the day before. I don't remember what I said, but I do remember that I picked the wrong answer. Oh well, I had a fifty percent chance of being correct.

This story illustrates Paul's wit in action. He was quick in repartee and had (still does) a way of using language cleverly. His lectures and his writings attest to this, but it is also true in the most casual conversation as well.

I left Chicago for Berkeley in 1959, and Halmos left Chicago for Michigan in 1961. Since then our paths have crossed occasionally at a meeting or at a lecture (now that he is nearby at Santa Clara), but we have never had as close a contact as in those days at Chicago. It's been my loss.

*Department of Mathematics*
*University of California*
*Berkeley, CA 94720*

# Reminiscences of Chicago in the Fifties

## Morris W. Hirsch

I came to the Chicago Math Department for graduate study in 1952 after two years of undergraduate study at Saint Lawrence University in upper New York State. I knew practically no mathematics, but a knowledge of mathematics did not seem to be a prerequisite! The University required only that I pass a special exam on general knowledge and intellectual ability.

Although Robert Hutchins had left a couple years earlier, Chicago was still under his influence. Hutchins' pathbreaking, iconoclastic ideas about education were in full force: to study original works ("great books"); that the purpose of university study is to develop the student's ability to think critically rather than to prepare the student for a profession; emphasis on intense class discussion, but grades based only on written examinations; the firm belief that much of high school is a waste of time; and downgrading of intercollegiate athletics (there was no football team). Hutchins' program created the most exciting intellectual atmosphere I've ever seen in a University. While his ideas probably had little effect on graduate math courses, they attracted first rate scholars, scientists, writers, and artists.

Hutchins strongly encouraged students to come to Chicago after only two years of high school; and this program was still in full swing. But while many undergraduates were very young, most students were in graduate school; and many students were veterans of World War II or the Korean War.

Besides academics, the Hyde Park community attracted writers, actors, and musicians of all kinds, as well as assorted crazies and weirdoes. Many graduates and dropouts from the University stayed in the community for years. As a result, Hyde Park contained a rich mixture of interesting people; and the intellectual and cultural environment was extraordinarily fertile and exciting.

The campus was not as isolated from the surrounding community as it is now; there were more bookstores, bars, restaurants, and coffee houses than there are now, and more cheap apartments. The streets were not as dangerous, although this changed for the worse during my six years as a student.

The University took its then traditional role *in loco parentis* very lightly; students were not required to live in dormitories; and warnings were posted in dormitories the day before bed-checks were made.

There was a great deal of folk music and jazz in Chicago. There was probably rock and roll too, but at the University of Chicago we (at least the people I knew) certainly would not listen to anything so lowbrow. I played for dances at the dormitory in a jazz band which included bassist William Dement, now a leading researcher on sleep and dreams, a young instructor named Homer Goldberg whom I've lost track of, myself on piano, and others I've forgotten.

A group of us founded the University of Chicago Folklore Society, sponsoring square dances, hootentannies, and concerts with Pete Seeger, Peggy Seeger, Big Bill Broonzy, Sonny Terry, Odetta, and others. The University Administration was worried about Pete Seeger, since he was under investigation by the Committee on Unamerican Activities of the House of Representatives. But he was allowed to sing on the campus.

Paul Robeson, the famous black actor, singer, and Communist, who was not allowed to leave the country, also gave a concert on the campus. That night many FBI agents improved their musical knowledge.

Every year a local cooperative school held a fundraising jazz concert; the master of ceremonies was the semanticist S.I. Hayakwawa (later President of San Francisco State College, and then Senator from California). I remember well a talk he gave at one of the concerts, in which he contrasted the earthy, realistic lyrics of folk blues with the self-pitying themes of the "kick me in the face because I love you school of popular songs, which Mr. Clancy Hayes will now illustrate with a song he wrote." Clancy did fine.

There were at least three theatre groups near the campus, all excellent. I spend the summer of 1954 working with the Compass Theatre, the forerun-

Moe Hirsch, 1954

Irving Segal, 1950

ner of Second City and many other improvisational groups, along with Mike Nichols, Elaine May, Barbara Harris, and other unknown actors. Nichols was an announcer on radio station WFMT, and for years ran a marvelous folk music program called "The Midnight Special" on which my friends and I occasionally played.

I roomed with five other graduate students in an amazingly cheap and dirty apartment at 5546 Ingleside Avenue called the GDI, which stood for "The God Damned Independents." Originally started by veterans in the late forties, it continued as an informal cooperative until about 1955. Two of my housemates, Ed Silverstein and Jerry Friedman, were physics graduate students. They invited me to the Physics Department's annual picnic where Enrico Fermi played fascinatedly with a yoyo. Jerry was recently awarded a Nobel Prize. I remember his telling us gleefully, upon getting his doctorate, that his first grade teacher had wanted to put him in a special class for the mentally retarded.

Ed and I liked to amuse our housemates by playing tricks on them. Every night for a week the two of us gambled in the kitchen, playing a flashy card game called "Scan," in which large sums of money rapidly changed hands. One evening when Ed thought I was winning too much he put the king and queen of spades on the table and cried "Grand Nullo!," sweeping all the cards and money to his side of the table. Our housemates were astounded,

Saunders Mac Lane, 1950


especially when I told them (truly) it was the first time I had ever seen a Grand Nullo. They became very interested in watching the action, and we tried to explain the game to them, urging them to play. But we didn't tell them that in fact there was no such game: we just made up the rules as we went along. When finally we admitted that the game was just a joke, our housemate Arthur was terribly disappointed to hear that the game wasn't real: "I was just beginning to catch on," he complained. He was a brilliant chemistry student, but strange in some ways.

It is perhaps obvious that I found plenty of activities more exciting than attending classes. My grades in the first two years, were, I regret to say, atrocious; I even managed to fail Marshall Stone's course in measure theory (it didn't help that the textbook was in French). Not only did I find extracurricular activities more interesting than classes, but for some reason many of my first classes were assigned to some of the worst lecturers in the department. They were either just about to retire, or else they were fresh Ph.D.s; but none of them was able to inspire me to study.

Honorable exceptions included Irving Segal's course in set theory (even though he never prepared or remembered the proofs of the equivalence of 12 forms of the Axiom of Choice); Shiing-Shen Chern's course in differential geometry; Paul Halmos's course in metric spaces, and Ed Spanier's course in topology. For the first time mathematics became not only interesting but exciting and challenging. I was inspired enough to study seriously for the Master's Exam, managing to pass it at a high enough level that I was permitted, if not encouraged, to continue in the Ph.D. program. At least one student who was asked to leave refused to do so, and today is quite eminent.

After I got serious about mathematics, I discovered there were some wonderful teachers on the faculty. I took inspiring topology courses with Ed Spanier, who later directed my doctoral work. His exceptionally clear and well-organized lectures perhaps were the reason I developed an early love for the subject. I took as many courses as I could with Spanier, including Topological Groups and Calculus of Variations in the Large.

I also made a point of taking all courses offered by Halmos. I even audited an undergraduate course he gave on (of all things) complex variables. One of my favorite memories is his reply when a student asked him how to compute a certain line integral. Halmos wrote it down, stared at it for not more than two seconds, and said firmly, "I never integrate in public!" A very useful remark. Halmos also taught me "topological algebra," which was functional analysis; and a version of mathematical logic.

Halmos was not only an inspired lecturer, he wrote some of the finest textbooks I've ever seen. Several generations of students have learned linear algebra from his "Finite Dimensional Vector Spaces" and measure theory from his "Measure Theory," two of his books that I learned a lot from.

From Kaplansky I learned field theory. I remember complaining to fellow student Eben Matlis that despite Kap's fine lectures (which all seemed to begin mysteriously with "Consider the following obvious identity"), I disliked the subject because there was nothing I could visualize. "But that's exactly why I like it" was the reply, "I can never understand those pictures the geometers are always drawing!" Kap also taught an interesting course in differential Galois theory. But one of the best courses was one he gave on homology theory, despite (or more likely, because of) the fact that it was far from his own research.

Stone taught a course — rather beyond me — on spectral theory, which began with Hilbert spaces over the quaternions, and ended with recent research that was completely over my head. His handwriting was terrible, but that didn't matter because he stood directly in front of what he wrote, spoke to it, and immediately erased it.

There is not enough space to review, or even list, all the first-rate courses I had. These included Chern's differential geometry, whose modern treatment of tensors fields I didn't properly appreciate until I taught the subject myself; O.F.G. Schilling's proseminar, which required students to learn advanced material on their own; Lashof's course on Lie groups. (When Schilling lent a student a book, as he often did, he always gave the firm instruction, "Please don't wrap herring in it!") From Saunders Mac Lane I remember inspiring seminars on fiber spaces and Eilenberg–MacLane spaces, which he modestly called "$K(\pi, n)$ spaces."

A remarkable number of the math students at Chicago in the fifties went on to become fine mathematicians. Some that I recall are: Hy Bass, Paul Cohen, Jerry Spanier, Larry Wos, Joe Wolf, Joshua Leslie, Robert Bonic, John Thompson, Steve Schanuel, Ed Posner, Harold Levine, Eben Matlis, Elon Lima, Guido Weiss, Ray Kunze, Woody Stinespring, Lester Dubins, Don Ornstein, Lennie Gross (one of my housemates), Arshag Hajian, Larry Glasser, Eli Stein, Anil Chowdury (now Nerode), Jack Feldman, Harold Widom — and there were many others. My colleague Rob Kirby came just after I left.

Mac Lane organized the annual Math Department beer parties, which were very jolly affairs. The students wrote skits and songs poking fun at the faculty, using what we considered ingenious puns like "Professor Sixpence" (for Schilling) and "Dr. Gropingduck" (for Grothendieck). One of the songs, I forget by whom, was to the tune of an official University anthem. I record it here for posterity:

> 'C' stands for C-star Algebra,
> 'H' stands for Hilbert Space,
> 'I' stands for Integration,
> A course that we cannot face;
> 'C' stands for Category,
> 'A' for Associative Law;
> Gee, but we're weary, of Galois Theory!
> O Orthonormal, Oh we're conformal,
> Oh what a school we're in!

Another deathless ditty, which I helped write, goes to the tune of "Glory, Glory Hallelujah":

### The Ballad of Bourbaki

> Analysts, topologists, geometers agree,
> When it comes to generality there's none like Bourbaki;
> One theorem by them will cover $N$ by you or me,
> Bourbaki goes marching on!

Chorus:

> Glory, glory hallelujah!
> Their generality will fool ya,
> They're axiomatically peculiah,
> Bourbaki goes marching on!

> To prove that two plus two is four, here is what they do:
> They prove that $a$ plus $b$ is $c$ when $a$ and $b$ are two;
> We know that $c$ is half of $d$, and $d$ is four times two—
> Bourbaki goes marching on!
> (Repeat chorus ad nauseam.)

After having imbibed sufficiently, Mac Lane would boom out his spirited rendition of "My name is Samuel Hall," with everyone joining on the refrain "Damn your eyes!"

Those parties were typical of a spirit of warm camaraderie in the math department at Chicago, which I have not experienced elsewhere. In my thirty years at Berkeley, for example, I remember only one skit performed at a Math Department party; and it was written not by the students but by a visiting lecturer (Mark Green). Very few faculty members show up at departmental picnics, or even the afternoon coffee hours.

The department was a relatively small place then, which may be part of the reason for the feeling of unity. In fact mathematics as a whole was a smaller community, much more isolated than it is today. You had to be a little crazy to go into mathematics. Now many more students go into math and science, and there are many more industrial and government jobs open to mathematicians. It is more like any other profession. I believe this is all to the good, but it also means that mathematicians no longer tend to see themselves as a very special community.

The song about Bourbaki brings up the fact that its (their? — Bourbaki's) grip on the department was absolute. It is very easy to describe the applied mathematics I was exposed to: None. The words were never uttered. Functional analysis was taught without mentioning differential equations. In Zygmund's course on differential equations we were exposed to the "heat equation," the "wave equation," and so forth, but I don't recall being taught their derivations from physical principles — they were just quaint names. Complex function theory was taught without benefit of the intuition available from electrical current flow, which I now think of as analogous to teaching calculus without mentioning velocity. Although there was a Committee on Mathematical Biology on the campus, I didn't learn of its existence until much later. Certainly I never heard a colloquium talk about any applied subject.

The extreme emphasis on abstract structure reflected the spirit of the times; no doubt similar conditions obtained at Princeton, Harvard, and elsewhere. And applications of mathematics in the fifties were not nearly as developed as in they are in the nineties: For example, there was no such thing as software; complexity theory didn't exist; differential equations theory was computational and *ad hoc;* there were no such things as "chaos" or "fractals." Still, we could have been offered cybernetics, probability, and information theory, or even classical mechanics. In fact Halmos, a few years ahead of his time, wrote the first book (at least in English) on Ergodic Theory.

This is not a criticism of Chicago; rather it is a critique of the dominant spirit in American academic mathematics in the fifties. In retrospect, however, I feel that I missed what should have been an important part of my education.

Halmos has different ideas about the value of Pure vs. Applied Mathematics, stated plainly by the title of his controversial article (in *Mathematics Tomorrow*, Lynn Steen ed.), "Applied Mathematics is Bad Mathematics." It is ironic that the first example he gives of mathematics that is "discrete, finite, pure," is the problem of tiling a square by distinct squares. The main tool in its solution are equations coming from Kirchhoff's laws of electrical circuits — applied math *par excellence*!

An important educational role was played by the many visitors passing through Eckhart Hall in the fifties. Perhaps they were temporary lecturers, or even merely colloquium speakers, but I vividly remember lectures, courses, and seminars by Armand Borel, Richard Palais, Christopher Zeeman, Bruce Reinhart, Robert Hermann, Serge Lang, Eldon Dyer, René Thom, Norman Steenrod, R.H. Bing, and many others.

Bing, Wallace, and Pitcher, 1955

I remember vividly a colloquium by Bing. Growing up in New York I had met very few southerners, and had unconsciously absorbed the common New York prejudice that they were ignorant rednecks. It shocked me to hear sophisticated mathematics expounded in a strong Texas drawl.

Bing was a bit of a joker. His talk dealt with embeddings in three-space of Cartesian products of a surface and an interval (showing that one of the surfaces must be tame). In the question period Thom asked, in his heavy French accent, if this could be generalized to fibrations (a fibration is a generalized product). Bing pretended to misunderstand him: "I don't know what you mean by *vibrations*" he replied, "this theorem has nothing to do with *vibrations!*" I have no doubt that R.H. knew very well what Thom meant.

At another colloquium a Japanese mathematician gave an interesting talk about the development of mathematics in Japan. He said its slow development was due to the jealousy and secrecy of competing schools. Students had to take an oath in blood not to reveal what they learned to outsiders. From the audience André Weil remarked, "I think that's a very good idea."

Weil ran an elite seminar much too advanced to me. I didn't have any course from him, nor can I remember ever speaking with him at Chicago.

When I was a post-doc at the Institute (that is, the Institute for Advanced Study, still *the* institute to me), I submitted my doctoral thesis to the *Annals of Mathematics*, of which he was editor. He riffled the pages a few times, found it wanting, and handed it back to me with the words: "We don't publish doctoral theses unless they solve the Riemann Hypothesis." I don't think that was strictly true, but at least I didn't have to wait six months for a rejection.

In Princeton Weil ran a current research seminar, which I did attend. At the organizing meeting he assigned topics and dates to all the speakers. When a young mathematician from Japan, assigned to speak in a few weeks, protested that his English was very bad, Weil told him firmly, "By November 15 you will learn English." And he did.

In order to earn some money I taught elementary mathematics at night at Illinois Institute of Technology. By and large these students were neither interested in nor prepared for further study in mathematics, but for various reasons needed another course — not the subject, just a record of passing the course. I have never forgotten one student who insisted on simplifying $(a + b)/a$ into $b$, explaining that he did this by "canceling" the $a$'s. When I showed him with several examples that this gives the wrong answer, he looked puzzled, and asked me, "Yes, I see that it's not true in those particular examples, but why isn't it true *in general*?" For 35 years I have been unable to think of a suitable reply. Probably Halmos would have known what to say.

Nathan Reingold was chair of the Math Department at IIT. He generously told us teachers, "I don't care whom you pass, but save the A's and B's for the good students," so my canceling student passed. Reingold used to consult on the side. He told me he once was offered a job to compute some very strange probabilities. When he asked what they were for, he was told they applied to gambling equipment. He was worried about accepting the job, since gambling was illegal, but was assured that in Chicago it was not against the law to manufacture or sell gambling equipment, only to use it. He was not told who bought the equipment.

The person who most strongly influenced my mathematical career was Steve Smale, who came to Chicago in 1956 directly after his Michigan Ph.D. (he was Raoul Bott's first student). I met him in Mexico City at the Symposium on Algebraic Topology; his work in the topology of smooth manifolds was one of the few things there I could understand. Back at Chicago I would go to his office every day to explain my much simpler proof of his result; and every day he would patiently explain why I didn't have a proof. Eventually he taught me to understand his work, which dealt with the then unnamed field of Differential Topology. With the help of his inspiring geometrical intuition, and Spanier's masterful teaching of algebraic topology, I eventually wrote a thesis.

By an amazing coincidence, Chern, Spanier, and Smale left Chicago forever and came to Berkeley at the same time I did. This was a stroke of good fortune for me, but it must have been a severe shock for Chicago.

While Chicago has continued its tradition of the highest mathematical excellence, I shall always look back on my experience in Eckhart Hall during the fifties as a wonderous Golden Age.

*Department of Mathematics*
*University of California*
*Berkeley, CA 94720*

# Reflections on Graduate Education

## Rob Kirby

In 1958 at the University of Chicago, Paul Halmos taught a course in general topology using Kelley's book of the same name. It was the first math course that I really enjoyed and worked for; it was partly Halmos's teaching and partly those wonderful problems in Kelley's book, e.g., the long one on the Cantor set. Furthermore, the course got me started in the direction of geometric topology.

Halmos was a stern teacher (most were sterner in those days). He would lock the classroom door shortly after the bell rang to discourage interruptions by latecomers. This classroom had another door in the back of the room which connected it with the back of the adjacent classroom. One fringe-type student came late one day, found the main door locked, and went through the other class and came in the back door. I don't remember Halmos's exact words, but he left no doubt as he threw the student out that this was the most grievous error the student had yet made in his short life.

Chicago had a number of excellent teachers in those days. Besides Halmos, I took courses from Mac Lane who radiated energy and love for mathematics, Kaplansky who was lucid and played Lasker (an interesting version of checkers) with me occasionally, A.A. Albert, who was ambidextrious and could fill a blackboard faster than anyone, writing with one hand when the other got tired (sometimes it seemed like he wrote with both simultaneously), but I missed Chern and Spanier who left for Berkeley about 1959.

My impression is that undergraduate math was less pressured and slower paced in the 1950's than it is now. For example, Albert's third year linear algebra course was much more leisurely than the second year linear algebra course we teach at Berkeley in large lectures. Similarly with multivariable calculus, which is more high-powered now. There's a sense now of so much material to be covered and so little time. Sometimes this trend is counteracted by innovations in the presentation of material. For example, I took multivariable calculus from Widder's old book, and it was for me an impenetrable maze of different formulas and notation each time the dimension of domain or range changed. A few years later, Nickerson, Spencer, and Steenrod turned up, and it was an eye-opener to discover that a differential, in any dimension, was just a linear transformation. I'm amazed that most students are still taught the old way. Similarly I didn't really understand continuity and epsilons and deltas until I took Halmos's topology course

Rob Kirby, 1969

and discovered that "the inverse image of $U$ must be open when $U$ is open" summed it up.

In 1961–62, Norman Steenrod visited Chicago and organized five lectures for graduate students on how to do research. He started the series, and was followed by Mac Lane, Kaplansky, Calderon, and Herstein (in some order). All of them told us seriously that you had to work hard to write a thesis. They gave many anecdotal examples from their own fields and careers. One addressed the question of how it was ever possible to learn enough to prove a theorem before the giants in the subject did it. He said that as you work and learn, you eventually know more about a perhaps very narrow topic than anyone else, and then the results begin to come. He acknowledged the difficulty of doing this with hot, competitive topics, but claimed that even in those fields there was much to be done that the luminaries didn't have time for. I don't remember much more, but in general the lectures were entertaining and sometimes illuminating, and I'm surprised it hasn't been done more often.

So the following is what I might say in such a lecture; it begins and is sprinkled with personal anecdotes which have influenced my ideas on graduate education.

As an undergraduate, I'd been far more interested in chess, poker, and almost any sport, than in the game of mathematics. I had little chance of getting into a good graduate school. However, I failed German and didn't get a B.S. in four years, so in my fifth year I took most of the graduate courses on which the Masters Exam (really a Ph.D. prelim) was based. With a B.S. I asked to be admitted to graduate school so as to take the Exam. They cautiously said yes if I got grades closer to B than C in the fall quarter. I got a B and a C (measure theory from Halmos and algebraic topology from Dyer) and a Pass, and no one told me to leave.

The Masters Exam could have four outcomes: you could pass with financial aid, pass without aid but with encouragement, pass with advice to pursue studies elsewhere, and fail. I got the third pass, but really liked Chicago and turned up the next year (1960) anyway.

Norman Steenrod, 1955

So that's how I snuck into graduate school. And ever since I've felt that, within reason, it is best to let students make the decision about whether to quit their math studies or not. Yes, they have to pass their exams, but it is research and teaching that we really care about, so rather than flunk them out, let's give them as much chance as possible for doing research towards a thesis. Generally, a student is a better judge of his capabilities than the faculty; in fact, a student will often underestimate himself because of the common syndrome of believing that everyone else is smarter, or at least more knowledgeable.

Of course, this sentiment of giving almost any warm body a chance runs afoul of the tradition that a Ph.D. advisor has the responsibility of giving the student a problem and helping him solve it. I believe that a student has to earn his advisor's time and help, with hard work at the very least. A weak student does not have the right to be led by the hand through a thesis.

I applied to teach at Roosevelt University in downtown Chicago, where a few other Chicago grad students also taught. Mac Lane wrote a nice letter and my application arrived in August when, I speculate, it turned up on the top of the pile of applications just when Roosevelt decided it needed to hire. I taught there 4 years, 9 units per term, mostly college algebra and trigonometry, but 14 units the last year, including calculus. Those 14 units took about 17 hours per week, including office hours, and seem considerably less now than teaching a "light" load at Berkeley (6 or 3 units) and advising graduate students. If one only has to teach, then 14 units for 32 weeks of the year is a light load. In this sense, TA's are very well paid; the problem is that they are not paid for their research, which of course may never exist.

The Chair at Roosevelt was a wonderful old codger named G.D. Gore, who had received his Ph.D. at Chicago in the early 1930's. More than once I heard his story about obtaining a Ph.D.; afterwards, he couldn't keep his foot on the ground, like the convict whose ball and chain were finally removed.

Chicago's Ph.D. qualifying exam was an oral exam on two topics. I chose homotopy theory from Hu's book and group theory from Hall's book (most of the former and some of the latter). I failed with a dismal performance. I recall being asked whether the projection of a 2 simplex onto one of its edges was a fiber map, and answering no after observing that the fibers were not homeomorphic and then being confused.

But I received excellent advice: start talking mathematics, come to tea, and become part of the mathematical community. I'd begun to study books, but it was an eye-opener to hang around other students and occasionally faculty and realize how much folklore existed, how valuable intuition could be, and how other people thought about mathematics. The involvement was vastly different when doing math with others than when passively listening to a professor in class. Tea was filled with math (and cake — everyone was charged a dollar per quarter which some luckless grad student, e.g., Lance Small, had to collect; and occasionally chess, sometimes with David Sanchez, then a newcomer from Cesari at Michigan, now head of Math. Sciences at NSF). I recall Steenrod describing the foliation of the 3 sphere by Hopf circles, thereby giving a picture of the Hopf map and turning a bit of homotopy theory into something geometrically intuitive.

I passed on my second try at the qualifying exam, but with a weaker performance in topology than in group theory, so the recommendation was that I not pursue research in that subject. However, George McCarty, a new Ph.D. and instructor, had written a thesis (with Richard Arens at UCLA) on homeotopy groups (homotopy groups of spaces of homeomorphisms of topological spaces). He'd been one of my oral examiners and, perhaps feeling isolated among the algebraic topologists at Chicago, encouraged me to read his thesis. That's how I got hooked by geometric topology.

Finding a thesis advisor was another matter. I was embarassed to ask the professor closest to geometric topology, Dick Lashof, for he'd seen my weak performances on my topology orals. I spoke to Mac Lane who had the reputation of requiring his students to come in once a week and lecture on what they'd been doing. I said that I was interested in McCarty's stuff and possibly a couple of other things closer to Mac Lane, and he told me to spend the summer thinking and picking a thesis topic.

This, I think, is good advice. Earlier a physicist had told me that one should pick an advisor that one could work with as a person, and then do whatever the advisor recommended for a thesis; if you didn't like the subject, well you could always switch fields after your thesis (imagine telling a fresh Ph.D., worried about publishing and a tenure-track job, to just change subjects and start anew). Instead, you should pick a subject that you love, for it's likely to be the main focus of your life for some time, and only then should you choose the best advisor you can.

So I asked Eldon Dyer, who had been on sabbatical during my orals. When asked, he didn't immediately answer for he didn't know me that well. I didn't bring it up again, but just stopped by to talk to him occasionally when I had a math question. After awhile, it seemed clear that I was, de facto, his student. Our primary meeting spot was on the handball court for a threeway cutthroat game with John Polking.

Dick Swan and Eldon Dyer, 1958

Around 1962, I became interested in the annulus conjecture, which states that the region bounded by two (locally flat) disjointly imbedded $(n-1)$ spheres in $n$ space is homeomorphic to an annulus [an $(n-1)$ sphere cross the unit interval]. One day I thought I had a proof and collected Dyer and MacLane for an audience. The mistake was at the end when I claimed a nested sequence of closed sets had a nonempty intersection (consider the sets $(0, 1/n]$ in the positive reals). Later Mac Lane kindly warned me that the annulus conjecture was a bit hard for a thesis topic. But later I proved a weaker form of the conjecture which used an additional hypothesis, and Dyer, an editor for Proc. AMS, accepted it. Neither of us knew that this result was folklore in the Bing school of topology, and I finally felt like a mathematician.

But the most important part of my graduate education was getting together every day with Walter Daum, a smart student with similar interests, to read the literature: Zeeman's notes on PL topology, and papers of Haefliger, Homma, Stallings, Smale, Mort Brown, Connell, Kister, and some of the Bing school. Ever since, I've felt that it was more important to have colleagues of similar ability to work with on an equal footing, than to have giants around casting large shadows. I think my success with my Ph.D. students is due to having enough of them to form a critical mass; they then learn from each other as I did with Daum. As for casting large shadows, one of my early students, Selman Akbulut, recently told me that he was gratified to meet me for he then realized that you could be a slow mathematician and yet a successful one (I think he'd met topologists like Dennis Sullivan and felt that if he had to shine like Dennis to be successful, then he was in trouble).

I finished my thesis (now outmoded and forgotten) and dreamed of going to the Institute for Advanced Study. But Dyer and Lowell Paige, then chair at UCLA, determined my destination over a beer at the winter meeting of the AMS. As far as I knew, at that time (1965) southern California had no geometric topologists (look at it now, the center of the universe). But Bob Brown, David Gillman, and Ned Staples had just arrived, and the four of us formed a lively group along with a number of visitors. It was a good environment for a mathematician like me to do my own work in, and I finally proved the annulus conjecture one evening in August 1968, while looking after my 4 month old son, Rolf.

Chern has often said, sometimes in reference to Chinese students who were worried about leaving a mathematical center and returning to a small, provincial institution, that it is ok to get away from the influence of the superstars and pursue your own work in comparative isolation. Of course, that worked for him and also for me. But other ways work for other people — listen to Bott describe the wonderful years he spent at the Institute for Advanced Study around 1950.

Let me finish with a few general remarks. Up through the 1950's the Ph.D. was supposed to take three years. You passed qualifying exams in the second year and then found an advisor who gave you a doable problem and told you what to read and what methods to use to solve it. But it was said that 75% of Ph.D.s never published anything beyond their thesis. But no wonder. Finding doable problems is half the job and figuring out what to use is much of the rest. Those Ph.D.s had not really done independent research yet (with exceptions of course).

So from the beginning, I have encouraged my students to find their own problems. It may take an extra year or two, but when they're finished they will not only have done honest research, but will have a backlog of problems which they've worked on, so that they can hit the ground running in their first job. My responsibility has been to help them get into promising subjects with a future, to nurture a small community of students who learn from and support each other, to organize seminars and give courses on recent developments, and to treat them like mathematicians who just happen to have not yet written their first paper.

Geometric topology had many accessible problems, and I think this procedure worked well for many years (perhaps I was also blessed with unusually good students). But I confess that it has become more difficult to advise students recently because the influence of Thurston, Donaldson, Jones, Witten, and others on low dimensional topology has made the subject far, far broader. Students seem to need more guidance than in the past (how, for example, is a student to find his way through topological quantum field theory on his own?), and at the same time it is getting harder to give guidance (who can keep up with hyperbolic geometry, gauge theory, and quantum invariants?). Perhaps this is just a matter of perspective, but I don't think so; mathematics and nearby areas interact much more today than at any time in my past.

One last piece of advice comes from an acquaintance who teaches piano. He tells his students not to set themselves up for disappointment. Don't focus on winning a certain competition; you may have an off day, a competitor may have the day of his life, the judges may not like your looks or your style, any number of things may to wrong that aren't really your fault. Instead, the piano teacher advises, concentrate on being the best pianist that you can be, and your career will usually fall into place satisfactorily. Ditto for mathematicians, who have gotten along well by downplaying prizes and other status symbols.

*Department of Mathematics*
*University of California*
*Berkeley, CA 94720*

# Michigan Years

## R.G. Douglas

Paul Halmos came to Michigan in the Fall of 1961, while I arrived a year later. He resigned his position at Michigan going to Indiana in 1969, the same year I moved to Stony Brook. Several years later we met at a colloquium he gave at the Belfer Graduate School at Yeshiva before the bean counters closed it down. Upon Leon Ehrenpreis's proffered introduction of me to Paul, Paul remarked, "Do I know Ron Douglas? Why I invented him!" While an obvious exaggeration, the statement contained a good deal of truth reflecting what had happened during our "Michigan years."

Whatever the circumstances of his departure from Chicago, Paul had been welcomed at Michigan and his respected position in Ann Arbor was clear to me when I arrived the following year. I had first met Paul the preceding year at a colloquium he gave at Tulane University during Mardi Gras. Pat Porcelli, my thesis advisor, had assembled a carload of colleagues and graduate students to make the pilgrimage from Baton Rouge. Paul was the first "celebrity mathematician" I had met. I recall he spoke on the invariant subspaces of the shift operator, a subject I knew nothing about

Ron Douglas, 1966

at the time. While T.H. Hildebrandt's visit to Louisiana State the following Fall had something to do with my being offered the first instructorship bearing his name at Michigan, it was Paul's presence in Ann Arbor which sealed my decision to accept it.

The Fall of 1962 was an exciting and a good time for mathematics. The United States had embarked on a large expansion of science and mathematics, especially at universities, in response to Sputnik. It was difficult to spend well all the money that was available and there was the feeling that everything was possible. I had had to choose between NSF and NDEA Fellowships for graduate study in a department that was rapidly developing and expanding in response to federal and state funding. Then I had the opportunity to start my professional life at one of the historic mathematics departments in one of the great universities in the United States. It was a heady time and much of the success which American mathematics has enjoyed in the decades since is based on that investment in education and people.

Although my thesis was not in operator theory, it was difficult for a young analyst at Michigan not to end up working in operator theory. Paul had just published his first "Ten Problems" and this had attracted much attention. Indeed I had just missed an informal summer program centered around Paul at which Harold Widom had presented his proof of the connectedness of the spectrum of Toeplitz operators. The proof was a tour de force, as was Paul's insight that such a result might be true. A perspicuous proof still eludes us. In any case one had to feel at the center of operator theory at Michigan.

Formal contact with Paul was somewhat limited my first year. However, Paul's views, professionalism, and enthusiasm for mathematics and life could be learned at lunch, where Paul held court each day in the Michigan Union. All were welcome to join the famous round table where the latest mathematical tidbits and gossip were shared. (Gossip to a mathematician is talking about everything but rigorous mathematics.) Although many contributed to this lively conversation, Paul's absence due to a traveling day, made it clear to all who the main character was. Paul has always had strong, but well-reasoned opinions on all aspects of mathematics and indeed on most everything else. But he has always taken the opinions of others seriously, although that did not necessarily mean changing his own.

Paul was an editor of the *Proceedings* that first year. Hence, when I had completed my first paper, I submitted it to Paul. It was a note on invariant means, a subject to which Porcelli had introduced me. I made an appointment one afternoon at Paul's office in Angell Hall and gave it to him. His only comment that I recall was that he doubted that I would ever receive a Fields Medal working on invariant means. But he took the paper and my first editorial experience began. After one referee equivocated and a second was rather negative, Paul asked me to rewrite the paper and resubmit it, which I did. I was devastated when a third referee wrote a scathing report

Moe Schreiber, 1962

which Paul forwarded to me. What to do? The reports left little doubt as to the character of my writing but not what was wrong. Somehow the subject came up at lunch and Moe Schreiber, true to his name, volunteered to assist me. After much red penciling and several rewrites, I learned how to write mathematics. Here Paul was catalyst, with Schreiber his former student playing the key role. However, the importance of clear, understandable writing was obvious around Paul who never demeaned effort put into writing or exposition.

Paul encouraged a sharing of responsibilities in such a way that one felt honored to be chosen. By the following summer, I had been asked to take over the scheduling of our seminar although the outside speakers were provided by Paul. It was a good experience for me to have reason to speak with the Michigan analysts about the mathematics they were doing and to encourage them to speak in the seminar. One person that could always be counted on was Allen Shields. He was always enthusiastic about mathematics and loved speaking both formally and informally about it. While I believe Allen's first love was the interface between function theory and functional analysis, in Ann Arbor this mixture was leavened with operator theory. Thus, Allen's view of operator theory was less parochial than that of Paul who has always been something of a purist. While Paul has worked in several fields, it has always been one at a time.

Ciprian Foias, 1965

My first doctoral student was Paul Muhly who finished the year I left Ann Arbor. Still, I was involved with Paul's students almost from the moment I arrived. Don Sarason finished his doctorate at the end of my first year at Michigan. Although we spoke several times, I did not get to know him very well until his return visits to Ann Arbor in later years. However, I was second reader for Lew Coburn the following year and talked and worked with all the Michigan students in operator theory from then on. In my second or third year, I recall the party Paul had for the graduate students in his Fall Operator Theory course. There was considerable surprise when he told them I would take over for the spring, since I was still younger than most of them.

Despite all of the activity in operator theory in Ann Arbor, I resisted working in this area for my first two or three years. For one thing, a large group, which included Arlen Brown, Carl Pearcy, and Allen Shields, had formed and I wasn't certain I wanted to compete with them. Indeed, Carl and Arlen had by then completed their famous characterization of commutators. However, two events sealed my fate. First, the visit of Bela Sz-Nagy to Ann Arbor encouraged the seminar at Michigan to work through his series of papers with Ciprian Foias studying contraction operators. Second, Louis de Branges and James Rovnyak announced that they had solved the invariant subspace problem and Paul, along with Henry Helson, Shizuo Kakutani, and Peter Lax, had been asked to referee it. Since this problem

had become the holy grail for operator theorists, there was much excitment
and I along with the others tried to work through the paper. This resulted
in numerous discussions with Paul and a steady correspondence with de
Branges. Although refereeing is usually done anonymously, for this paper
the process became public. Eventually Paul suggested that I replace him as
a referee. After much work over a year or two, it was agreed that there was
a serious gap in the proof. Since their approach and that of Sz-Nagy and
Foias were two ways of looking at the same topic, when the dust settled
I had become an authority on an important part of operator theory. This
mastery, plus ideas and programs which I developed during my visit at the
Institute for Advanced Study, has stood me in good stead since. I have
taken the subject which Paul had helped shape, single operator theory, but
cast it in a broader context choosing to emphasize its connections with the
rest of mathematics.

Teaching and communicating mathematics is very important to Paul.
While in the sixties it was common to discuss students and teaching at
Ann Arbor, both undergraduate and graduate, a young instructor had to
be impressed at Paul's seriousness of purpose. Students were awed by the
famous mathematician who took their pictures during the first week of class
and called them by name the second. His lectures were as carefully prepared
as were his seminars and colloquium talks. Moreover, he was interested in
content, curriculum, and teaching methods, although he chose to think
about such things mostly on his own.

There was much for a young instructor to learn, and although I owe much
to many people in Ann Arbor during those years, I owe a special debt to
Paul.

*College of Arts and Sciences*
*SUNY at Stony Brook*
*Stony Brook, NY 11794*

# Walking and Talking with Halmos†

## Lawrence J. Wallen

Writing about the departed is a far less thorny proposition than dealing with the living. This is especially true when one's subject has been variously mentor, colleague, collaborator, boss, father confessor, and just plain friend. Was such and such a comment confidential? What about letters? The difficulties are apparent. Nevertheless, against every instinct, I decided to set down something of my long, if lacunary, relationship with Paul Halmos.

Mathematicians, like, say, violinists, want to read about their stars but are treated, alas, with precious few opportunities. Most of us were weaned on Bell's *Men of Mathematics* which made wonderful reading even if some of the history was suspect. Constance Reid proved that we'll gobble up tidbits about even the most uncharismatic of men provided he proved enough hard theorems and has a space named for him. Ulam showed that one doesn't even need a space if adventures are promised. The moral seems to be that we're starved for gossip, insights, and revelations about our heroes. Halmos, of course, realized this fact and moreover, had the wit to know that one photograph (and a couple of words) brought names to life. This little effort should be viewed then as a public service in this direction. A caveat is in order. Anything that looks like "see p. 306," refers to the Automathography.

I met Halmos, vicariously, in a little book with a ghastly orange cover, called, of course, *Finite-Dimensional Vector Spaces*, and I guess many of my aging colleagues made his acquaintance the same way. The book was a revelation, sheer magic. The intimations I'd gotten that mathematics could be beautiful were finally realized. It read like a short story one wished were longer. It's a little sad that in making a real text of it, a little of the magic seems to have disappeared. Not all though. Who else would talk of "zeroish" transformations?

In the summer of 1955, the University of Chicago staged an extravaganza of functional analysis. This sort of thing may be commonplace today but was not so then. I'd already, after a year of graduate study at MIT, been seduced by the subject and the line-up was all-star caliber, Mackey, Irving Segal, Kaplansky, and Halmos. Sad to say, the first two were well beyond

---

†This little piece is dedicated to the memory of Allen Shields whom Paul and I loved and who left us far too soon.

me and the third too algebraic for my tastes. It remained for Halmos and ergodic theory to salvage my summer and savings.

I can't say I really remember my feelings on first putting a face to the orange cover, though I do remember being surprised. He was a lot bigger than I'd imagined and the moustache was a surprise (why?). The voice and delivery are a lot more vivid since they haven't changed in the ensuing 3.5 decades. Many words and phrases were clipped, others almost drawled and the timbre was pleasantly resonant. In fact, the total effect was rather musical. The vestigial Hungarian accent Paul bemoans was so slight as to seem just a mannerism, at least to my tin ear. The lectures proceeded at a no-nonsense but unhurried pace.

Two deviations from the normal flow come to mind even today. The lectures were attended by a lot of future stars, the most luminous of whom was the vociferous Grothendieck. Occasionally, something or other caused some buzz in the audience, especially among the cognoscenti. "There's too much static" was the invariable, slightly menacing rebuke from the front. It always worked. I plagiarized this snappy reproof often and got only blank stares. Something must have been missing. The other memory involves a rare excursion into whimsy. Paul had written a paper titled "In general, a measure preserving transformation is mixing." A little later, Rohlin published "A general measure preserving transformation is not mixing" (no contradiction, see p. 112). Paul said he always wanted to write yet another

Larry Wallen, 1968

paper with the title, "In general, a measure preserving transformation IS TOO mixing."

Meeting the protagonist was a bit of a project and accompanied with lots of trepidations. An informal poll of listeners concerning the Halmos personality gave a mean of "intimidating." Unconvinced, and deciding that the way to a mathematician's heart was through his mathematics, I accosted him at tea one afternoon with something irresistibly clever like "here's a generalization of. . .". The reply was, I now realize, typical. "I've been there. It doesn't fly." This reads curt but there was a smile and a sort of benign tone, not in the least intimidating. By the way, it didn't fly.

Kolmogorov, 1965

Regarding the metaphors above, they're pretty typical of Paul's conversation, spoken and written. Often, they're variations on old saws, inserted at odd places, and they make communication a little more fun. A recent letter contained the following, "I tried to reconstruct your proof of. . . . I failed. I've got the melody but forget the words."

Late that summer, I met Virginia (Ginger) Halmos, a striking woman I recall thinking. Any picture of Paul that omits Ginger is grossly incomplete. In the first place, she's crucial to keeping the entropy of the Halmos household improbably small and in keeping Paul and the cats hale and hearty. This, of course, doesn't define Ginger. She's a woman of remarkable intelligence with a fine wit that not everybody is privy to. She's the ecological Halmos who fishes foundering lizards from the pool and worries about wetlands. She still can be seen riding her bike in the perilous environs of San Jose and not infrequently sports a Band Aid from a minor contretemps. Paul frequently, and not without cause, frets about her safety.

Anyway, Ergodic Theory, did make the summer. It was a beautiful set of lectures and nobody seemed to mind that Kolmogorov hadn't reinvented entropy yet.

A dozen years passed before I saw Paul again. I knew he was doing something called algebraic logic and recall hoping this was a transient abberation. Judging from the care Paul lavished on his account of the logic years (pp. 202–216), this period must have been a high water mark in his career. He was, after all, doing what he liked most and did best, making connections between heretofore separate disciplines, making precision from fuzziness and just plain understanding. Moreover, the field has a proximity to his second love, linguistics, which, say, ergodic theory doesn't.

In 1967, I came by a year at Ann Arbor, and reacquaintance with Paul was of high priority. The moustache had grown a luxuriant beard with the usual effect of trading youth for distinction. After possibly one amenity, we were immersed in invariant spaces. This was just a year after the famous nonstandard proof by Bernstein and Robinson that polynomially compact operators have invariant subspaces. Paul standardized the proof to (almost) everybody's relief. Much later, I asked him if he had ever REALLY worked on the GENERAL invariant subspace problem. By GENERAL I meant no assumptions on the operator, like, say, quasi-nilpotence, but what I meant by REALLY was vague, even to me. The answer was, as always, arresting. "I've never stopped working on it." There was, of course, an active operator theory seminar, Michigan being at the time a mecca for the subject, and Paul gave talks on his then new notion of quasi-triangularity. Though this seemed an appealing and natural idea, no one really would have guessed how important it would prove to be. Indeed, operator theory was poised for the great leap forward of the seventies and beyond, and certainly ideas (and equally importantly, questions) of Halmos pointed out some of the most fruitful directions. Subnormality, dilations, and quasi-triangularity don't exhaust the list.

Michigan would have seemed an idyllic situation for Paul. There were good friends, good colleagues, good students. There was an active social life, a fine home, and a terrific Old English sheep dog (Bertrand Russell by name). There was good walking and there was even that miserable climate (damp cold winters and damp hot summers) on which he seems to thrive best (but see p. 346). Consequently when I heard rumors about his leaving, I dismissed them as smoke without fire.

But there was fire. It was spring and I was potentially unemployed. Paul told me one day that he was going to head the department at Hawaii and would I come. Had he mentioned some place remotely reasonable like, say, Maine, the reply wold have been an instant affirmative. But Hawaii wasn't a real place, it was a Michener construct. Nevertheless, in the end, the prospect of working with Halmos was too attractive to refuse. By September, we were the only two operator theorists 2500 miles from anywhere and so was born the Pacific Basin Two Man Seminar (or T.M.S.).

The essential ingredients of any T.M.S. are congeniality and constancy. The importance of the first is clear but Paul gets high wisdom points for realizing how crucial is the second. We tried 7:00 A.M. and 4:30 P.M. and it was no contest. By 4:30 fatigue was evident; what we really wanted by then was a beer. So every Tuesday at 6:30 A.M. I wended my way to Paul's Diamond Head home which came complete with pool, lanai, and burglars.

There was never a formal agenda and utter silence was a possibility but somehow it never happened. Usually, urgency will win the day, and at the time, I was urgently trying to find the structure of one-parameter semigroups of partial isometries. Paul's interest in continuous semigroups was about zero but he certainly knew a lot about partial isometries. In particular, he knew how poorly they behaved when multiplied. So, almost immediately he turned my nice analytic problem with its delicious promise of direct integrals into a purely algebraic one. What, he asked, is the structure of the discrete analog, namely positive powers? He was in his element and we solved this fairly quickly. The transition back to the continuous case was my baby and not easy but was ultimately accomplished.

When it came to writing up the discrete paper, not one word was mine. Most of us detest putting a fait accompli in a finished form, but with Paul it was a labor of love and importance and he has a genius for it. He christened our atoms "truncated shifts" (descriptive if not poetic, and the name seems to have stuck), and fashioned a little diagram that made a cumbersome, index-filled proof transparent.

"Behind every analytic truth is an algebraic or combinatorial one" is an oft repeated (and purposely exaggerated) Halmos contention. Though counter-examples abound, there is more than a grain of truth here, as I found with my semigroups. Paul is essentially an algebraist-combinatorialist. He seems happiest rooting about in matrices, and even now has on his desk a huge notebook ominously labelled "Bad Matrices." Recently, he showed me an analytically virtuosic measure-theoretic counter-example, which prompted me to call him what he's always calling me, a "crypto-analyst." He denied it and claimed that the idea was firmly rooted in discrete sets. It may seem a little insane to claim that the author of *Measure Theory* is not "essentially" an analyst, yet very possibly, the gem of the book is a purely set-theoretic result, the monotone class theorem. In the final analysis, essences may be too elusive for mathematicians and ought to be left to philosophers.

The next semester, Allen Shields joined our dual seminar. Normally, three's a crowd in seminars as well as on dates. When there are only two, you've got to pull your weight or collapse is inevitable. With three, slough-

Larry Wallen and Allen Shields, 1969

ing off is possible. Allen, however, was such a wonderful mathematician and person that he could never be anything but a plus.

By the fall of 1969, the ranks of Pacific operators had diminished by half. Paul resigned as chairman for the reasons of p. 161 plus possibly a few more. Hawaii was possibly the worst place for him to try out chairing (though he was a very good chairman). The condition known as polynesian paralysis (mañana plus) no doubt irked one of mankind's most punctual members. Moreover, and this is not as frivolous as it sounds, Honolulu is poor walking country. Being an urban mountain top, a serious walk entails either lots of CO inhalation or lots of vertical travelling, neither of which Paul liked.

Over the next five years there were conferences, meetings, letters, and sometimes succinct communiques which read like — Are you coming to the meeting? I'm free for a beer Friday between six and seven. Meetings were always hectic times for Paul, involved as he was with AMS and editorial business. An hour off was a luxury. In 1974–75, I went to Bloomington, the new operator center, for a sabbatical year. The T.M.S. resumed without missing many beats, dominated to some degree by Paul's preoccupation with integral operators. A couple of incidents are unaccountably vivid.

Programmable calculators were coming into vogue and Paul, who loves gadgets, naturally got one at a staggering cost. At one session, I was having unusual difficulty concentrating. There seemed to be a sort of insistent noise in the room which was finally traced to an unfamiliar small black object. Exasperated at my ignorance, he informed me that IT was factoring $N$. For once I prevailed by declaring that either IT went or I went.

The other incident involved an intellectural gaffe on my part. I began the session by suggesting we look at a certain "situation." The retort was immediate. "Pose a problem, ask a question, but don't give me situations. It'll probably be the wrong problem but it's a place to start." This credo pretty well embodies his research and pedagogical modus operandi. He once proposed starting a real variables course with the question: Is there a continuous nondecreasing function whose graph lies in the unit square and which has length two? (I don't know whether or not he actually did.) The lesson exists at all levels. Don't start by finding the area under $f(x)$, do it for $x^2$. The conceptual difficulties are already present in the special case, and the student needn't worry about which $f$ the lecturer is talking about.

When Paul decided it would be fun and profitable to live a long time, he went about it in a characteristically systematic way. He got a guru, none other than the redoubtable Nathan Pritikin. Pritikin's ideas were pretty straightforward. Namely, we eat too much (and badly) and move around too little. The situation can be rectified by doing less of one (and better) and more of the other. For years now both Halmoses have been faithfully following Dr. Pritikin's diet with only the occasional lapse at post-colloquium dinners and affairs-of-state banquets. In various university eating establishments, Paul could be seen with his spartan lunch, surrounded by gluttonous colleagues. I once asked him if he were ever hungry. The reply was memorably laconic. "Almost always." Well, Pritikin wasn't for sissies and, at any rate, the diet seems to have worked.

The other half of Pritikin, the moving part is more interesting. Paul had already been a walker, and ultimately became one of the discipline's most enthusiastic and vigorous practitioners. Of course, the way to ensure one's getting the minimum daily dose of one hour (equals four miles) is to live two miles from the office. Paul claims (and there's no reason for doubt) that he found the San Jose house by drawing a (Euclidean) circle of radius two miles, centered at O'Connor Hall and scouting the perimeter. Not unreasonable. One can expect a reasonable vacancy in 12 city miles. Once, a colleague told me he'd seen a man looking a lot like Halmos striding at an improbable speed around L.A. airport. Was this possible? Not only possible but totally unsurprising.

I accompanied Paul on lots of walks in and out of Bloomington that year. There were problems however. Since $V$(elocity) $= L$(ength of stride) $\times f$(requency) I was in trouble, as my $L$ was a lot less than his $L$. The only solution was to up my $f$. Initially, I tried jogging ahead, then back, but this ploy ruined conversation. Another problem that surfaced was my intolerable habit of occasionally slowing down to observe the birdlife. Paul had, out of indifference, fashioned a fittingly Boolean taxonomy. "There are eagles and there are sparrows," he declared with finality. If I pointed out a red-tailed hawk, it became a small eagle. Anything smaller, he claimed not to be able to see. The only dent I ever put in his systematics was the owl-in-residence at Santa Clara. He probably didn't consider it a real bird.

Despite this, the walks were wonderful. Sometimes we talked, sometimes we didn't. Serious math talk was difficult as I had to concentrate on $f$. One walk was memorable, but it occurred seven years later. Paul was planning to visit Jim Williams in the hospital and asked if I would join him, and if so, at what time. I said, sure, ninish would be fine. I realized the faux pas immediately but the deed was done. "I don't like ishes." Well, ishes are no worse than isms (but see paragraph three). We started at nine, very sharp. On the way, Paul was occupied in convincing me that the dimension of a certain set was $N$. Preoccupied with maintenance of $f$, I nodded in that way that certain lost students do, but was secretly unconvinced. That afternoon, during a very Baroque recital, I managed three proofs that the dimension was not $N$. Feeling that he shouldn't retire deluded, I called him that evening. Admitting the error of his ways he noted, rather wistfully, that in his dozen or so years at I.U., nobody had ever called him for a mathematical reason.

Paul does have exacting standards of punctuality. My own track record is pretty good but there are tales of missed lunch dates by mildly dilatory elements (admittedly hearsay). Of course, everybody has an occasional lapse. A colleague of mine, the proverbial unimpeachable source, swears that Paul was once a half hour late for a dinner and was profusely apologetic. Certainly possible, even understandable, but on the other hand, maybe the others were half an hour early. When those watches which lose maybe a minute every century first emerged, Paul was the first on his block to own one. At a Wabashian post-seminar dinner, Charles Putnam (of the beautiful inequality) showed up with his own super-watch. Comparing notes and times Paul declared Putnam's watch to be incorrect. I can't remember how many seconds were involved, but Charley was upset. What I've never resolved is how it was decided just who was right. Maybe Paul's reputation won the day.

One more walking anecdote reveals a very different side of the Halmos character. It concerns the so-called dog problem. Practically every household in Bloomington has at least one, and the farther away from town, the larger and more irascible the beast. I'd already had some biking confrontations and had adopted the policy of arming myself with a pocket full of stones. Since dogs don't like walkers much more than bikers, I decided to continue the policy. Paul, and even more so, Ginger, were appalled. In fact, his tactic was not only more humane, but more effective. He armed himself with dog biscuits. Every dog on the route had become his personal friend. He even knew their names (at least he called each something they didn't object to).

In the course of our conversations, stationary or otherwise, we at least touched on most topics and, generally speaking, there was harmony. There were, however, a few exceptions. On music (p. 30), which was important to us both, we were definitely at loggerheads. Paul is a self-proclaimed, Bach to Mozart reactionary, while I dote on all those people he loathes, Bartok, Hindemith, and Schoenberg. Although this is a matter of taste, each of us is firmly convinced the other is dead wrong. As a historical note, the same Warren Ambrose who introduced Paul to Bach, showed me Webern almost a generation later. Of course we both love detective novels. Paul's ideal is that impeccable logician, philologist, and beer connoisseur, Nero Wolfe. Mine is the plodding, intuitive, beer connoiseur Maigret. Paul has an op-art Vasarely in his living room while I go for misty Monets. Strangely enough, we're still on speaking terms.

For a period in the eighties, Santa Clara University was the seat of the only two readable mathematical journals in the U.S. Paul was editor of the Monthly and Jerry Alexanderson his opposite number for Mathematics Magazine. The Monthly job could easily have been a drain, but Paul, with his genius for organization, his certain knowledge of what he wanted, and more important, how to get it, seemed to have enviable stores of energy left (see p. 379). He taught courses at Santa Clara and Stanford, attended seminars and colloquia at Stanford, Berkeley, and other places, worked on the picture book, gave far flung lectures, and, of course, mathematized. Incidentally, Jerry wasn't exactly sedentary either.

In 1986–87, a little jaded with Paradise, Cotten (my much better half) and I happened to live about two miles from the Halmos house. Sheer serendipity; I couldn't afford to draw circles. On lots of Saturdays, the drill went like this: (a) Paul races to my place, (b) we both race back to his, (c) we talk around mathematics, (d) I'm stranded at his place. One such morning, I hear the familiar Gestapo knock, Paul strides in appropriately winded, and unceremoniously mutters TRIBO. ORBIT say I. There's no mystery. The morning paper has a little anagram game that one normally breezes through. Occasionally, a wiring problem surfaces but ORBIT didn't happen to be mine. With Paul, you always had to be ready for word problems he'd spring out of the blue. Was the preferred pronunciation SKISM or SHISM? I lost; it's neither. Find a word containing the ordered quintriplet AEIOU, all of whose other letters are consonants (pretty easy).

He got his revenge a little later. We were off to a colloquium and I confessed to having had a little trouble with CAFOIS. The talk was plain awful. The speaker scrupulously ignored all the well-known (common-sensical) Halmos formulas for giving an intelligible lecture. We were anesthetized with 40 minutes of banal calculations. When release seemed imminent, the dreaded rhetorical question surfaced, "Would anyone mind staying an additional few minutes?" I fully expected Paul to say "yes," but decorum won. Outside, instead of the diatribe I'd expected, Paul said merely, "What a FIASCO!"

Late in the eighties, the inevitable proved to be just that. The walking regimen was important, both physically and psychologically. I wondered, and fretted about the effect of some hobbling injury, say, tendonitis or shin splints, something like that. Actuality was more serious. Surgery was required on both legs. The convalescence was a protracted period of pain and, worse, doubt. This was the only time most of us, who knew Paul as thoroughly upbeat, ever saw him in the doldrums. The story has a happy ending. Tackling his physiotherapy with characteristic tenacity, he's back threatening the safety of his slow-moving fellow humans.

Though I'd sat through lots of Paul's research level lectures, I'd never witnessed his performance with undergraduates. Some months ago, I requested and received permission to do just that (having promised not to butt in). It was the second meeting of a beginning function theory course and he prepared for it as meticulously as he would for a Stanford colloquium. He'd taken pictures of the students, put names to faces, and managed with only a couple of juxtapositions, to make everyone feel a little more important. I learned what a power series was and thought idly that, all in all, it was probably healthy that the kids didn't realize they were being taught by a legend.

Even though this vignette is anecdotal and doesn't stress the Mathematician Halmos, mathematics is surely the heart of Paul Halmos. I remember being in one of those periods of dissatisfaction that mathematicians are particularly prone to. We were on a walk at some meeting and I was berating mathematics as being irrelevant, trivial, an ill-disguised ego trip, and other uncomplimentary things. He heard me out, waited till I'd run out of steam, and then said only "Mathematics is my religion."

The above is certainly not the whole truth, but it is, as well as I can recall, nothing but the truth. Warren Ambrose, or Ginger or worse, Paul, may wonder just who is this guy I've been talking about. So be it. Finally, I want to thank Cotten for her merciless and uncompromising criticism. Even more finally, I'd like to thank Paul Halmos for letting me sit in.

*Department of Mathematics*
*University of Hawaii*
*Honolulu, HI 96822*

# Being Professional

## John H. Ewing

It's impossible to *reminisce* about events that happened within the past 20 years; reminiscences describe events that are forgotten by everyone except (possibly) the teller. But here are *observations* about Paul Halmos during his Indiana years. They are observations as I might have recorded them a few years after arriving in Bloomington in 1973. I wish I had. I came fresh from a two-year instructorship at Dartmouth. I knew plenty about cohomology operations, about $H$ spaces, and (I thought) about writing papers; I knew little about being a mathematician. I suspect most young mathematicians begin their careers in similar circumstances. Here's what I learned by watching.

Paul Halmos is an imposing figure in the department, not because he writes the most papers but because he influences every part of mathematical life. He is an editor, a member of countless professional committees, a writer of textbooks, a master teacher, a researcher. He seems to do it all, and I wonder what magic allows him to do so many things well.

Paul belongs to no university or college committees (except for a group of Distinguished Professors who select other Distinguished Professors). Although he is always interested in departmental affairs, he spends little time talking about university politics. He watches dispassionately as deans come and go. We pay administrators to do their job, he says, so we ought to let them do it. If they're bad, fire them.

On the other hand, Paul belongs to both the Society and the Association, sits on committees and boards for both, and attends national meetings, summer and winter. He is a member of editoral boards for several book series and devotes a great deal of professional energy to mathematics publishing. Both at meetings and in Bloomington, he regularly meets with editors for Springer-Verlag, Klaus Peters, and (a little later) Walter Kaufmann-Bühler.

Every part of departmental life is important to Paul. Nothing is insignificant, from having names and addresses of new faculty members on the first day of the new semester to having the right supplies in the office. Paul insists on lunch each day at 11:30 sharp. He walks to several doors, banging on them with his cane and usually not waiting for a response. A large group of people follows him to lunch in the cafeteria, where the food is awful but the conversation is lively. (Paul buys only a container of butter milk and brings the rest of his Spartan lunch — lettuce and pita.) Much department business is transacted at lunch, but there is mathematics as well. What are the possible values of pi (for any convex body)? What sums can be obtained from changing signs in a convergent sequence? What if we consider

Max Zorn, 1983 (photo by Jens Zorn)


the finite case ...? But suppose the dimension is ...? A steady stream of
puzzles and problems and tidbits. Sometimes, people talk about a single
topic for days or weeks. Occasionally, someone writes a paper based on
this discussion, but that seems unimportant. The younger faculty also find
out about mathematical culture and history: Is the latest counterexample
to the invariant subspace problem correct? Who's a good candidate for
the next Fields Medal? Who were the big shots in the 50's? We have a
department of around 40 and nearly half the faculty takes part in these
lunches; Paul is the driving force both in establishing the tradition and
keeping it alive.

Daily teas are equally important to Paul. When he is here, he seldom
misses a tea. He talks with old friends (especially Max Zorn), but he also
makes an effort to circulate, to talk with visitors and graduate students and
the physicists (who share our tea room). When he meets someone new, he
often starts with the phrase, "What's your game?" This usually leaves the
recipient mumbling and embarrassed; that doesn't stop Paul from repeating the initiation to others. He will often remind speakers (before their
talks) that 55 minutes is the ideal length of a talk. This is only partially
successful. Paul often brings his camera to colloquium teas and snaps pictures of virtually all guests. I notice that everyone feigns embarrassment
but feels flattered. For Paul, taking someone's picture is a way of shaking
hands, of making people feel comfortable.

Many other mathematical-social occasions are important to Paul. He and Ginger often have parties for visitors. He values (and never misses) departmental picnics, where he seeks out young faculty and graduate students as well as their families. He begins a tradition of having a large buffet at the beginning of each academic year to meet all new faculty. For Paul, mathematics does not only mean writing papers; mathematics is an essential part of his life, a part he enjoys.

Mathematics is not the only part of his life. He loves language and words; he loves music (some of it); he loves the environment of a major university. Paul and Ginger often have dinners to which they invite the intellectual elite of the community, famous musicians and artists and historians. Many, but not all, have some Hungarian connection. Occasionally, Paul and Ginger

Alice and Klaus Peters, 1972

Walter Kaufmann-Bühler, 1975

invite young faculty members to share the experience. In contrast, Paul plays poker with a group of car dealers and truck drivers, and seems to feel equally at ease in their company. He insists he wins over the long run. Paul is an avid walker and almost everyone in town knows him by sight, if not by name: He's the man with the cane who walks rapidly down High Street (or Maxwell or Jordan or ..., depending on where the person lives). Paul and Ginger both love animals of every kind. They have cats who (with justification) think of themselves as people. Paul carries dog biscuits at all times and is on good terms with every misanthropic dog in town.

Both Paul and Ginger are less comfortable with children. When they come to my house for dinner the first time, Paul tries to coax my wary dog to him by lying on the floor. My young children mistake him for a Grandfather and scamper across the floor, ready to play. "Take them away!" he shouts. The children learn to respect Paul; the dog becomes his fast friend.

Paul has strong opinions on almost everything. He's not dogmatic — he listens and (occasionally) changes his mind — but he's an iconoclast. He

Paul Halmos, 1973

challenges the obvious at lunch, at parties, and in department meetings. Grants? Not necessary for mathematics. Democracy? Not a good thing in a university. Qualifying exams? Give the student less time, not more, to complete them. Department meetings are the most explosive because people take them so seriously. When discussing whether to make an offer to a young mathematician, Paul waits until the end and says he's against it: This one doesn't smell right. (Guaranteed to make junior faculty scream about a "lack of procedure.") Should we make a big offer to a middle-aged hotshot? Absolutely, he says, this will be the best mathematician in the department. (Guaranteed to make at least one or two senior faculty angry in *any* department.) His opinions are not always predictable, but they are always definite, and supported by a succinct rationale. One thing becomes clear: Some people in the department wait for Paul to speak and then automatically take the opposite point of view. Diplomacy is not one of his virtues.

Paul has equally strong opinions about mathematics. Some mathematics is good; some is bad. He has strong opinions about articles, about books, about talks: Each is good or bad. He even has strong opinions about opinions — it's important to have them. Forming opinions and supporting them is part of being a mathematician.

Good mathematics is not the same as Paul's mathematics. He works in operator theory, but he relishes working on a problem in algebra or combinatorics or hard-analysis. He has a passion for mathematics, and when he begins to think about a problem, small or large, he immerses himself in it. He calls to say he thought of something during his morning walk: What about ... ? He wants to find out about mathematics he doesn't know, and he sometimes asks a young colleague to explain a topic, without much success. Paul believes all mathematics can be explained (to other mathematicians) while walking — no paper, no blackboard. His younger colleagues disagree and always head for the blackboard or start to scribble definitions on a napkin.

Paul is also passionate about his own mathematics, just like the rest of us. I see him in animated discussion with his student, Sunder. He attends the Wabash Seminar (mainly functional analysis) faithfully. He works on papers and goes to conferences. But Indiana is known for its large group of operator theorists, and in this respect Paul, while a leader, is one of many. I gradually realize that Paul is admired more by those outside the department, and he is admired not so much for his great mathematics as for being a great mathematician.

What's the magic that allows him to do so many things well? Paul is a great writer, but he is great both because he has natural talent *and* he works extraordinarily hard at writing. Paul devotes immense amounts of time to every enterprise. When he teaches a course, he starts to prepare a year or more in advance. When he gives a talk, he writes and revises and rehearses for weeks. When he edits the book reviews for the Bulletin, he spends months to establish a functioning system, worrying about how to word each post card and form letter. When Walter Kaufmann-Bühler visits Bloomington to hold an editorial meeting, Paul spends many hours working through each file, going over every detail of every project.

Details are important to Paul. He is fastidious about having the right word for an article, about being certain that letters are sent, about filing papers in the right order (back to front), about keeping track of finances. This makes him demanding on secretaries, and he manages to find exceptional ones who accept his demands; it's more than natural selection. He is equally demanding of chairmen (What is my teaching schedule the year after next?) and editors (Tell me precisely how you establish that price?) and young colleagues (Didn't you mean to say "just as" rather than "like"?). He is always punctual and makes a joke of measuring tardiness in seconds, not minutes; it's only partly a joke. He is disciplined: When he works on a major project — a book or an article — he allocates a certain portion of time each day. He is demanding of everyone, including himself.

A passion for mathematics, hard work, attention to detail — Paul calls it "being professional." This is Paul's magic recipe for success: You must have talent, but you also must be professional.

As far as I know, no department has ever offered a seminar in how to ply our trade. There are books and pamphlets on teaching — there are even articles on how to write research papers — but we learn most of our trade by watching others. Many young mathematicians have learned (and continue to learn) by watching Paul Halmos. I did. Of course, we don't all carry dog biscuits in our pockets; we don't go around banging on doors with our canes; and we don't share all of Paul's strong opinions about mathematics and the world. But we know what it means to be professional.

*Department of Mathematics*
*Indiana University*
*Bloomington, IN 47405*

# Reflections of a Department Chair

## G.L. Alexanderson

In eary 1984 I received a telephone call from Paul Halmos — that was before he converted to e-mail — during which he said, among other things, that he would like to be someplace with more sunny days. The Bloomington winter seemed long. A telephone call from Paul was not unusual; I was coeditor of the Problem Section of the *Monthly,* for which Paul was at that time editor-in-chief. He asked whether I happened to know of any open positions in California. I responded that I didn't, but I would think about it. When I had, I called him and asked him whether he might consider Santa Clara, a place I suspect he knew little about at that stage. But I pointed out one salient fact about Santa Clara — we do see the sun a lot. I raised the question of his joining us at Santa Clara with some hesitation because, though we may have good weather, Santa Clara is not the kind of institution at which Paul had spent his career: the Institute for Advanced Study, Syracuse, Chicago, Michigan, Hawaii, UC Santa Barbara, and Indiana. Our department is undergraduate, with approximately 20 faculty.

Paul's knowledge of Santa Clara was enhanced by our flying him out to take a look at the campus. I don't know that we had particularly good weather while he was here, but it may have looked greener than the midwest, so it wasn't long before we were into serious negotiations. Since Paul never leaves things to chance, the conversations ranged not only through the critical issues like salary and secretarial assistance but also to details like the number of feet of bookshelf space in an office. The result was that the following January Paul joined the faculty of Santa Clara. I cannot speak for Paul, but from my experience the arrangement has been a splendid success. He was full-time with a half-time teaching load here while continuing to edit the *Monthly.* Subsequently he has dropped down to half-time, possible under our phased retirement scheme.

From his arrival on campus he has participated fully in the life of the Department. He participates along with all the other senior faculty in promotion, tenure, and hiring questions. He seldom misses a departmental colloquium. He sees his students. He has lunch with colleagues regularly at the Faculty Club. And in the evenings Paul and his wife, Virginia, are seen at concerts and lectures on campus, as well as at all sorts of functions put on by the mathematics students: their picnics, pizza nights, trivia bowls, whatever. (In this area he's much more conscientious about participating in student-faculty activities than some of the rest of us.) He has become a

Gerald Alexanderson, 1989

very visible member of the faculty and is widely known throughout the university community.

Chairs at institutions similar to ours have asked me how having Paul in the Department has worked out. With an eye to trying something like this themselves, they want to know what it's like to bring a very distinguished mathematician from a research university to an undergraduate department in what is essentially a liberal arts college environment. My responses have no doubt encouraged similar moves. But, of course, there's only one Paul Halmos. He has not only had a distinguished career in research but has long had an international reputation as a brilliant mathematical expositor. His concern for conveying mathematical ideas to others makes it easier for him to step into a department largely concerned with teaching.

Our administration has been delighted to have Paul on the faculty. Soon after he arrived on campus, his review of the year in mathematics for the *Chronicle of Higher Education* appeared. That's something that administrators read and seeing the name of the university in this context made an impression on them. Mathematical achievements are not always recognized by professional academic administrators, but to them appearing in the *Chronicle of Higher Education* matters.

Of course, from a chair's point of view, it's not always easy working with someone as well-organized as Paul. I know that I have consistently disappointed him by not knowing his class schedule one or two years in advance. He likes to start preparing his classes well before the beginning of the course. And those who have observed his appointment book are aware that he knows several years in advance, it seems, just where he's going to be when. I'm lucky to know where I'll be next week.

Paul is a great hit with students. I have tried not to waste his talents on students who don't enjoy mathematics, so he has been teaching primarily majors. But his student evaluations reach stratospheric heights. He has taught a variety of courses here: linear algebra, set theory, topology, abstract algebra, complex variables, and problem solving. His classes are challenging — he often uses a modified Moore method — so students expect to have to work and to be alert during class. They never know when they'll be put on the spot. They also know that he'll know who they are. One of his first classes here was rather startled when he arrived for the first meeting with his camera — those who know Paul know he's never far from his camera — to take pictures of each student so that he could learn the names more quickly. The good students know, of course, that if they do well it will definitely be to their benefit. A positive letter to a graduate school signed by Paul R. Halmos will do wonders (and has). His being here has been a great opportunity for the students. I recall a few years back when he was teaching a course in set theory and Garrett Birkhoff happened to be visiting the campus (as, I believe, a Phi Beta Kappa visiting lecturer). Birkhoff gave a guest lecture in Paul's class and the discussion that developed between Halmos, Birkhoff, and our own senior foundations person, Karel de Bouvère, was fascinating to observe. I don't know how many students appreciated what they were witnessing, but I did.

Paul's being in the Department has also provided us with wider contacts for colloquium speakers and sabbatical visitors than we would normally have. The Department has certainly benefited from this. Of course, visitors of all sorts have to be prepared for Paul's ever-present camera.

So what has Paul done since coming to a small liberal arts department in the West? He has accumulated a couple more honorary degrees: Kalamazoo College and the University of Waterloo. He has served both on the Executive Committee of the Council of the AMS and the Executive Committee of the MAA. He still edits four series for Springer-Verlag. He has given talks from Auckland to Caltech to Tel Aviv and a lot of places in between. Two books have appeared: his "automathography" entitled *I Want To Be a Mathematician*, published by Springer-Verlag and subsequently by the MAA, and *I Have a Photographic Memory*, his picture book for the AMS. And his most recent research paper (coauthored with José Barría of our department) appeared in 1990 with another scheduled to appear in 1991. One of the high points of the 75th anniversary meetings of the MAA in Columbus in the summer of 1990 was Paul's talk, "Has Progress in Mathematics Slowed Down?," with the acclaimed follow-up article in the *Monthly*.

Has Paul Halmos slowed down since leaving the heady environment of

a major research university? Not obviously. There are surely things that
he misses in not being at an institution that is at the center of things.
But while here he has taught functional analysis at Stanford from time to
time and he spent the fall of 1989 at the Mathematical Sciences Research
Institute in Berkeley. He gave the after-dinner address at the AMS dinner
at the annual meetings in San Francisco in January 1991. And he still takes
his four to six mile walk in the morning before coming to the office. If he's
slowed down, it's a pace some of us would like to accelerate to.

My advice to chairs looking for an infusion of excitement into an undergraduate department is: find a Paul Halmos and hire him for your department with dispatch.

*Department of Mathematics*
*Santa Clara University*
*Santa Clara, CA 95053*

# Some Mathematics

*If a man's wit be wandering, let him study
mathematics; for in demonstrations, if his wit be
called away never so little, he must begin again.*
—Lord Bacon

*Mathematicians are like Frenchmen: whatever you
say to them they translate into their own language
and forthwith it is something entirely different.*
—Goethe

# Paul Halmos and the Progress of Operator Theory

## John B. Conway

John Conway, 1977

For most of the mathematical public the name Paul Halmos is synonymous with excellent mathematical exposition. Having been Paul's colleague for many years, both as a faculty member at Indiana University and as a researcher in operator theory, I feel that I know many other aspects of his mathematical personality. To begin with, he is totally dedicated to mathematics and scholarship. He works harder than most mathematicians I know. His reputation as a lecturer is in no small way due to his hard work which results in careful preparation and great attention to detail.

Another part of his approach to research is a very careful analysis of what is happening. With the advent of a new mathematical discovery Paul studies the work and tries to decipher what makes it all come together. Sure we all do this and we try to get our graduate students to do the same. But Paul brings something extra to the process. He is constantly asking questions and abstracting the ideas so that they can be placed in a broader framework. This has resulted in a truly significant contribution that extends well beyond the theorems he has proved.

The facet of Paul Halmos's mathematical personality that I will discuss here is his contribution as a research mathematician in the theory of operators on Hilbert space, and specifically his role as a shaper of the direction of research over the past 40 years. To be sure Paul has a number of papers and theorems that anyone would be proud to call his own. But the thing that has always struck me about his work is the extraordinary number of topics and problems that are dominant themes in the current research of today and that have their origin in his work. Over the years Paul has demonstrated an uncanny ability to extract crucial properties from a given mathematical entity and lay it open before his colleagues in such a manner that there is a universal inclination to look and explore further. I hope I will be able to adequately illustrate this in the course of this article.

Paul has posed many questions that have attracted the attention of many outstanding mathematicians. I am not going to survey all of these, but rather I will focus only on those which have resulted not just in a solution but in the creation of a body of mathematics, a subarea of operator theory.

Let's begin with the first of his papers that concerns operators on a Hilbert space [12]. Herein lie two ideas, the ramifications of which are prevalent today: dilation and extension. An operator $T$ on a Hilbert space $\mathcal{K}$ is said to be a *dilation* of an operator $A$ on a Hilbert space $\mathcal{H}$ provided $\mathcal{H}$ is a subspace of $\mathcal{K}$ and $A = PT \mid \mathcal{H}$, where $P$ is the orthogonal projection of $\mathcal{K}$ onto $\mathcal{H}$ and the vertical line denotes restriction. $T$ is an *extension* of $A$ provided that the subspace $\mathcal{H}$ is invariant for $T$, that is, $T\mathcal{H} \subseteq \mathcal{H}$. Thus $T$ is an extension of $A$ if $T$ is a dilation of $A$ that satisfies $A = T \mid \mathcal{H}$ (no $P$ required).

Why? The idea behind studying dilations and extensions is to study the properties of the operator on the smaller space (the operator $A$ in the above definitions) by using the properties of the dilation or extension. Now a little mathematical maturity on the part of the reader induces him to question whether anything in this generality can possibly have any value. To a certain extent this is correct, but some results are given below which show that the mature view here is a jaded one. Specifically, arbitrary operators have most interesting dilations, though to get interesting and useful extensions, attention must be restricted to special classes of operators.

Here are some examples of each concept at work. In fact these are probably the examples that inspired Halmos to study these ideas. Let $m$ be normalized Lebesgue measure on the unit circle in the complex plane, $\partial \mathbb{D} = \{z : |z| = 1\}$. Let $H^2$ denote the classical Hardy space consisting of the functions $h$ in $L^2(m)$ with vanishing negative Fourier coefficients: $\hat{h}(n) \equiv \int h(z)\bar{z}^n dm(z) = 0$ for $n < 0$. This can also be described as the space of analytic functions $h$ on the open unit disk $\mathbb{D}$ such that the coefficients in their Taylor expansion are square summable: $h(z) = \sum_n a_n z^n$ and $\sum_n |a_n|^2 < \infty$. (For those unfamiliar with these matters, the Taylor coefficients of the function analytic on the disk are the same as the Fourier coefficients of the function on the circle.) If $\phi$ is a bounded function on the circle, then there is a natural operator $M_\phi$ defined on $L^2(m)$, the multiplication operator: $M_\phi f = \phi f$ for all $f$ in $L^2(m)$. There is also the classical Toeplitz operator $T_\phi$ defined on $H^2 : T_\phi h = P(\phi h)$ for all $h$ in $H^2$, where $P$ denotes the orthogonal projection of $L^2(m)$ onto $H^2$. Clearly $M_\phi$ is a dilation of $T_\phi$. If, in addition, $\phi$ has vanishing negative Fourier coefficients (and, as it turns out, corresponds to a bounded analytic function on $\mathbb{D}$) then $H^2$ is invariant under $M_\phi$ so that $M_\phi$ is an extension of $T_\phi$. When $\phi$ is a bounded analytic function the operator $T_\phi$ is called an analytic Toeplitz operator.

In this paper Halmos showed that if $A$ is any contraction, then $A$ has a unitary dilation. The proof is trivial. In fact, let $\mathcal{K} = \mathcal{H} \oplus \mathcal{H}$ and let $B$ and

$C$ be the positive square roots of $1 - AA^*$ and $1 - A^*A$, respectively. If

$$U = \begin{bmatrix} A & B \\ C & -A^* \end{bmatrix}$$

acting on $\mathcal{K}$, then it is easy algebra to check that $UU^* = U^*U = 1$ and so $U$ is the promised unitary dilation. No big deal (the paper has more to it than this), though this does show that arbitrary operators (every operator is a scalar multiple of a contraction) can have very nice dilations. Now the voice of maturity asserts that dilations cannot possibly be useful. But there is a fine line between maturity and cynicism.

This observation that every contraction has a unitary dilation prompted B. Sz-Nagy [20] to explore this notion of a dilation and led him to what is today called the Sz-Nagy Dilation Theorem. Namely, every contraction $A$ has a power unitary dilation $U$. That is, there is a unitary dilation $U$ of $A$ such that $A^n = PU^n \mid \mathcal{H}$ for all positive integers $n$ (refer back to the definition of a dilation for the notation). In other words, there is one unitary dilation of $A$ such that every positive power of the unitary operator is a dilation of the corresponding power of the contraction. Thus began a series of discoveries that culminated in the book of Sz-Nagy and Foias [21] in which a model for an arbitrary contraction is established, developed, and exploited. For one thing if a certain minimality condition is imposed on the unitary dilation, then it is unique. Moreover assume that $A$ is what is called a *completely nonunitary* contraction, that is, $A$ cannot be written as $B \oplus V$ for some unitary operator $V$. In this case the minimal unitary dilation has a spectral measure that is absolutely continuous with respect to the Lebesgue measure $m$ on $\partial \mathbb{D}$. This permits the application of Fourier analysis to the study of operator theory with remarkable consequences.

Thus the small observation of Halmos has grown to a major subfield of operator theory: the Sz-Nagy–Foias model theory. The intrigued reader can consult their book [21] for further developments as well as additional historical notes on this subject.

In addition Halmos's 1950 paper defined what are today called subnormal and hyponormal operators. Say that an operator $S$ is *subnormal* if it has a normal extension; call $S$ *hyponormal* if $S^*S - SS^*$ is a positive operator. It is not hard to see that every subnormal operator is hyponormal. In fact if $N$ is a normal extension of $S$, then writing $N$ as a $2 \times 2$ operator matrix relative to the decomposition $\mathcal{K} = \mathcal{H} \oplus \mathcal{H}^\perp$ yields that

$$N = \begin{bmatrix} S & X \\ 0 & Y \end{bmatrix}$$

for some operators $X$ and $Y$. Now $N^*N = NN^*$ (the definition of a normal operator), and doing the necessary matrix arithmetic we have, among other things, that $S^*S - SS^* = XX^*$ which is a positive operator. Hence $S$ must be a hyponormal operator.

Some examples of subnormal operators are the analytic Toeplitz operators defined above; the multiplication operators $M_\phi$ are their normal extensions. It is not hard to see that there are hyponormal operators which are not subnormal. Indeed the reader can find such an example in [12].

In [12], Halmos also gives an internal characterization of subnormal operators; that is, a characterization that relies only on the internal workings of the operator on the host Hilbert space $\mathcal{H}$ and not on any external object such as an extension.

Later one of Paul's students, Joseph Bram, wrote his thesis on subnormal operators [7]. Here Halmos's internal characterization of subnormal operators was improved and many additional properties explored. Today, subnormal operators form one of the success stories in operator theory [9], attracting the attention of many experts and students. Similar events followed the concept of hyponormal operators and here also results have become so numerous that the topic has commanded its own treatise [19].

Why are subnormal operators and hyponormal operators important and why have they attracted so much attention? Why does any area of mathematics attract attention? I can't answer this last question, though it seems to me that a common thread to all those parts of mathematics that have their loyal followers is that they have an abundance of examples and make contact with other parts of mathematics. This is certainly true of both subnormal and hyponormal operators. Subnormal operators are intimately connected with analytic functions and particularly with the theory of rational approximation. One of the principal sources of examples of hyponormal operators is from the theory of singular integrals. Both of these parts of operator theory make heavy use of the results and techniques of these other branches of analysis. In addition the operator theory has succeeded in suggesting interesting questions and problems about these parts of analysis.

Another milestone in the advancement of operator theory came when Halmos [13], in the process of studying the invariant subspace problem, introduced the idea of a quasitriangular operator. An operator $T$ is *quasitriangular* if there is an increasing sequence of finite rank projections $\{P_n\}$ such that $P_n \to 1$ in the strong topology and $\|TP_n - P_nTP_n\| \to 0$. To better understand the definition, first realize that in light of the fact that the finite rank projections $\{P_n\}$ are increasing (that is, their ranges are increasing), the requirement that $P_n \to 1$ in the strong operator topology means nothing more than that the union of the ranges of these projections is a dense subset of $\mathcal{H}$. If the condition that $\{TP_n - P_nTP_n\}$ converges to 0 is replaced by the condition that $TP_n - P_nTP_n = 0$ for all $n$, then $T$ is said to be a *triangular operator*. Let's explore this terminology a bit.

If $P$ is a projection and $TP - PTP = 0$, it is an easy exercise for the reader to verify that the range of $P$ is an invariant subspace for $T$. If $\mathcal{H}$ is finite dimensional and $T$ has an upper triangular matrix representation with respect to the basis $\{e_1, \ldots, e_m\}$, then each of the spaces $\mathcal{L}_n$ spanned by $\{e_1, \ldots, e_n\}$ is invariant for $T$. So if $P_n$ is the projection of $\mathcal{H}$ onto $\mathcal{L}_n$,

$TP_n - P_nTP_n = 0$ and $P_n \to 1$ (in fact, the sequence is eventually equal to the identity).

The same reasoning, as well as the converse, applies in the infinite dimensional case. If $\{P_n\}$ is as in the definition of a quasitriangular operator and $TP_n - P_nTP_n = 0$ for all $n$, an orthonormal basis for $\mathcal{H}$ can be found by taking the union of appropriate bases for each of the spaces $(P_n\mathcal{H}) \cap (P_{n-1}\mathcal{H})^\perp$ $(P_0 = 0)$ such that relative to this basis the resulting infinite matrix representation of $T$ is triangular. Hence the definition.

For infinite dimensional spaces the most appropriate analog of an operator on a finite dimensional space is the concept of a compact operator. An operator $T$ is *compact* if $\{Th \colon \|h\| \leq 1\}$ has compact closure. Just as every operator on a finite dimensional space has a triangular matrix representation, every compact operator is quasitriangular. (In fact there are compact operators that are not triangular. So the price paid for being infinite dimensional is that we must settle for the sequence $\{TP_n - P_nTP_n\}$ being asymptotically 0 rather than being identically 0.)

This last fact was implicit in a paper of Aronszajn and Smith [6] where they showed that every compact operator has a nontrivial invariant subspace. It was Halmos's efforts to fully understand this result of Aronszajn and Smith that led him to abstract the concept of quasitriangularity. After the introduction this seems a natural enough concept. After all the triangular representation of an operator on a finite dimensional space is an excellent tool for their study; perhaps the concept of quasitriangularity can be similarly useful.

There are many examples of quasitriangular operators besides the compact ones. For example, every normal operator is quasitriangular. Moreover, if $T$ is quasitriangular and $K$ is any compact operator, then $T + K$ is quasitriangular. One of the most famous operators, however, fails to be quasitriangular. If $\mathcal{H} = \ell^2$, the space of square summable sequences, and $S \colon \ell^2 \to \ell^2$ is defined by $S(a_0, a_1, \ldots) = (0, a_0, a_1, \ldots)$, then $S$ is called the unilateral shift. [The unilateral shift is also the analytic Toeplitz operator where for the function $\phi$ we take $\phi(z) = z$.] It turns out that the unilateral shift is not quasitriangular although its adjoint is.

In [10], R.G. Douglas and Carl Pearcy showed that if $T$ is an operator for which there is a complex number $\lambda$ such that $T - \lambda$ is semi-Fredholm with a negative Fredholm index, then $T$ is not quasitriangular. This squares with the example of the unilateral shift since this operator is a Fredholm operator with index $-1$. (An operator $A$ is a *semi-Fredholm operator* provided the range of $A$, $\operatorname{ran} A$, is closed and at least one of the spaces $(\operatorname{ran} A)^\perp$ and $\ker A$ is finite dimensional. The operator is called a *Fredholm operator* if both of the spaces $(\operatorname{ran} A)^\perp$ and $\ker A$ are finite dimensional. When we have a semi-Fredholm operator $A$ the *index* of $A$ is defined by $\operatorname{ind} A \equiv \dim[\ker A] - \dim[(\operatorname{ran} A)^\perp]$. The unilateral shift has index $-1$ because it has no kernel—it is an isometry—and $(\operatorname{ran} S)^\perp$ is spanned by the vector $(1, 0, 0, \ldots)$. Here and later we will have need to refer to semi-Fredholm operators and the Fredholm index. The reader who is not familiar with this topic will have to suffer a bit as an explanation of this important idea is too lengthy to insert in this article. Such a discussion can be found in most of the standard works on operator theory, for example, see [8]. What is used most here, and in most other places where the Fredholm index is applied, are the values of $\text{ind}(A - \lambda)$ for the complex numbers $\lambda$ where the operator $A - \lambda$ is semi-Fredholm. The complex numbers $\lambda$ for which $A - \lambda$ is semi-Fredholm and $\text{ind}(A - \lambda) \neq 0$ of necessity belong to the spectrum of $A$; the exact values of $\text{ind}(A - \lambda)$ provide additional spectral information about the operator. An elementary exercise for the reader is to show that when the underlying Hilbert space is finite dimensional the only possible value of the index is 0. Thus the Fredholm index is a concept whose use is restricted to operators on infinite dimensional spaces.)

Later, in the last of a sequence of four papers, Apostol, Foias, and Voiculescu [2] proved the converse of this result of Douglas and Pearcy. Thus an operator $T$ is quasitriangular if and only if for each complex number $\lambda$ such that $T - \lambda$ is semi-Fredholm we have that $\text{ind}(T - \lambda) \geq 0$. (See [11] for an excellent expository account of these matters.) This furnishes a characterization of quasitriangular operators in terms of their spectral properties alone.

Now at the end of his paper Halmos [13] posed the question, "Does every quasitriangular operator have a nontrivial invariant subspace?" In light of the Apostol, Foias, and Voiculescu result this question is seen to be equivalent to the general (celebrated) question of whether every operator on Hilbert space has an invariant subspace. Indeed, if $T - \lambda$ is semi-Fredholm and $0 > \text{ind}(T - \lambda) = \dim[\ker(T - \lambda)] - \dim[\text{ran}(T - \lambda)]$, then the range of $T - \lambda$ is seen to be a nontrivial invariant subspace for $T$. [Incidentally, if the adjoint of $T$, $T^*$, is not quasitriangular, then $T$ also has a nontrivial invariant subspace. In fact here $0 > \text{ind}(T - \lambda)^* = -\text{ind}(T - \lambda)$ so that $\ker(T - \lambda)$ is a nontrivial invariant subspace.] Halmos comments in his paper after posing this question that there is reason to believe that the general question is equivalent to his, but I doubt that he ever suspected that such a dramatic result would bear him out.

The irony here has always struck me. A concept that evolved from a positive invariant subspace result (compact operators have invariant subspaces) has proved to be the intractable case for the invariant subspace problem. Thus the general invariant subspace problem is equivalent to showing that *biquasitriangular* operators (those operators $T$ such that both $T$ and $T^*$ are quasitriangular) have invariant subspaces.

With the reader's indulgence, this might be a good place to consider the importance of the invariant subspace problem. Why is it important? In fact I think this problem, like most of the great problems in mathematics, is important because mathematicians have decided it is important. Unlike The Riemann Hypothesis, there are no important applications awaiting its solution. It is just that through the years the best operator theorists of the time, starting with John von Neumann, have all regarded this problem as interesting and important and have devoted some portion of their life to its solution.

Another direction of research initiated by Halmos and attracting increasing numbers of contributors today is the area of nonabelian approximation. This is the term applied to approximation in $\mathcal{B}(\mathcal{H})$, which has a strong claim to being a nonabelian version of the space of continuous functions on a compact Hausdorff space. Here we always assume $\mathcal{H}$ to be separable in order to avoid technicalities (and trivialities). If I am given a set of operators, which operators belong to its closure? [Here "closure" refers to the topology defined by the norm on $\mathcal{B}(\mathcal{H})$; it is possible to talk of approximation in the so-called strong and weak operator topology, but this is not what will be considered here.]

One of the first such results appeared in [14] when Paul proved that the set of irreducible operators is a dense $G_\delta$. [An operator $T$ is *irreducible* if there are no nontrivial reducing subspaces; that is, if there are no closed subspaces other than $(0)$ and $\mathcal{H}$ that are invariant for both $T$ and $T^*$.] In another paper on this topic (see [16]) he considered the distance from an arbitrary operator to the set of positive operators.

But the real beginning of nonabelian approximation theory came in 1970 when Paul delivered ten lectures in a conference at Texas Christian University sponsored by the Conference Board of the Mathematical Sciences. These lectures were published that same year [15]. This list of ten problems has had an enormous affect on operator theory and probably deserves an expository article of its own. In fact Halmos himself has written such an article [17] on the tenth anniversary of his lectures. Another follow-up article is fully justified.

Question 7 in his famous list of 10 problems was "Is every quasinilpotent operator (the definition is given below) the norm limit of nilpotent ones?" Question 8: "Is every operator the norm limit of reducible ones?"

This last question seems like a so-what question. It is a natural thing to ask after his result that the irreducible operators are dense, but it is also the type that often elicits the reaction, "Who cares?" I have often been mystified by such a reaction. Different questions, of course, have varying levels of appeal to different mathematicians so I understand any individual person declining to work on a specific problem. But clearly the poser of the question cares. More importantly, it seems to me that only answers can be summarily dismissed. The value of a question cannot be ascertained until it is answered.

In fact this "so what" question elicited some beautiful mathematics that has had far ranging consequences for operator theory as well as the theory of $C^*$ algebras. This eighth question was answered in the affirmative by Dan Voiculescu [23] when he proved his celebrated noncommutative Weyl–von Neumann Theorem on the representations of separable $C^*$ algebras.

Let $\mathcal{A}$ be a separable $C^*$ algebra contained in $\mathcal{B}(\mathcal{H})$ (remember that $\mathcal{H}$ is assumed separable) and containing the identity operator. A *representation* of $\mathcal{A}$ is a *-homomorphism $\rho\colon \mathcal{A} \to \mathcal{B}(\mathcal{K})$ for some Hilbert space $\mathcal{K}$ such that $\rho(1) = 1$. Two representations $(\rho_1, \mathcal{K}_1)$ and $(\rho_2, \mathcal{K}_2)$ are *approximately equivalent* if there is a sequence of unitary operators $U_n\colon \mathcal{K}_1 \to \mathcal{K}_2$ such that for each $A$ in $\mathcal{A}$:

$$\|\rho_1(A) - U_n^* \rho_2(A) U_n\| \to 0$$

and

$$\rho_1(A) - U_n^* \rho_2(A) U_n$$

is compact. It is a result of Voiculescu that if a sequence of unitary operators can be found such that the first of these two conditions is satisfied, then a sequence of unitaries can be found such that both conditions are satisfied and the representations are approximately equivalent (see Theorem 1.5 of [23]). In other words, two representations are approximately equivalent if a sequence of unitary operators can be found such that the first condition is satisfied.

Here are examples of two approximately equivalent representations. Let $T$ be any operator in $\mathcal{B}(\mathcal{H})$ and let $\mathcal{A}$ be the $C^*$-algebra generated by $T$. Thus $\mathcal{A}$ is the closure $\{p(T, T^*)\colon p$ is a polynomial in the two noncommuting operators $T$ and $T^*\}$. Define $\rho\colon \mathcal{A} \to \mathcal{B}(\mathcal{H})$ by $\rho(A) = A$ for all $A$ in $\mathcal{A}$; yes, that's right, $\rho$ is the identity representation. Now let $T_1$ be an operator in the closure of $\{U^*TU\colon U$ is a unitary operator$\}$. So there is a sequence $\{U_n\}$ of unitaries such that $U_n^*TU_n \to T_1$. It is easy to check that $U_n^* p(T, T^*)U_n \to p(T_1, T_1^*)$ for every polynomial $p$. Thus $\|p(T_1, T_1^*)\| = \lim \|U_n^* p(T, T^*)U_n\| = \|p(T, T^*)\|$. It follows that there is a representation $\rho_1\colon \mathcal{A} \to \mathcal{B}(\mathcal{H})$ such that $\rho_1[p(T, T^*)] = p(T_1, T_1^*)$ for every polynomial $p$. In particular, $\rho_1(T) = T_1$.

Note that the converse of this construction is valid. If $\mathcal{A}$ is the $C^*$ algebra generated by the operator $T$, $\rho\colon \mathcal{A} \to \mathcal{B}(\mathcal{H})$ is the identity representation, and $\rho_1\colon \mathcal{A} \to \mathcal{B}(\mathcal{H})$ is an approximately equivalent representation, then the operator $T_1 = \rho_1(T)$ belongs to the closure of the unitary orbit of $T$.

A representation $\rho\colon \mathcal{A} \to \mathcal{B}(\mathcal{K})$ is *irreducible* if there is no nontrivial subspace of $\mathcal{K}$ that reduces $\rho(A)$ for every operator $A$ in $\mathcal{A}$. If $\mathcal{A}$ is the $C^*$ algebra generated by the single operator $T$, then $\rho$ is an irreducible representation if and only if $\rho(T)$ is an irreducible operator. Voiculescu's noncommutative Weyl–von Neumann Theorem asserts that every representation of a separable $C^*$ algebra is approximately equivalent to the infinite direct sum of irreducible representations. (Actually this is a corollary of Voiculescu's main theorem.) Note that this says that every representation is approximately equivalent to a representation that is reducible. (Incidentally, in simple terms the original Weyl–von Neumann Theorem states that every self-adjoint operator is approximately unitarily equivalent to a diagonal self-adjoint operator—that is, a self-adjoint operator whose eigenvectors span the space.)

Now let's apply Voiculescu's Theorem to the identity representation $\rho: \mathcal{A} \to \mathcal{B}(\mathcal{H})$ of the $C^*$ algebra generated by the arbitrary operator $T$. Thus $\rho$ is approximately equivalent to a representation $\rho_1: \mathcal{A} \to \mathcal{B}(\mathcal{H})$ that is the infinite direct sum of irreducible representations. Putting $T_1 = \rho_1(T)$ and letting $\{U_n\}$ be the sequence of unitaries such that $U_n^* \rho_1(A)U_n \to \rho(A) = A$ for all $A$ in $\mathcal{A}$, we get that $U_n^* T_1 U_n \to T$ and each of the operators $U_n^* T_1 U_n$ is decidedly reducible.

This solves Halmos's Problem 8 from [15]: the reducible operators are norm dense in $\mathcal{B}(\mathcal{H})$. Voiculescu clearly attributes his work [23] to his efforts to answer this question of Halmos. In the meantime Voiculescu's Theorem has been used by many who work in $C^*$ algebras and is considered one of the better results in this field in recent times.

Now let's return to Halmos's Question 7 from [15], "Is every quasinilpotent operator the norm limit of nilpotent ones?" A *quasinilpotent operator* is an operator $T$ whose spectrum is the singleton $\{0\}$; this is equivalent to the condition that $\|T^n\|^{1/n} \to 0$. A *nilpotent* operator is one that satisfies the equation $T^n = 0$ for some integer $n$. Again this seems like an innocent and naive question, but it triggered a sequence of papers that have produced truly significant mathematics. Halmos's Question 7 also has an affirmative answer as was shown by Apostol and Voiculescu [5].

Quasinilpotent operators have a special interest for operator theorists. They seem very mysterious. Part of the interest stems from the invariant subspace problem. If $T$ is an operator and its spectrum is disconnected, then it is easy to show (by using the Riesz functional calculus learned in a first course in functional analysis) that $T$ has a nontrivial invariant subspace. Thus it suffices to consider the case where the spectrum of $T$ is connected. The simplest case of a connected set in the plane, or anywhere else, is a singleton. So why not assume that $T$ has as its spectrum the singleton $\{0\}$, that is, why not assume that $T$ is quasinilpotent. In fact it is an unsolved problem as to whether every quasinilpotent operator has a nontrivial invariant subspace. In fact it would not be too surprising if this question were equivalent to the entire invariant subspace question.

A number of papers on nonabelian approximation theory then began to flow from the Romanian operator theory group (which has since become, for the most part, a group of Americans) and other mathematicians. Voiculescu [22] characterized the closure of the algebraic operators which can also be used to answer Halmos's question. Another result of some significance surfaced when Apostol and Morrel [4] characterized the closure of the set of all operators whose spectrum is contained in an arbitrary but fixed subset of the plane. Actually they succeeded in giving what amounts to an approximate Jordan canonical form for operators on an infinite dimensional space. The word "canonical" gets lost in the change to infinite dimensions; there is nothing unique about their model.

Apostol and Morrel called their approximate Jordan forms "simple models." From the point of view of the experienced operator theorist the use of the word "simple" is accurate, but the novice might think this word inappropriate. The point is that there are some basic building blocks that can be combined to approximate any operator. Let's look at these blocks.

Let $\Omega$ be a finitely connected region in the complex plane such that its boundary $\Gamma$ is a finite collection of analytic Jordan curves; call such a region a *Cauchy region*. Let $H^2(\Omega)$ be the closure in the $L^2$ space of arc length measure on $\Gamma$ of the space of rational functions with poles off the closure of $\Omega$. Define the operator $A = A_\Omega$ on $H^2(\Omega)$ as multiplication by the independent variable: $(Af)(z) = zf(z)$ for all $z$ in $\Omega$ and all $f$ in $H^2(\Omega)$. The operators $A_\Omega$ for the Cauchy regions $\Omega$ form one type of building block. The second type of building block is found by taking the adjoints of the operators $A_\Omega$; let $B_\Omega = A_\Omega^*$.

The last type of block is formed by letting $C$ be a normal operator with finite spectrum. An easy application of The Spectral Theorem says that if $\lambda_1, \ldots, \lambda_k$ are the distinct points in the spectrum of $C$ and $P_1, \ldots, P_k$ are the orthogonal projections onto the corresponding eigenspaces, then $P_i P_j = 0$ for $i \neq j$ and

$$C = \bigoplus_{i=1}^{k} \lambda_i P_i.$$

These three types of operators are the simple building blocks.

Finally, a piece of notation. If $T$ is any operator acting on the Hilbert space $\mathcal{H}$ and $1 \leq n \leq \infty$, let $\mathcal{H}^{(n)}$ be the direct sum of $\mathcal{H}$ with itself $n$ times and denote by $T^{(n)}$ the operator on $\mathcal{H}^{(n)}$ defined by $T^{(n)}(h_1, h_2, \ldots) = (Th_1, Th_2, \ldots)$.

The result of Apostol and Morrel is that for any operator $T$ and any $\varepsilon > 0$, there are Cauchy regions $\Omega_1, \ldots, \Omega_n, \Lambda_1, \ldots, \Lambda_m$, positive extended integers $p_1, \ldots, p_n, q_1, \ldots, q_m$, a normal operator with finite spectrum $C$, and an operator $S$ similar to

$$\bigoplus_{i=1}^{n} A_{\Omega_i}^{(p_i)} \oplus \bigoplus_{j=1}^{m} B_{\Lambda_j}^{(q_j)} \oplus C$$

such that $\|T - S\| < \varepsilon$.

The reason for the operators of the type $A_\Omega$ and $B_\Lambda$ is to take care of the fact that there may be complex numbers $\lambda$ such that $T - \lambda$ is semi-Fredholm with $\mathrm{ind}(T - \lambda)$ either negative or positive, respectively. If $\mathrm{ind}(T - \lambda)$ never has a nonzero value where it is defined, then $T$ can be approximated by an operator similar to a normal operator with finite spectrum. In the finite dimensional case the Fredholm index is always 0 so that this result does not recapture the standard theorem on Jordan forms. In this case all it says is that any operator on a finite dimensional space can be approximated by an operator similar to a normal operator. It is an amusing exercise to give a direct proof of this fact.

Unlike Jordan forms in finite dimensional spaces, this result of Apostol and Morrel cannot be used to characterize operators that are approximately similar. What does this mean? In fact a result of great significance both for nonabelian approximation theory as well as for operator theory in general is the characterization of the closure of the similarity orbit of an operator by Apostol, Herrero, and Voiculescu [3]. That is, for any operator $T$ they identify all the operators that belong to the closure of the set $\{RTR^{-1}:$ $R$ is invertible$\}$. This characterization is given solely in terms of various parts of the spectrum, the values of the Fredholm index, and other spectral properties of the operator. It is certainly a candidate for the most important result in nonabelian approximation. In finite dimensional spaces, Jordan canonical forms give complete similarity invariants for any operator. This work [3] can be thought of as approximate similarity invariants just as the work [4] can be thought of as approximate Jordan forms.

If the above discussion does not suffice to convince the reader that nonabelian approximation theory has progressed to a subarea of operator theory, then the existence of a monograph (see [1] and [18]) can also be offered in evidence.

Over the years Paul Halmos has asked many questions. Solutions of his problems have almost always resulted in interesting mathematics and often in the establishment of reputations. A problem from Halmos is something to be taken seriously. The examples given here are ample testimony to this.

There aren't many people who can give a talk or write an article the principal aim of which is to pose problems. For most of those who can make such a presentation there aren't many who are willing to sit and listen. Paul's success in both aspects of such a venture is a testament to his insight, his sense of taste, the force of his personality, and his ability to present the problems in an intriguing and thought provoking way.

There are other ways in which he has influenced operator theory. Each area of mathematics has a personality and style. There is no doubt that the style of operator theory in the United States has been greatly influenced by Paul Halmos. It has always seemed to me that written and oral exposition in operator theory is somewhat better than in many other areas; operator theorists attend meetings of the society regularly, thus fostering greater exchange and *esprit de corps* (for a relatively small area we seem to have an inordinate number of conferences). I attribute many of these traits to Paul's style and ardour for mathematics and operator theory. We have all been enriched by it.

*Happy Birthday, Paul!*

# REFERENCES

1. C. Apostol, L. Fialkow, D. Herrero, and D. Voiculescu, *Approximation of Hilbert Space Operators,* Pitman Publishing Co., London, 1984.

2. C. Apostol, C. Foias, and D. Voiculescu, "Some results on non-quasi-triangular operators, IV," *Rev. Roumaine Math. Pures Appl.* **18** (1973), 487–514.

3. C. Apostol, D.A. Herrero, and D. Voiculescu, "The closure of the similarity orbit of a Hilbert space operator," *Bull. Am. Math. Soc.* **6** (1982), 421–426.

4. C. Apostol and B.B. Morrel, "On uniform approximation of operators by simple models," *Indiana Univ. Math. J.* **26** (1977), 427–442.

5. C. Apostol and D. Voiculescu, "On a problem of Halmos," *Rev. Roumaine Math. Pures Appl.* **19** (1974), 283–284.

6. N. Aronszajn and K.T. Smith, "Invariant subspaces of completely continuous operators," *Ann. Math.* **60** (1954), 345–350.

7. J. Bram, "Subnormal operators," *Duke Math. J.* **22** (1955), 75–94.

8. J.B. Conway, *A Course in Functional Analysis*, Second edition, Springer-Verlag, New York, 1990.

9. J.B. Conway, *The Theory of Subnormal Operators*, Amer. Math. Soc. Math. Surveys, Vol. 36, Providence, 1991.

10. R.G. Douglas and C. Pearcy, "A note on quasitriangular operators," *Duke Math. J.* **37** (1970), 177–188.

11. R.G. Douglas and C. Pearcy, "Invariant subspaces of non-quasitriangular operators," *Proc. Conf. Operator Theory*, Springer Lectures Notes, Springer-Verlag, Berlin, 1973, Vol. 345.

12. P.R. Halmos, "Normal dilations and extensions of operators," *Summa Bras. Math.* **2** (1950), 125–134.

13. P.R. Halmos, "Quasitriangular operators," *Acta Sci. Math. (Szeged)* **29** (1968), 283–293.

14. P.R. Halmos, "Irreducible operators," *Michigan Math. J.* **15** (1968), 215–233.

15. P.R. Halmos, "Ten problems in Hilbert space," *Bull. Am. Math. Soc.* **76** (1970), 887–933.

16. P.R. Halmos, "Positive approximants of operators," *Indiana Math. J.* **21** (1972), 951–960.

17. P.R. Halmos, "Ten years in Hilbert space," *Integral Equations Operator Theory* **2** (1979), 529–564.

18. D.A. Herrero, *Approximation of Hilbert Space Operators*, I, Pitman Publishing Co., London, 1982.

19. M. Martin and M. Putinar, *Lectures on Hyponormal Operators*, Birkhäuser, Basel, 1982.

20. B. Sz-Nagy, "Sur les contractions de l'espace de Hilbert," *Acta Sci. Math. (Szeged)* **15** (1953), 87–52.

21. B. Sz-Nagy and C. Foias, *Harmonic Analysis of Operators on Hilbert Space*, North-Holland, Amsterdam, 1970.

22. D. Voiculescu, "Norm-limits of algebraic operators," *Rev. Roumaine Math. Pures Appl.* **19** (1974), 371–378.

23. D. Voiculescu, "A non-commutative Weyl–von Neumann Theorem," *Rev. Roumaine Math. Pures Appl.* **21** (1976), 97–113.

*Department of Mathematics*
*University of Tennessee*
*Knoxville, TN 37996*

# Characteristic Classes and Characteristic Forms

## Shiing-shen Chern

Chern, 1959

## Stiefel–Whitney and Pontrjagin Classes

A vector bundle is a family of vector spaces, real or complex, which is parametrized by a manifold, such that (1) it is locally a product, and (2) the linear structure on the vector spaces (called fibers) has a meaning. More precisely, this can be described by a diagram:

$$V_q \xrightarrow{\ j\ } E$$
$$\pi \downarrow \qquad \downarrow \pi \quad , \qquad (1)$$
$$x \ \longrightarrow \ M$$

where $M$ is a manifold and $x \in M$. The bundle $E$ consists of the vector spaces $V_q$ which are inverse images under $\pi$ of all points $x \in M$. We suppose all spaces and maps to be smooth. Conditions (1) and (2) can be expressed as follows: (1) Every point $x \in M$ has a neighborhood $U$ such that $\pi^{-1}(U)$ is a product $U \times Y$ ($Y$ is the $q$-dimensional vector space) and thus has the local coordinates $(x, y_U)$, with $x \in U$, $y_U \in Y$. (2) When the neighborhoods $U$ and $V$ have a nonempty intersection, the coordinates $y_U$ and $y_V$ are related by a linear transformation:

$$y_U g_{UV}(x) = y_V, \qquad (2)$$

where we write the action on $Y$ by $GL(q)$ [$= GL(q; R)$ or $GL(q; C)$] to the right.

An immediate question is: Is such a local product structure always a product globally? This is not an easy question. Its answer is provided by the following theorem:

**Theorem (Poincaré–Hopf).** *Let $M$ be a compact orientable manifold and $s(x)$, $x \in M$ be a generic vector field on $M$. Then*

$$\Sigma \ (zeroes \ of \ s) = \chi(M), \qquad (3)$$

*where $\chi(M)$ is the Euler–Poincaré characteristic of $M$.*

We shall call a map

$$s: M \longrightarrow E \tag{4}$$

a section, if it satisfies the condition $\pi \circ s =$ identity. A product bundle has sections which are nowhere zero. Hence the theorem of Poincaré–Hopf says that if $\chi(M) \neq 0$, the tangent bundle of $M$ is not a product.

But formula (3) says more. It can be used to define $\chi(M)$ in terms of the left-hand side, i.e., in terms of a generic vector field. More generally, let $s_1, \ldots, s_k$ be $k$ vector fields on $M$ in general position. The points where they are linearly dependent form a $(k-1)$-dimensional set. It is tempting to make it a cycle and consider its homology class. This was done by E. Stiefel in his doctoral thesis in 1936. At the same time Hassler Whitney took a broader viewpoint and studied general sphere bundles. He also noticed that, as the "obstructions" are defined by conditions, the invariants should be cohomology classes. Without going into the fine points of algebraic topology such as local coefficients, the upshot is the introduction of the Stiefel–Whitney characteristic classes for a real vector bundle:

$$W^i \in H^i(M, \mathbb{Z}_2), \quad 1 \leqq i < q \tag{5a}$$

with the Euler class

$$W^q \in H^q(M, \mathbb{Z}). \tag{5b}$$

The restriction to coefficients $\mathbb{Z}_2$ in relation (5a) is dictated by the topological properties of the Stiefel manifolds, viz, the presence of torsion.

Whitney noticed the importance of the tautological bundle

$$\pi_0: \longrightarrow Gr(q, N), \tag{6}$$

where $Gr(q, N)$ is the Grassmann manifold of all the $q$-dimensional linear spaces through the origin in $R^{q+N}$ or $C_{q+N}$, as the case may be, and the fibers are the spaces themselves. This bundle is now called universal, because if $N$ is sufficiently large, any bundle can be induced by a mapping

$$f: M \longrightarrow Gr(q, N).$$

Again under the assumption that $N$ is large, Pontrjagin observed that the mapping $f$ is defined up to a homotopy. The characteristic classes are naturally defined to be the elements in the image

$$f^* H^*[Gr(q, N)] \subset H^*(M).$$

From this viewpoint the crucial problem is the homology of the Grassmann manifolds. This was studied in a definitive way by C. Ehresmann, beginning in his thesis in 1934 (see [3]). It turns out that with the exception of the Euler class in relation (5b), for real vector bundles the integral characteristic classes are of dimension $4k$. They are called the Pontrjagin classes and are denoted by $p_k(E) \in H^{4k}(M; \mathbb{Z})$. For reference to results in this section we recommend the classical book [4].

# Characteristic Classes in Terms of Curvature

It was a trivial observation, and a stroke of luck, when I saw in 1944 that the situation for complex vector bundles is far simpler, because most of the classical complex spaces, such as the complex Grassmann manifolds, the complex Stiefel manifolds, etc., have no torsion. In particular, for a $q$-dimensional complex vector bundle, by generalizing the construction of the Stiefel–Whitney classes, the consideration of $q - s + 1$ sections in general position leads to the Chern classes $c_s(E) \in H^{2s}(M; \mathbb{Z})$, with $1 \leq s \leq q$, which are cohomology classes with integer coefficients.

These classes have the important property that they are expressible in the sense of the de Rham theory by exterior differential forms constructed geometrically. The basic idea is that of a connection or absolute differentiation. The sections (4) form a vector space which we denote by $\Gamma(E)$. A connection is a map

$$D: \Gamma(E) \to \Gamma(E \otimes T^*M), \qquad (7)$$

i.e., a $\Gamma(E)$-valued linear differential form, satisfying the following conditions:

$$D(s_1 + s_2) = Ds_1 + Ds_2, \qquad (8)$$

$$D[h(x)s] = dh \otimes s + hDs, \qquad (9)$$

where $s_1, s_2, s \in \Gamma(E)$ and $h(x)$, with $x \in M$, is a function on $M$. Such a connection always exists, even in the real analytic category.

A connection has local invariants in the form of "curvature," measuring the noncommutativity of the differentiation so defined. To describe it we call a frame an ordered set of $q$ sections $s_i$, with $1 \leq i \leq q$, such that $s_1 \wedge \cdots \wedge s_q \neq 0$. In a neighborhood $U$ we consider a frame field $s_i(x)$, with $x \in U$. Then we can write

$$Ds_i = \Sigma \omega_i^j \otimes s_j, \quad 1 \leq i, j \leq q. \qquad (10)$$

The matrix

$$\omega = (\omega_i^j) \qquad (11)$$

is a $(q \times q)$ matrix of one-forms and is called the connection matrix. Since any section is a linear combination of the sections of a frame, the connection $D$ is completely determined by $\omega$. In an obvious matrix notation we write Eq. (10) as

$$Ds = \omega s. \qquad (12)$$

Let

$$s^1(x) = g(x)s(x), \quad x \in U \qquad (13)$$

be a change of the frame field, with $g(x)$ being a nonsingular $(q \times q)$ matrix of functions on $U$. By differentiating Eq. (13) and using the conditions (8) and (9) we get

$$dg + g\omega = \omega^1 g, \qquad (14)$$

where $\omega^1$ is the connection matrix relative to the frame field $s^1(x)$.

Taking the exterior derivative of Eq. (14), we get

$$g\Omega = \Omega^1 g, \tag{15}$$

where

$$\Omega = d\omega - \omega \wedge \omega \tag{16}$$

and $\Omega^1$ has a similar expression in terms of $\omega^1$. By definition $\Omega$ is a $(q \times q)$ matrix of two-forms and is called the curvature matrix.

Equation (15) motivates us to consider the determinant

$$\det\left(1 + \frac{i}{2\pi}\Omega\right) = 1 + c_1(\Omega) + \cdots + c_q(\Omega), \tag{17}$$

so that $c_i(\Omega)$ is a $2i$-form on $U$, with $1 \leq i \leq q$, independent of the choice of the frame field. We take a covering $\{U_\alpha\}$ of $M$ and define $c_i(\Omega)$ in each member of the covering. Equations (15) and (17) show that they agree in the intersection of any two neighborhoods of the covering. Hence $c_i(\Omega)$ is a $2i$-form in $M$. Moreover, it is closed. By the de Rham theory it defines an element $\{c_i(\Omega)\} \in H^{2i}(M; \mathbb{R})$.

A fundamental result, which is not hard to prove, says that $\{c_i(\Omega)\}$ is exactly the Chern class $c_i(E)$ introduced above. The form $c_i(\Omega)$ is called a Chern form. This analytical representation of $c_i(E)$, by curvature, has important developments. For details see [1] and [2].

The Pontrjagin classes can be expressed in terms of the Chern classes through the complexification of the real vector bundle.

## Transgression

When a characteristic form is pulled back in a bundle, it could happen that it becomes an exact form. Such a process is called transgression. It is clearly of great importance.

My favorite example is the Gauss–Bonnet formula. Let $M$ be a two-dimensional oriented Riemannian manifold. To a unit tangent vector $e_1$ at $x \in M$ there is a unique unit tangent vector $e_2 = e_1^\perp$, so that $xe_1e_2$ is a frame. Let $\omega_1, \omega_2$ be its dual coframe. In the unit tangent bundle

$$\pi \colon E \to M \tag{18}$$

the forms $\omega_1, \omega_2$ are well-defined. In fact, $\omega_1 \wedge \omega_2$ is in $M$ and is the element of area. There is a uniquely defined one-form $\omega_{12}$, the connection form, which satisfies the conditions

$$d\omega_1 = \omega_{12} \wedge \omega_2, \quad d\omega_2 = \omega_1 \wedge \omega_{12}, \tag{19}$$

expressing the fact that the connection has no torsion.

By taking the exterior derivatives of conditions (19), we find

$$d\omega_{12} = -K\omega_1 \wedge \omega_2, \tag{20}$$

where $K$ is the Gaussian curvature.

Equation (20) is of great importance. By Gauss–Bonnet $\frac{1}{2\pi}\{K\omega_1 \wedge \omega_2\}$ represents the Euler class. Now Eq. (20) says that its pull-back in $E$ can be further "integrated," giving the connection form. The equation gives the most natural proof of the Gauss–Bonnet formula, which generalizes to high dimensions. It has many other applications.

Further transgression phenomena occur in the principal bundle or frame bundle. In Eq. (13) we let $g$ run over all nonsingular $(q \times q)$ matrices, i.e., we consider all frames over $M$. The frame bundle $P$ so obtained has the local coordinates $(x, g)$. In $P$, the connection form, to be called $\varphi$, is well-defined. By Eq. (14) it has the local expression

$$\varphi = dgg^{-1} + g\omega g^{-1}. \tag{21}$$

The curvature form

$$\Phi = d\varphi - \varphi \wedge \varphi = g\Omega g^{-1} \tag{22}$$

is also well-defined in $P$.

By Eq. (17) we have

$$c_1(\Phi) = \frac{i}{2\pi}\mathrm{Tr}\,\Phi.$$

Since $\mathrm{Tr}(\varphi \wedge \varphi) = 0$, we get from Eq. (22),

$$c_1(\Phi) = \frac{i}{2\pi}d\,\mathrm{Tr}\,\varphi. \tag{23}$$

In other words, in $P$ the first Chern form can be written explicitly as an exact form.

Similarly, Eq. (17) gives

$$c_2(\Phi) = \left(\frac{i}{2\pi}\right)^2 \{c_1(\Phi)^2 - \mathrm{Tr}(\Phi \wedge \Phi)\}. \tag{24}$$

Exterior differentiation of Eq. (22) gives the Bianchi identity

$$d\Phi = \varphi \wedge \Phi - \Phi \wedge \varphi. \tag{25}$$

It follows that

$$d\{\mathrm{Tr}(\varphi \wedge \varphi \wedge \varphi)\} = 3\,\mathrm{Tr}(\varphi \wedge \varphi \wedge \Phi)$$
$$d\,\mathrm{Tr}(\varphi \wedge \Phi) = -\mathrm{Tr}(\varphi \wedge \varphi \wedge \Phi) + \mathrm{Tr}(\varphi \wedge \Phi).$$

We put

$$CS(\varphi) = +\frac{1}{3}\mathrm{Tr}(\varphi \wedge \varphi \wedge \varphi) + \mathrm{Tr}(\varphi \wedge \Phi). \tag{26}$$

Then we have
$$dCS(\varphi) = \text{Tr}(\Phi \wedge \Phi). \tag{27}$$

The form in Eq. (26) is called the Chern–Simons form. It transgresses the second Chern form in the frame bundle. This is a three-form defined without reference to a metric in $M$. For dimension reason it is closed if $M$ is of dimension 3.

The Chern–Simons form has played an important role in recent developments in theoretical physics (see [6]).

## Holomorphic Line Bundles and the Nevanlinna Theory

We consider the case $q = 1$, where the fiber is the complex line and the group $GL(1; C)$ is abelian. The universal bundle (6) becomes

$$\pi_0 : C_{N+1} \to Gr(1, N) = P_N(C), \tag{28}$$

the latter being the complex projective space of dimension $N$. Denote by $z_0, z_1, \ldots, z_n$ the coordinates in $C_{N+1}$. Restricting the bundle to the unit sphere

$$z_0 \bar{z}_0 + \ldots + z_N \bar{z}_N = 1,$$

we get the Hopf fibering. For $N = 2$ this gives the famous Hopf map $\pi_0 : S^3 \to P_1(C) = S^2$, the first example of a continuous map from a space to one of lower dimension, which is not homotopic to the constant map.

To study the geometry of the universal line bundle (28) we introduce in $C_{N+1}$ the hermitian inner product

$$(Z, W) = z_0 \bar{w}_0 + \cdots + z_N \bar{w}_N, \tag{29}$$

where
$$Z = (z_0, \ldots, z_N), \quad W = (w_0, \ldots, w_N) \in C_{N+1}. \tag{30}$$

Let $Z_0 \in C_{N+1}$ and let $Z_0, Z_1, \ldots, Z_N$ be a unitary frame such that

$$(Z_A, Z_B) = \delta_{A\bar{B}}, \quad 0 \leqq B, A, C \leqq N. \tag{31}$$

The family of all unitary frames can be identified with the group $U(N+1)$. We can write

$$dZ_A = \Sigma \omega_{A\bar{B}} Z_B, \tag{32}$$

where $\omega_{A\bar{B}}$ are the Maurer–Cartan forms of $U(N + 1)$. As a consequence of Eq. (31) they satisfy the relations

$$\omega_{A\bar{B}} + \omega_{\bar{B}A} = 0, \quad \omega_{\bar{B}A} = \bar{\omega}_{B\bar{A}}. \tag{33}$$

Exterior differentiation of Eq. (32) gives the Maurer–Cartan equations

$$d\omega_{A\overline{B}} = \Sigma\omega_{A\overline{C}} \wedge \omega_{C\overline{B}}. \tag{34}$$

We suppose $Z_0 \in C_{N+1}$ and the projection $\pi_0$ in Eq. (28) be defined by sending it to the point of $P_N$ with $Z_0$ as its homogeneous coordinate vector. By Eq. (32) with $A = 0$, a connection in the bundle (28) can be defined by

$$DZ_0 = \omega_{0\overline{0}}Z_0. \tag{35}$$

By our general theory its curvature is, by Eq. (34),

$$\Omega := d\omega_{0\overline{0}} = \Sigma\omega_{0\overline{k}} \wedge \omega_{k\overline{0}} = -\Sigma\omega_{0\overline{k}} \wedge \omega_{\overline{0}k}. \tag{36}$$

This gives a transgression in the line bundle, because the right-hand side is a two-form in $P_N$.

On the other hand, $P_N$ has the Kähler metric

$$ds^2 = \Sigma\omega_{0\overline{k}}\omega_{\overline{0}k}. \tag{37}$$

Its Kähler form is

$$\frac{i}{2}\Sigma\omega_{0\overline{k}} \wedge \omega_{\overline{0}k} = \frac{1}{2i}\Omega. \tag{38}$$

Integrating over a complex projective line, we find

$$\int_{P_1} \frac{1}{2\pi i}\Omega = 1.$$

We can say that the Poincaré dual $*\{\frac{1}{2\pi i}\omega\}$ is the hyperplane $P_{N-1}$ in $P_N$. It follows that if an algebraic curve in $P_N$ is given by a holomorphic map

$$f: M_1 \rightarrow P_N,$$

where $M_1$ is a compact Riemann surface, we have

$$\frac{1}{2\pi i} \int_{f(M_1)} \Omega = \text{Area}[f(M_1)] = \text{order}. \tag{39}$$

In other words the order of an algebraic curve is identified with the area of the image of $M_1$ under $f$.

The transgression formula (36) allows this identification to be carried further. If $M_1$ is a compact Riemann surface with boundary and $n[f(M_1) \cap P_{N-1}]$ is the number of points of intersection of the image $f(M_1)$ with a generic hyperplane $P_{N-1}$, a simple application of Stokes Theorem to Eq. (36) expresses the difference $n[f(M_1) \cap P_{N-1}] - \text{area}[f(M_1)]$ as an integral over the boundary. The boundary term involves the connection form $\omega_{0\overline{0}}$.

In the holomorphic category the connection form $\omega_{0\bar{0}}$ can be "integrated," so that Eq. (36) is sharpened to a double transgression formula. In fact, let $Z \in C_{N+1} - \{0\}$ and

$$Z_0 = \frac{Z}{|Z|}, \quad |Z|^2 = (Z, Z).$$

We have

$$
\begin{aligned}
\omega_{0\bar{0}} &= (DZ_0, Z_0) = (dZ_0, Z_0) \\
&= \frac{1}{2|Z|^2}\{(dZ, Z) - (Z, dZ)\} = (\partial - \bar{\partial})\log|Z|.
\end{aligned}
\tag{40}
$$

It follows that
$$\Omega = d(\partial - \bar{\partial})\log|Z| = -2\partial\bar{\partial}\log|Z|. \tag{41}$$

This double transgression formula is the key to the so-called first main theorem in the Nevanlinna theory.

The deep result in the Nevanlinna theory is contained in the second main theorem. From our viewpoint it is simply the double transgression formula applied to the canonical bundle. It deals with the question of the relation between intrinsic invariants and mapping invariants and is a generalization of the Riemann–Hurwitz and Plücker formulas.

Recently the Nevanlinna theory gains attention because it can be seen as the limiting case of algebraic number theory. In fact, the second main theorem, properly generalized to arithmetic geometry, contains Roth's theorem on diophantine approximation and Faltings' theorem on the Mordell conjecture (see [5]).

## REFERENCES

1. S. Chern, *Complex Manifolds without Potential Theory*, Second edition, Springer-Verlag, New York, 1979.

2. S. Chern, Vector bundles with a connection, in *Global Differential Geometry*, Studies in Mathematics, Mathematical Association of America, 1989, Vol. 27, pp. 1–26.

3. C. Ehresmann, Sur la topologie de certains espaces homogènes, *Annals Math.* **35** (1934), 396–443.

4. N. Steenrod, *The Topology of Fibre Bundles*, Princeton University Press, Princeton, 1951.

5. Paul Vojta, *Diophantine Approximation and Value Distribution Theory*, Lecture Notes in Mathematics, Springer-Verlag, New York, 1987, Vol. 1239.

6. E. Witten, Quantum field theory and the Jones polynomial, in *Braid Group, Knot Theory, and Statistical Mechanics,* edited by C.N. Yang and M.L. Ke, World Scientific, Singapore, 1989, pp. 239–329.

*Mathematical Sciences Research Institute*
*1000 Centennial Drive*
*Berkeley, CA 94720*

# Equivalents of the Invariant Subspace Problem[†]

## Peter Rosenthal

Peter Rosenthal, 1969

Paul Halmos has shaped much of modern operator theory through his published research, insightful questions, and beautiful expositions. The invariant subspace problem was not originally due to Halmos (von Neumann showed that compact operators have invariant subspaces back in the 1930's [2]), but he played a vital role in arousing interest in the problem through his lectures and writings, and in suggesting the investigation of special cases that turned out to prove to be very fruitful.

In particular, the question of the existence of invariant subspaces for polynomially compact operators was publicized by Halmos [21] and answered via nonstandard analysis by Bernstein and Robinson [6]. Halmos's translation of the proof into standard analysis [18] and isolation of the concept of quasitriangularity [19] led to the deep work of Apostol, Foias, and Voiculescu on quasitriangular operators ([1], also see [14]), and undoubtedly also stimulated Lomonosov's very different approach [23] to a much more general result.

Similarly, Halmos's definition of subnormal operators [20] and question about their having invariant subspaces [21] led to a general study of subnormal operators [10], to Scott Brown's deep theorem on existence of invariant subspaces [7], and to further interesting generalizations [8].

Thus Halmos has had significant influence on work on the invariant subspace problem, and it seems appropriate to have some discussion of it in a volume celebrating the 75th anniversary of his birth.

This is an expository paper addressed to nonspecialists, but we also hope that something in it might suggest something that may lead to further progress.

---

[†]To Paul Halmos with admiration, gratitude, and affection.

We present several reformulations of the invariant subspace problem. Most of the results are very elementary; only one is a little more sophisticated. However, each of the reformulations looks quite different from the original problem, at least on the surface, and thus appears to offer at least a slight potential of providing a useful point of view.

The invariant subspace problem concerns bounded (that is, continuous) linear operators on complex, infinite-dimensional, separable Hilbert space (which is, up to isomorphism, the space of all square-summable sequences of complex numbers); it should be noted that there are counterexamples to the corresponding problems on Frechet spaces [3] and on Banach spaces ([15], [32], [33]; also see [4]).

To describe the problem we fix some standard terminology. A *subspace* of a Hilbert space is a subset that is closed in the topological sense as well as with respect to the vector operations. A subspace is *invariant* under an operator if it is taken into itself by the operator. The *trivial* subspaces are the entire space and the space consisting of the zero vector alone, each of which are clearly invariant under every linear operator. The *invariant subspace problem* is the question whether every bounded linear operator on Hilbert space has a nontrivial invariant subspace.

There are many special cases in which it is known that operators have nontrivial invariant subspaces. In particular, a recent theorem of Brown, Chevreau, and Pearcy states that every operator of norm one whose spectrum includes the unit circle has nontrivial invariant subspaces ([8]; also see [5]). Also, Lomonosov's theorem states that every operator that commutes with an operator that is not a multiple of the identity that itself commutes with a compact operator other than 0 has a nontrivial invariant subspace ([23]; see also [4] or [31]). While it is known that Lomonosov's theorem does not cover all operators [17], the only operators that are known not to satisfy its hypothesis all have obvious invariant subspaces.

There are a number of other invariant subspace theorems (see [31]), but no solution to the general problem is in sight.

We begin with a discussion of universal operators, which are particular operators with the property that certain information about their invariant subspaces would solve the general problem. Two universal operators are presented: Rota's classical example of the backwards shift of infinite multiplicity [34] and a more recent example [28] of a universal composition operator. We then mention a very elementary reformulation of the invariant subspace problem, which goes back to Aronszajn and Smith [2], as a problem about solving a quadratic equation in linear operators. A geometric equivalent due to Nordgren, Radjavi, and the author [26], in terms of direct-sum decompositions of Hilbert spaces, is also described in detail. A reformulation in the language of "weak resolvents" of an operator is also discussed.

Outlines of the proofs of the most elementary results are given; filling in the details of the outlines would not be difficult for the average mathematical reader.

## Universal Operators

An operator is said to be *universal* if every bounded linear operator on separable Hilbert space is similar to a multiple of a part of it; i.e., $T$ is universal if for every bounded linear $A$ there is a scalar $c$ and an invariant subspace $\mathcal{M}$ of $T$ such that $A$ is similar to the restriction of $cT$ to $\mathcal{M}$.

We say that an operator $T$ *has property* $\mathcal{S}$ if whenever $\mathcal{M}$ is an infinite-dimensional invariant subspace of $T$ there is another nontrivial invariant subspace $\mathcal{N}$ of $T$ which is properly contained in $\mathcal{M}$. For any fixed universal operator $T$, the invariant subspace problem is equivalent to the statement: $T$ has property $\mathcal{S}$. For if every operator has nontrivial invariant subspaces then the restriction of $T$ to $\mathcal{M}$ does. Conversely, if a universal operator $T$ has property $\mathcal{S}$ and $A$ is any operator, then $A$ is similar to the restriction of $cT$ to an invariant subspace $\mathcal{M}$ for some $c$, which implies that $A$ has nontrivial invariant subspaces since similarity preserves existence of invariant subspaces.

It may seem surprising that there exist universal operators. The first (and still best-known and most important) example is due to Rota [34]. The *backwards unilateral shift of (countably) infinite multiplicity* is the operator $T$ defined as follows. Let $\mathcal{H}$ be any separable infinite-dimensional Hilbert space, and let $\mathcal{K}$ be the direct sum of countably many copies of $\mathcal{H}$. That is, $\mathcal{K}$ is the set of all sequences $(x_1, x_2, x_3, \ldots)$ with the $x_i$ in $\mathcal{H}$ and $\sum_{i=1}^{\infty} \|x_i\|^2$ finite. Then $T$ is defined on $\mathcal{H}$ by

$$T(x_1, x_2, x_3, \ldots) = (x_2, x_3, \ldots).$$

**Theorem** ([34]). *The backwards unilateral shift of infinite multiplicity is universal.*

**Outline of proof:** Let $T$ be the backwards shift of infinite multiplicity. Given any operator $A$ on $\mathcal{H}$, let $c$ be a positive number greater than the norm of $A$, and $s = \frac{1}{c}$. If $Y$ is the mapping of $\mathcal{H}$ into $\mathcal{K}$ defined by

$$Yx = [x, sAx, (sA)^2 x, (sA)^3 x, \ldots]$$

then $YAx = cTYx$, and $A$ is similar to the restriction of $cT$ to the range of $Y$.

Thus the invariant subspace problem is equivalent to the statement that the unilateral shift of infinite multiplicity has property $\mathcal{S}$. This is equivalent to a certain factorization problem for operator-valued analytic functions ([22], [31], p. 56). There was a period (about 1960 to 1980) during which a great deal of work was done from this point of view, but in recent years it seems that few people have been attempting to solve the invariant subspace problem using this approach.

S.R. Caradus realized that Rota's example could be greatly generalized.

**Theorem** ([9]). *If $T$ is onto and has an infinite-dimensional null space, then $T$ is universal.*

The proof of Caradus's Theorem is not very different from that of Rota's Theorem. Its utility for finding invariant subspaces depends upon finding universal operators that have some additional structure that might make it possible to determine if they have property $S$. An example containing some analytic structure is the following.

Let $\mathcal{H}^2$ denote the usual Hardy space; i.e., $\mathcal{H}^2$ is the space of all functions analytic in the unit disk which have square-summable power series coefficients. Define the function $\phi$ by

$$\phi(z) = \frac{z - 1/2}{1 - (1/2)z}.$$

Then $\phi$ is a hyperbolic disk automorphism. The function $\phi$ induces a composition operator $C_\phi$ on $\mathcal{H}^2$ defined as follows: for $f \in \mathcal{H}^2$, $(C_\phi f)(z) = f[\phi(z)]$. (The study of composition operators involves an interesting interplay between function theory and operator theory — see [25] and [11] for excellent surveys of the subject.) The operator $C_\phi$ has spectrum consisting of a nondegenerate annulus [24].

Nordgren, Wintrobe, and the author used a decomposition theorem for such operators [28] and Caradus' Theorem to prove the following.

**Theorem** ([28]). *If $C_\phi$ is as above and $\lambda$ is any point in the interior of the spectrum of $C_\phi$, then $C_\phi - \lambda$ is universal.*

Since $C_\phi$ and $C_\phi - \lambda$ have the same invariant subspaces, it follows that every operator has nontrivial invariant subspaces if and only if $C_\phi$ has property $S$. This can be reformulated as follows, where "span" means "smallest subspace containing."

**Theorem** ([28] and [29]). *For $\phi(z) = \frac{z-(1/2)}{1-(1/2)z}$, the invariant subspace problem is equivalent to the assertion that whenever $f$ is a nonconstant function in $\mathcal{H}^2$ there exists a function $g$ in $\mathcal{H}^2$ that is not identically $0$ such that the span of*

$$\{C_\phi^n g : n = 0, 1, 2, \dots\}$$

*is a proper subspace of the span of*

$$\{C_\phi^n f : n = 0, 1, 2, \dots\}.$$

This reformulation is a very concrete problem about analytic functions. Also, the proof of the equivalence involves some nontrivial analysis. Whether the answer to the invariant subspace question is affirmative or negative, it seems possible that this version of the problem might provide a good approach. To date, however, it has not led to any further insight into the problem.

It seems that one of the problems in using the backwards shift approach to the invariant subspace problem is the fact that the backwards shift does not have a simple analytic expression, although its adjoint, the forwards shift, is simply multiplication by the independent variable on a vector-valued $\mathcal{H}^2$ space. The composition operator, on the other hand, does have a simple expression. However, composition is a much more complex operation than multiplication.

## Solving a Quadratic Equation

For a given operator $A$, how many operators satisfy the quadratic equation $AX = XAX$?

**Theorem ([2]).** *The equation $AX = XAX$ has a solution other than $X = 0$ and $X = I$ if and only if the operator $A$ has a nontrivial invariant subspace.*

**Outline of proof:** It is very well known and easy to verify that a subspace is invariant under $A$ if and only if the projection $P$ onto the subspace satisfies $AP = PAP$. Thus it need only be shown that the existence of a nontrivial solution $X$ (that need not be a projection) yields an invariant subspace. Given such an $X$, the subspace $\{f : Xf = f\}$ is invariant under $A$, and, since this subspace contains the range of $AX$, it is not hard to see that it is nontrivial unless $A = 0$ or $X = 0$ or $I$.

If every operator has a nontrivial invariant subspace then every operator has an infinite number of invariant subspaces (simply use the fact that restrictions and quotients of a given operator would also have invariant subspaces). Thus the following holds.

**Theorem.** *Every operator has a nontrivial invariant subspace if and only if for every operator $A$ the equation $AX = XAX$ has an infinite number of solutions $X$.*

One might hope to use this observation and known techniques for solving equations to try to prove the existence of invariant subspaces. An investigation of this was begun years ago by Daughtry [12], but it appears to be very difficult to get any substantial results.

# Direct-Sum Decompositions and Invariant Subspaces

The following seems to be a problem about the geometry of Hilbert space. Suppose that we are given two different direct-sum decompositions of a separable infinite-dimensional Hilbert space:

$$\mathcal{H} = \mathcal{K} + \mathcal{L} = \mathcal{M} + \mathcal{N}$$

(we are assuming that $\mathcal{K} \cap \mathcal{L} = \mathcal{M} \cap \mathcal{N} = \{0\}$, but we are not assuming that the decompositions are orthogonal). The problem is: do there necessarily exist equal nontrivial subdecompositions, in the sense that there are subspaces $\mathcal{K}_0$, $\mathcal{L}_0$, $\mathcal{M}_0$, and $\mathcal{N}_0$ of $\mathcal{K}$, $\mathcal{L}$, $\mathcal{M}$, and $\mathcal{N}$, respectively, such that $\mathcal{K}_0 + \mathcal{L}_0 = \mathcal{M}_0 + \mathcal{N}_0$ and $\mathcal{K}_0 + \mathcal{L}_0$ is not $\{0\}$ or $\mathcal{H}$? Nordgren, Radjavi, and the author [26] showed that this is also equivalent to the invariant subspace problem, as follows.

**Theorem** ([26]). *Every pair of direct-sum decompositions of Hilbert space has nontrivial equal subdecompositions if and only if every operator has a nontrivial invariant subspace.*

To see this it is convenient to introduce another reformulation.

If $P$ is the projection on $\mathcal{K}$ along $\mathcal{L}$ and $Q$ is the projection on $\mathcal{M}$ along $\mathcal{N}$ then the existence of equal subdecompositions is easily seen to be equivalent to the existence of common invariant subspaces for $P$ and $Q$. Thus the above is equivalent to the following.

**Theorem** ([26]). *Every operator has a nontrivial invariant subspace if and only if every pair of idempotent operators has a common nontrivial invariant subspace.*

**Outline of proof.** If $A$ is any operator on a Hilbert space $\mathcal{H}$, then

$$P = \begin{pmatrix} A & A \\ I - A & I - A \end{pmatrix}$$

is an idempotent on $\mathcal{H} \oplus \mathcal{H}$. If $Q$ is defined by

$$Q = \begin{pmatrix} I & 0 \\ 0 & 0 \end{pmatrix}$$

then it is not hard to see that $P$ and $Q$ have a common nontrivial invariant subspace if and only if $A$ has a nontrivial invariant subspace.

Although the proof of this equivalence is very easy and elementary, it seems that the equivalent formulation should be useful. It is easily seen that it suffices to restrict to the case where both idempotents have infinite-dimensional ranges and null spaces. The fact that any two such idempotents are similar then allows us to fix one of the idempotents to be any given one, while the other is arbitrary. For example, it might be productive to let $P$ be the orthogonal projection of $\mathcal{L}^2$ of the circle onto $\mathcal{H}^2$, and try to find a $Q$ that has no nontrivial invariant subspace in common with $P$. Such would provide a counterexample to the invariant subspace problem.

It should be mentioned that Chandler Davis [13] showed that there do exist three idempotents (even orthogonal projections) that have no common nontrivial invariant subspaces. Also, Theorem 4.2 has been generalized [30] to pairs of operators satisfying quadratic equations other than $X^2 = X$.

**Theorem** ([30]). *Fix polynomials p and q of degree two. Then every operator has a nontrivial invariant subspace if and only if every pair A, B of operators satisfying $p(A) = q(B) = 0$ has a nontrivial common invariant subspace.*

# Weak Resolvents

If $A$ is a bounded linear operator, a function of the form $z \to [(z-A)^{-1}f, g]$, for vectors $f$ and $g$, is said to be a *weak resolvent* of the operator $A$ [27]. If $f$ is in an invariant subspace of $A$ and $g$ is orthogonal to that subspace, then clearly the function $z \to [(z - A)^{-1}f, g]$ is identically 0 when the modulus of $z$ is greater than the norm of $A$ (consider the Neumann series). This establishes half of the following.

**Theorem** ([27]). *The operator A has a nontrivial invariant subspace if and only if there exist nonzero vectors f and g such that the function $z \to [(z - A)^{-1}f, g]$ is meromorphic on $\{z: |z| > \|A\|\}$.*

**Outline of proof.** Given such a meromorphic function, its behavior near infinity shows that it must in fact be a rational function near infinity. Integrating the product of the denominator and the weak resolvent around a large circle and using the Riesz functional calculus gives the result.

In the finite-dimensional case, every weak resolvent is a rational function. In that sense this theorem is a natural generalization of the fact that finite-dimensional operators (on spaces of dimension greater than 1) have nontrivial invariant subspaces. However, it appears unlikely that this theorem will have any real applications. (The set of all weak resolvents of an operator, on the other hand, appears to be a useful similarity invariant — see [27] and [16].)

# Conclusions

For a number of years, Halmos has expressed the view that there is probably a counterexample to the invariant subspace problem, and has gone so far as to make a substantial wager with the eternally optimistic Heydar Radjavi. Read's construction of an operator on $\ell^1$ without nontrivial invariant subspaces seems, in my view, to increase the probability that Halmos will win the bet. It would be interesting if one of the reformulations mentioned above pointed the direction to a counterexample.

## REFERENCES

1. C. Apostol, C. Foias, and D. Voiculescu, Some results on nonquasitriangular operators IV, *Rev. Roumaine Math. Pures Appl.* **18** (1973), 487–514.

2. N. Aronszajn and K.T. Smith, Invariant subspaces of completely continuous operators, *Ann. Math.* **60** (1954), 345–350.

3. A. Atzmon, An operator without invariant subspaces on a nuclear Frechet space, *Ann. Math.* **117** (1983), 669–694.

4. B. Beauzamy, *Introduction to Operator Theory and Invariant Subspaces,* North-Holland, Amsterdam, 1988.

5. H. Bercovici, Notes on invariant subspaces, *Bull. A.M.S. (N.S.)* **23** (1990), 1–33.

6. A.R. Bernstein and A. Robinson, Solution of an invariant subspace problem of K.T. Smith and P.R. Halmos, *Pacific J. Math.* **16** (1966), 421–431.

7. S. Brown, Invariant subspaces for subnormal operators, *Int. Equations Operator Theory* **1** (1978), 310–333.

8. S. Brown, B. Chevreau, and C. Pearcy, On the structure of contraction operators II, *J. Funct. Anal.* **76** (1988), 30–55.

9. S.R. Caradus, Universal operators and invariant subspaces, *Proc. Am. Math. Soc.* **23** (1969), 526–527.

10. J.B. Conway, *Subnormal Operators,* Pitman, Boston, 1981.

11. C. Cowen, Composition operators on Hilbert spaces of analytic functions: A status report, in Operator Theory, Operator Algebras and Applications, *Proc. Sym. Pure Math.* **51**, Part I, AMS, Providence, 1990, pp. 131–145.

12. J. Daughtry, Operator equations and invariant subspaces, Ph.D. dissertation, University of Virginia, 1973.

13. C. Davis, Generators of the ring of bounded operators, *Proc. Am. Math. Soc.* **6** (1955), 970–972.

14. R.G. Douglas and C. Pearcy, Invariant subspaces of nonquasitriangular operators, in *Proceedings of a Conference on Operator Theory,* Lecture Notes in Math. 345, Springer-Verlag, Berlin, 1973, pp. 13–17.

15. P. Enflo, On the invariant subspace problem in Banach spaces, *Acta Math.* **158** (1987), 213–313.

16. C.K. Fong, E.A. Nordgren, H. Radjavi, and P. Rosenthal, Weak resolvents of linear operators II, *Indiana Univ. Math. J.* **39** (1990), 67–83.

17. D.W. Hadwin, E.A. Nordgren, H. Radjavi, and P. Rosenthal, An operator not satisfying Lomonosov's hypothesis, *J. Funct. Anal.* **38** (1980), 410–415.

18. P.R. Halmos, Invariant subspaces of polynomially compact operators, *Pacific J. Math.* **16** (1966), 433–437.

19. P.R. Halmos, Quasitriangular operators, *Acta Sci. Math. (Szeged)* **29** (1968), 283–293.

20. P.R. Halmos, Normal dilations and extensions of operators, *Summa Brasiliensis Math.* **II** (1950), 125–134.

21. P.R. Halmos, A glimpse into Hilbert space, in *Lectures on Modern Mathematics,* edited by T. Saaty, J. Wiley, New York, 1963, pp. 1–22.

22. H. Helson, *Lectures on Invariant Subspaces,* Academic Press, New York, 1964.

23. V. Lomonosov, Invariant subspaces for operators commuting with compact operators, *Funcktional Anal. Prilozen* **7** (1973), 55–56 (Russian); *Funct. Anal. Appl.* **7** (1973), 213–214 (English).

24. E.A. Nordgren, Composition operators, *Can. J. Math.* **20** (1968), 442–449.

25. E.A. Nordgren, Composition operators in Hilbert spaces, in *Hilbert Space Operators,* Lecture Notes in Math 693, Springer-Verlag, Berlin, 1978, pp. 37–63.

26. E.A. Nordgren, H. Radjavi, and P. Rosenthal, A geometric equivalent of the invariant subspace problem, *Proc. Am. Math. Soc.* **61** (1976), 66–69.

27. E.A. Nordgren, H. Radjavi, and P. Rosenthal, Weak resolvents of linear operators, *Indiana Univ. Math. J.* **36** (1987), 913–934.

28. E.A. Nordgren, P. Rosenthal, and F.S. Wintrobe, Invertible composition operators on $H^2$, *J. Funct. Anal.* **73** (1987), 324–344.

29. E.A. Nordgren, P. Rosenthal, and F.S. Wintrobe, Composition operators and the invariant subspace problem, *Proc. R. Sci. Can. VI* (1984), 279–283.

30. E.A. Nordgren, M. Radjabalipour, H. Radjavi, and P. Rosenthal, Quadratic operators and invariant subspaces, *Studia Math.* LXXXVIII (1988), 263–268.

31. H. Radjavi and P. Rosenthal, *Invariant Subspaces*, Springer-Verlag, Berlin, 1973.

32. C.J. Read, A solution to the invariant subspace problem, *Bull. London Math. Soc.* **16** (1984), 337–401.

33. C.J. Read, A short proof concerning the invariant subspace problem, *J. London Math. Soc.* **34** (1986), 335–348.

34. G.-C. Rota, On models for linear operators, *Common Pure Appl. Math.* **13** (1960), 469–472.

*Department of Mathematics*
*University of Toronto*
*Toronto, Ontario M5S 1A4*
*CANADA*

# Probability vs. Measure

## J.L. Doob

Joe Doob, 1980

Since Paul Halmos' 1950 book *Measure Theory* is still one of the few textbooks of measure theory to include some probability theory, it is appropriate in this volume to discuss the relation between these two subjects. Throughout this discussion "probability" means "mathematical probability." Any coin tossed will be unambiguously mathematical.

Analysis is too sacred a domain for its adherents to grant citizenship to a new subject without critical examination. In fact there was initial resistance to accepting measure theory and still more to accepting probability. Conversely, some probabilists still resist accepting measure theory, and a few resist accepting any refined mathematics. It is now taken for granted that every analysis student study measure theory, but the subject, formally inaugurated in 1902 by Lebesgue's thesis, remained somewhat suspect as late as the thirties.

The mathematical community is no more eager than other communities to accept what it considers alien ideas. When Saks lectured in Cambridge (Massachusetts) on what is now called the Vitali–Hahn–Saks theorem there was professional grumbling at what was considered the extremely abstract nature of the topic. Saks, in the introduction to his book *Theory of the Integral* quotes Poincaré's complaint aimed at the introduction of irregular functions into analysis: "Formerly when one invented a new function it was for some practical purpose, but nowadays they are invented with the explicit purpose of making invalid the reasoning of our fathers, and nothing else will ever come of their invention." By 1930 the grumbling Cambridge professors had accepted Lebesgue's measure theory, but why go further? Surely Vitali–Hahn–Saks abstractions had no useful purpose! Repeating this complaint in a newer context, many probabilists have thought, and some still do, that the absorption of probability by measure theory has no useful purpose. Kac, in his autobiography written in 1984, quotes with indications of approval a complaint by physicists Uhlenbeck and Wong, referring to measure theoretic refinements of probability "...it seems to us that these investigations have not helped in the solution of problems of direct physical interest." But the rigor achieved by this absorption is what made probability into respectable mathematics, thereby justifying past work as well as making progress possible into new unexpected areas.

By the time Kolmogorov published his 1933 monograph *Fundamental Concepts of Probability,* Lebesgue's measure theory had been accepted, but Kolmogorov's formalization of probability required measures on abstract spaces, a level of mathematical sophistication not yet incorporated into the background of every analyst. It was tactful of Kolmogorov, in showing that measure theory is the basis of probability, to state explicitly in the early pages of the monograph the identifications

(random variable) = (measurable function on a measure space)

(expectation of a random variable) = (integral of the function)

yet later in the monograph to define the left-hand sides and other measure theoretic concepts in probability language and to prove measure theorems in this language without remarking that he was repeating standard material. Kolmogorov's concession to cultural lag is perpetuated by the feeling held by many that even though probability calculations are identical with measure theory calculations, still probability is not a specialization of measure theory. This feeling is fostered by the fact that probability theory has preserved its special dialect, the terms expectation, conditional expectation, sample sequence (the sequence of values of a sequence of measurable functions for a specified value of the argument) and so on. This special slang should not be denigrated by mathematicians, because the ideas behind this slang have been and remain fruitful even in abstract research. Lévy, one of the greatest probabilists of this century, never wholly accepted probability as measure theory. Instead, he used intuitive notions derived from a necessarily imprecise version of nonmathematical probability to construct his probability creations.

Just as analysts were not eager to accept probability, many probabilists were not eager to be swallowed up by measure theory, and applied probabilists deplored the introduction of sophisticated mathematics. Thus there has been spirited discussion of the role of measure theory in probability along with discussions of the relation between probability and empirical observations. Although some of this discussion has no useful purpose it awakens the spiritual descendants of the medieval scholastics, who have seriously discussed what happens when a coin is tossed infinitely often, their coins apparently moving in some limbo between mathematics and the real world. The phrase "infinite sequence of coin tosses" can of course be given a mathematical interpretation, involving a measure on an infinite product space, and the properties of this space are not in doubt.

In spite of the cynicism of the preceding paragraph it may be enlightening to characterize the cultural place of probability in measure theory. Certainly probability language provides glamor. For example, probability language replaces the Birkhoff ergodic theorem by the more colorful *Strong Law of Large Numbers for Stationary Stochastic Processes*! A cultural point is that it is rare in nonprobabilistic measure theory to explicitly evaluate integrals which are not Riemann integrals. Non-Riemann integrals appear even in trivial probabilistic contexts. For example, consider the expected number of heads in two throws of a fair coin. The formulation is code for the following mathematical problem. Consider the space of the four pairs $(h, h)$, $(h, t)$, $(t, t)$, $(t, h)$ and define a measure on the space by assigning measure $\frac{1}{4}$ to each pair. The desired expectation is the integral with respect to this measure of the function defined on each pair as the number of $h$'s in the pair. This trivial example illustrates another cultural difference between probability and nonprobabilistic measure theory, namely the frequent use of integer valued functions in probability theory. A less trivial example of this usage is the following. Let $x_1, x_2, \ldots$ be measurable functions from a measurable space $X$ into a measurable space $Y$, and let $A$ be a measurable subset of $Y$. If, for each point $w$ of $X$, $z(w)$ is the minimum value of $j$ for which $x_j(w)$ is a point of $A$, or $z(w) = +\infty$ if $x_j(w)$ is not in $A$ for any value of $j$, then $z$ is an extended integral valued measurable function, the "first entry time of $A$ by the sequence $x$.", and if $z$ is so defined one can be reasonably sure that the writer is a probabilist. A nonprobabilist would be more likely to treat the decomposition of $X$ determined by the values of $z$ without explicitly defining $z$. A further cultural distinction between probabilistic and nonprobabilistic writing is that only probabilists are likely to use explicitly the fact (half the Borel–Cantelli theorem) that if $B_1, B_2, \ldots$ is a sequence of measurable sets of a measure space the sum of whose measures is finite, then lim sup $B_n$, i.e., the set of points in infinitely many members of the sequence $B$, has measure 0. This is a convenient fact to apply for example in an almost everywhere convergence proof, but is rarely used explicitly by nonprobabilists, perhaps for lack of a familiar name! Until measure theory texts routinely use such convenient techniques common in probability the psychological integration of probability in measure theory will remain incomplete.

The two foregoing examples are typical of the techniques of probability suggested by its nonmathematical context and not yet adopted by non-probabilists. More serious is the question of what distinguishes probability problems from nonprobabilistic measure problems. Kolmogorov suggested that the key difference is the stress on the independence concept by probabilists, in mathematical language the stress on product measures. There is, however, less stress on independence now than there was in 1933. In the spirit of his remark one would add now the concepts of conditional probability and conditional expectation, first defined in full generality in Kolmogorov's monograph. To justify the addition, one need only refer to all the research on Markov processes and martingale theory.

But the essential fact is that the historical nonmathematical context of probability has led to measure theoretic research in directions peculiar to probability. For example, this context has made probabilists less interested in orthogonality than in independence which, in a reasonable sense, is a sharpening of orthogonality. The context of probability is likely to be a function $(t, w) \rightarrow f(t, w)$ of a pair $(t, w)$ into a measurable space. Here $w$ varies in a measure space and $t$ is commonly a set of integers or of real numbers. Probabilists study (a) for fixed $t$ the distribution of the function $w \rightarrow f(t, w)$, supposed measurable, and (b) properties of the "sample function" $t \rightarrow f(t, w)$ for fixed $w$, in particular properties valid for almost all $w$. Study (a) leads to such topics as the central limit theorem, and study (b) to such topics as strong laws of large numbers, properties of Brownian motion paths, martingale convergence, and so on. Both (a) and (b) lead to interesting mathematics, with applications to classical nonmathematical probability contexts as well as to pure mathematics outside probability theory. It is significant that many basic probability results have been inspired by the historical nonmathematical context of probability rather than by the historical natural development of nonprobabilistic measure theory.

Some probabilists consider measure theory boring, to be avoided if at all possible, but even they should recognize that they are doing measure theory, possibly camouflaged by probabilistic terminology, even though it is the nonmathematical probability context that appeals to them. Some probabilists consider that a proof using sample functions is probabilistic and therefore more admirable than a proof using distributions of random variables, transition probabilities and so on, which they describe as a proof done by analysis. Thus one probabilist wrote that by applying his integral, Itô was able to study multidimensional diffusion "with purely probabilistic techniques, an improvement over the analytic methods of Feller." This writer has chosen to evict probability from measure theory and measure theory from analysis! Another probabilist has written "I believe that probability theory is more closely related to analysis, physics, and statistics than to measure theory as such." This writer excluded measure theory from analysis. His "probability" is not quite the mathematical probability discussed in this paper but rather the probabilistic study of certain nonmathematical topics, a study preferably carried through without the delicate measure theoretic considerations required in some applications, for example, those involving the Itô integral.

In conclusion, probability is subsumed under measure theory but probability has its own terminology and special point of view derived from its nonmathematical origin. Further development of the subject will be influenced, as is all of mathematics both by the natural continuation of its past development and by applications inside and outside mathematics. One aspect of probability is intrinsically heavily dependent on subtle measure theoretic concepts but a second no less important aspect does not need such subtleties. Adherents of each aspect are human and therefore scorn adherents of the other.

*Department of Mathematics*
*University of Illinois*
*Urbana, IL 61801*

# New Hilbert Spaces from Old[†]

## Donald Sarason

Don Sarason, 1985

It is a cliché that Hilbert spaces are often useful in connection with problems whose initial formulations do not mention them. A familiar case in point is the classical moment problem [3]. It is also a cliché that, in attacking problems about Hilbert spaces and Hilbert space operators, one is often led to construct new Hilbert spaces. Occasionally, one of these new Hilbert spaces originally lies hidden, lurking in the shadows, and, once revealed, makes possible an unexpectedly elegant solution.

My aim in this article is to illustrate the preceding remarks through a discussion of unitary dilations, a subject originated by Paul Halmos. The first section contains the basic facts about unitary dilations, including the existence theorem of B. Sz.-Nagy. The second section concerns one of the central results of the subject, the lifting theorem of Sz.-Nagy and C. Foias. An elegant proof of the lifting theorem due to R. Arocena will be presented.

In what follows, all Hilbert spaces mentioned are assumed to be complex and separable. All operators are assumed to be linear and bounded. The symbol $I$ will denote the identity operator; the space on which it acts will be clear from the context.

## Unitary Dilations

If $a$ is a complex number of modulus at most 1, then $a$ can be made the $(1,1)$ entry of a two-by-two unitary matrix, for example, the matrix

$$\begin{bmatrix} a & (1-|a|^2)^{1/2} \\ (1-|a|^2)^{1/2} & -\bar{a} \end{bmatrix}.$$

---

[†]To Paul Halmos, for caring about mathematics and about people.

In one of his first papers on operator theory [5], Halmos observed that the preceding construction, if suitably interpreted, applies not only when $a$ is a number but also when it is a Hilbert space contraction. Namely, if $T$ is an operator of norm at most 1 on the Hilbert space $H$, then (as will be explained shortly) the two-by-two operator matrix

$$U = \begin{bmatrix} T & (I - TT^*)^{1/2} \\ (I - T^*T)^{1/2} & -T^* \end{bmatrix}$$

defines a unitary operator on the direct sum $H \oplus H$. (The square roots here are the positive square roots of the given positive operators.) If one identifies $H$ with the first coordinate space in $H \oplus H$, one sees that one can produce the action of $T$ on a vector in $H$ by first applying $U$ to the vector and then projecting back onto $H$. In Halmos's terminology, $U$ is a dilation of $T$, and $T$ is a compression of $U$.

To prove that $U$ is unitary one verifies by direct calculation that $U^*U$ and $UU^*$ both equal the identity on $H \oplus H$. The verification uses the equality $T^*(I - TT^*)^{1/2} = (I - T^*T)^{1/2}T^*$, which Halmos deduced starting from the obvious relation $T^*(I - TT^*) = (I - T^*T)T^*$. From the latter relation one infers that $T^*p(I - TT^*) = p(I - T^*T)T^*$ for every polynomial $p$. If $(p_n)_1^\infty$ is a sequence of polynomials converging uniformly in the interval $[0, 1]$ to the square-root function, then $p_n(I - TT^*)$ tends in norm to $(I - TT^*)^{1/2}$ and $p_n(I - T^*T)$ tends in norm to $(I - T^*T)^{1/2}$. In the limit one thus obtains the desired equality.

Shortly after Halmos's paper appeared Sz.-Nagy [11] extended his ideas. Suppose $T$ is a contraction operator on the Hilbert space $H$. According to the theorem of Sz.-Nagy, there is a unitary operator $U$ acting on a Hilbert space $H'$ containing $H$ as a subspace such that, not only is $U$ a dilation of $T$, but also $U^n$ is a dilation of $T^n$ for every positive integer $n$. The Sz.-Nagy dilation is more intimately related to $T$ than the one of Halmos. It has been referred to as a strong dilation, and as a power dilation; usually, though, now that its study has developed into a branch of mathematics in its own right [12], it is called simply a unitary dilation of $T$. The dilation of Sz.-Nagy can be taken to be minimal, that is, not admitting a decomposition into the direct sum of two unitary operators the first of which is also a dilation of $T$. Such a minimal dilation is essentially unique and so is called *the* unitary dilation of $T$.

A few simple examples will serve to illustrate Sz.-Nagy's theorem. Let $L^2$ denote the usual Lebesgue space relative to normalized Lebesgue measure on the unit circle. It has a natural orthonormal basis, the Fourier basis, consisting of the functions $e_n(z) = z^n$ ($n \in \mathbb{Z}$). Let $L_+^2$ (usually called $H^2$) be the closed linear span in $L^2$ of the functions $e_n$ with $n \geq 0$ and $L_-^2$ the span of the functions $e_n$ with $n < 0$. The bilateral shift is the operator $S$ on $L^2$ of multiplication by the independent variable $[(Sf)(z) = zf(z)]$. The subspace $L_+^2$ is invariant under $S$; the restriction of $S$ to $L_+^2$ is called the unilateral shift and will be denoted by $S_+$. Let $S_-$ denote the compression of

$S$ to $L^2_-$, or, what amounts to the same thing, the adjoint of the restriction of $S^*$ to $L^2_-$; it is unitarily equivalent to $S^*_+$, the so-called backward shift. It is then very easy to verify that $S$ is the minimal Sz.-Nagy dilation of $S_+$ and $S_-$. It is also the minimal Sz.-Nagy dilation of the zero operator on the one-dimensional subspace spanned by the basis vector $e_0$. (Even when the original contraction $T$ acts on a finite-dimensional space, its Sz.-Nagy dilation acts on an infinite-dimensional one, except in the trivial case where $T$ is a finite-dimensional unitary operator.)

If one sanguinely assumes the Sz.-Nagy dilation exists and proceeds to analyze it, one learns from the analysis how to construct it. Thus, suppose $T$ is a contraction on the Hilbert space $H$ and $U$ is its minimal Sz.-Nagy dilation, acting on the Hilbert space $H'$ containing $H$ as a subspace. The orthogonal projection in $H'$ with range $H$ will be denoted by $P$. Let $H^+$ be the closed invariant subspace of $U$ generated by $H$; it is the closed subspace of $H'$ spanned by the vectors $U^n x$ with $x$ in $H$ and $n = 0, 1, 2, \ldots$. Let $H_0^+ = H^+ \ominus H$, the orthogonal complement of $H$ in $H^+$.

It is asserted that the subspace $H_0^+$ is invariant under $U$. To verify this it is enough to check that the vector $U(I-P)U^n x$ is orthogonal to $H$ whenever $x$ is in $H$ and $n = 1, 2, \ldots$. [This is so because the vectors $(I - P)U^n x$, with $x$ and $n$ as described, span $H_0^+$.] If $y$ is in $H$ then the inner product $\langle U(I - P)U^n x, y \rangle$ is the difference of two terms. Because $x = Px$ and $y = Py$, those terms can be written as $\langle PU^{n+1}Px, y \rangle$ and $\langle PUPU^n Px, y \rangle$. But each of the two preceding inner products equals $\langle T^{n+1}x, y \rangle$, so $U(I-P)U^n x$ is orthogonal to $H$, as desired.

The operator $U$ thus maps $H_0^+$ isometrically into itself. Moreover, the restriction $U \mid H_0^+$ must be a pure isometry, that is, it has no invariant subspace on which it acts as a unitary operator. (This follows because $H^+$ is the smallest $U$-invariant subspace containing $H$.) The purity of the isometry $U \mid H_0^+$ means it is a shift. Namely, if we let $L^+ = H_0^+ \ominus UH_0^+$, then we can write $H_0^+$ as the orthogonal direct sum of the subspaces $U^n L^+$, $n = 0, 1, 2, \ldots$. (Halmos's paper [6] contains a lucid discussion of these matters. The subspace $L^+$ is called by him a wandering subspace of the operator $U$.)

The subspace $H_0^- = H' \ominus H^+$ is invariant under $U^*$, being the orthogonal complement of the $U$-invariant subspace $H^+$. From the minimality of the dilation $U$ one infers that the restriction $U^* \mid H_0^-$ is a pure isometry, so that $H_0^-$ is the orthogonal direct sum of the subspaces $U^{*n}L^-$, $n = 0, 1, 2, \ldots$, where $L^- = H_0^- \ominus U^* H_0^-$. The $U$ invariance of the subspace $H_0^+$ implies that the subspace $H^- = H' \ominus H_0^+$ is $U^*$ invariant; it is the $U^*$-invariant subspace generated by $H$ and is the orthogonal direct sum of $H$ and $H_0^-$. The whole space $H'$ is the orthogonal direct sum of $H$, $H_0^+$, and $H_0^-$.

We introduce the defect operators $D_T = (I - T^*T)^{1/2}$ and $D_{T^*} = (I - TT^*)^{1/2}$, and the corresponding defect subspaces $\mathcal{D}_T$ and $\mathcal{D}_{T^*}$, the closures of the ranges of $D_T$ and $D_{T^*}$, respectively. If $x$ is a vector in $H$ then, as one easily sees, the vector $Ux - Tx$ lies in the subspace $L^+$; the square of its norm equals $\|x\|^2 - \|Tx\|^2$, which is the same as $\|D_T x\|^2$. We thus have a natural isometry of the subspace $\mathcal{D}_T$ into the subspace $L^+$, defined on vectors in the range of $D_T$ by $D_T x \mapsto Ux - Tx$. The range of this isometry is in fact all of $L^+$. Indeed, one can obtain a set of vectors that spans $L^+$ by projecting onto $L^+$ all of the vectors $U^n x$ with $x$ in $H$ and $n = 1, 2, \dots$. But

$$U^n x = T^n x + \sum_{k=1}^{n} U^{n-k}(UT^{k-1}x - T^k x).$$

The first term on the right is in $H$, and the $k$th term in the sum is in $U^{n-k}L^+$ and so is orthogonal to $L^+$ for $k = 1, \dots, n - 1$. The projection of $U^n x$ onto $L^+$ thus equals $UT^{n-1}x - T^n x$, showing that the vectors $Ux - Tx$ with $x$ in $H$ already span $L^+$. The isometry defined above thus maps $\mathcal{D}_T$ onto $L^+$ and so provides a natural identification between those two spaces.

By similar reasoning, there is a natural isometry of the subspace $\mathcal{D}_{T^*}$ onto the subspace $L^-$; it is defined on vectors in the range of $D_{T^*}$ by $D_{T^*} x \mapsto U^* x - T^* x$.

We now have a fairly good picture of the action of the operator $U$. On the subspace $H_0^+$ it acts as a shift, sending each of the mutually orthogonal subspaces $U^n L^+$ ($n = 0, 1, 2, \dots$) to the next one. On the subspace $H_0^-$ it acts as a backward shift, sending each of the mutually orthogonal subspaces $U^{*n}L^-$ ($n = 1, 2, \dots$) to the preceding one, and mapping $L^-$ itself into the orthogonal complement of $H_0^-$. If $x$ is in $H$, then the difference $Ux - Tx$ lies in $L^+$ and has the same norm as does $D_T x$. To fill out the picture, we need to understand better how $U$ acts on the subspace $L^-$.

The vectors $U^* x - T^* x$ with $x$ in $H$ form a dense subspace of $L^-$. The image of $U^* x - T^* x$ under $U$ is $x - UT^* x$, which is the difference between $x - TT^* x$ and $UT^* x - TT^* x$. The first of the two preceding terms equals $D_{T^*}D_{T^*}x$; it is, in other words, the image under $D^{T^*}$ of the vector in $\mathcal{D}_{T^*}$ that corresponds to $U^* x - T^* x$ under the natural isometry between $L^-$ and $\mathcal{D}_{T^*}$. The second term, $UT^* x - TT^* x$, lies in $L^+$; under the natural isometry between $L^+$ and $\mathcal{D}_T$ it corresponds to the vector $D_T T^* x$, which is the same as $T^* D_{T^*} x$ [by the equality $(I - T^*T)^{1/2}T^* = T^*(I - TT^*)^{1/2}$, mentioned earlier]. To summarize and paraphrase: the image under $U$ of a vector in $L^-$ has two components, one in $H$ and one in $L^+$; if the original vector corresponds to the vector $y$ in $\mathcal{D}_{T^*}$ under the isometry between $L^-$ and $\mathcal{D}_{T^*}$, then the component of the image in $H$ is $D_{T^*}y$, and the component of the image in $L^+$ corresponds, under the isometry between $L^+$ and $\mathcal{D}_T$, to the vector $-T^* y$.

While the discussion above was premised on the existence of the minimal Sz.-Nagy dilation of the contraction $T$, the information obtained can be interpreted as a set of instructions for the construction of the dilation. Following these instructions, we let $H_n$ be a copy of the defect space $\mathcal{D}_T$ for $n = 1, 2, \dots$ and a copy of the defect space $\mathcal{D}_{T^*}$ for $n = -1, -2, \dots$. We also let $H_0 = H$, and we let $H'$ be the direct sum of all the spaces $H_n$. We define the operator $U$ on $H'$ by letting it take the vector $\sum_{-\infty}^{\infty} \oplus x_n$ to the vector $\sum_{-\infty}^{\infty} \oplus y_n$ given by

$$y_n = \begin{cases} x_{n-1} & \text{for } n \leq -1 \text{ or } n \geq 2 \\ Tx_0 + D_{T^*}x_{-1} & \text{for } n = 0 \\ D_T x_0 - T^* x_{-1} & \text{for } n = 1 \end{cases}$$

A straightforward verification shows that this $U$ is unitary and that it is the minimal Sz.-Nagy dilation of $T$. The uniqueness of the dilation is established by the preceding analysis. The analysis shows that a unitary equivalence between two contractions extends to a unitary equivalence between their minimal Sz.-Nagy dilations. (This is the precise statement of the uniqueness part of Sz.-Nagy's theorem.)

The construction just given can be found in the book of Sz.-Nagy and Foias [12]. It refines a proof of Sz.-Nagy's theorem due to J.J. Schäffer [10]. The original proof of Sz.-Nagy was less direct.

## Lifting Theorem

The lifting theorem of Sz.-Nagy and Foias says that any operator commuting with a Hilbert space contraction $T$ is the compression of an operator commuting with the minimal unitary dilation of $T$. More generally, it describes the operators that intertwine two contractions.

Suppose $T_1$ and $T_2$ are contractions on the respective Hilbert spaces $H_1$ and $H_2$. For $j = 1, 2$ let $U_j$ be the minimal unitary dilation of $T_j$, acting on the Hilbert space $H'_j$ containing $H_j$ as a subspace. The orthogonal projection in $H'_j$ with range $H_j$ will be denoted by $P_j$, and the other notations from the analysis in the first section will be carried over, with the appropriate subscripts appended. (Thus, in place of $H^+$ we now have $H_j^+$, etc.) The inner product in $H'_j$ will be denoted by $\langle \cdot, \cdot \rangle_j$.

Suppose $X$ is an operator from $H_1$ to $H_2$ that intertwines $T_1$ and $T_2$, in other words, that satisfies $XT_1 = T_2 X$. The lifting theorem then states that there is an operator $Y$ from $H'_1$ to $H'_2$ satisfying the following conditions:

(i) $YU_1 = U_2 Y$;

(ii) $X = P_2 Y \mid H_1$;

(iii) $YH_1^+ \subset H_2^+$, and $YH_{1,0}^+ \subset H_{2,0}^+$;

(iv) $\|Y\| = \|X\|$.

In the case where $T_1 = T_2$, condition (iii) guarantees along with condition (ii) that $Y$ is a dilation of $X$ in the sense of Sz.-Nagy.

The very special case of the lifting theorem where $T_1 = S_+$ and $T_2 = S_-$ is especially interesting. In this case we can take $H'_1 = H'_2 = L^2$ and

$U_1 = U_2 = S$. Since now $H_{1,0}^+$ is trivial and $H_2^+$ is all of $L^2$, condition (iii) becomes automatic.

An operator from $L_+^2$ to $L_-^2$ that intertwines $S_+$ and $S_-$ is called a Hankel operator. The matrix of such an operator, relative to the usual orthonormal bases for $L_+^2$ and $L_-^2$, has constant cross diagonals, a property that characterizes Hankel operators. The entry on the $n$th cross diagonal is the Fourier coefficient with index $-n$ of the image under the operator of the constant function 1.

If $X$ is a Hankel operator then, according to the lifting theorem, $X$ can be obtained from an operator $Y$ on $L^2$ that commutes with $S$ and has the same norm as does $X$; to apply $X$ to a vector in $L_+^2$, one first applies $Y$ and then projects onto $L_-^2$. It is an elementary result that the operators on $L^2$ that commute with the bilateral shift $S$ are just the multiplication operators induced by the functions in $L^\infty$ of the unit circle, and that the norm of such a multiplication operator equals the essential supremum norm of the inducing function. Thus, a Hankel operator $X$ equals a multiplication operator followed by a projection. The function that induces the multiplication operator must clearly have for its Fourier coefficient of index $-n$ ($n = 1, 2, \ldots$) the entry on the $n$th cross diagonal of the matrix for $X$. The upshot is the following theorem of Z. Nehari [7]: Let $c_1, c_2, c_3, \ldots$ be complex numbers. Then the matrix

$$\begin{bmatrix} c_1 & c_2 & c_3 & \cdots \\ c_2 & c_3 & & \cdots \\ c_3 & & & \cdots \\ & & & \end{bmatrix}$$

induces a bounded operator on $\ell^2$ if and only if there is a function $f$ in $L^\infty$ of the unit circle such that $c_n = \hat{f}(-n)$ (the Fourier coefficient of $f$ with index $-n$) for all positive integers $n$. In the case of boundedness there is such an $f$ with $\|f\|_\infty$ equal to the norm of the operator.

Nehari's theorem solves the following one-sided trigonometric moment problem, now usually called the Nehari interpolation problem: Given complex numbers $c_1, c_2, \ldots$, when does there exist a function $f$ in the unit ball of $L^\infty$ such that $\hat{f}(-n) = c_n$ for all positive integers $n$? The answer, according to Nehari's theorem, is that such an $f$ exists if and only if the matrix above has norm at most 1 as an operator on $\ell^2$. The Nehari problem contains as special cases the classical interpolation problems of Nevanlinna–Pick and Carathéorody–Fejér; one will find more details on this in [9]. The general lifting theorem can be applied to a host of interpolation problems, many of which are discussed in the book of M. Rosenblum and J. Rovnyak [8].

V.M. Adamyan, D.Z. Arov, and M.G. Krein [2] discovered an approach to Hankel operators and the Nehari problem that is analogous to the operator-theoretic approach to the classical moment problem. In one respect, though, it is more subtle than the latter approach, for it involves a Hilbert space that initially lies hidden.

We owe to Arocena [4] the realization that the Adamyan–Arov–Krein approach can be adapted to yield a proof of the Sz.-Nagy–Foias theorem. This article ends with a presentation of Arocena's proof.

Without loss of generality we assume the operator $X$ that intertwines the contractions $T_1$ and $T_2$ is of unit norm. We form the algebraic direct sum of the two subspaces $H_1^+$ and $H_2^-$, which we denote by $K_0$. To avoid confusion between orthogonal direct sum and algebraic direct sum, we use the symbol $\dotplus$ to denote the latter: $K_0 = H_1^+ \dotplus H_2^-$. On $K_0$ we introduce an inner product, denoted by $\langle \cdot, \cdot \rangle_0$, by setting

$$\langle x_1 \dotplus x_2, y_1 \dotplus y_2 \rangle_0 = \langle x_1, y_1 \rangle_1 + \langle x_2, y_2 \rangle_2$$
$$+ \langle X P_1 x_1, y_2 \rangle_2 + \langle x_2, X P_1 y_1 \rangle_2.$$

The assumption that $\|X\| = 1$ guarantees that this inner product is positive semidefinite:

$$\langle x_1 \dotplus x_2, x_1 \dotplus x_2 \rangle_0 = \|x_1\|_1^2 + \|x_2\|_2^2 + 2 \operatorname{Re} \langle X P_1 x_1, x_2 \rangle_2$$
$$\geq \|x_1\|_1^2 + \|x_2\|_2^2 - 2\|x_1\|_1 \|x_2\|_2 \geq 0.$$

The space $K_0$ is a so-called pre-Hilbert space. We can produce a Hilbert space from it in two steps. First we form the quotient space $K_0/N$, where $N$ is the subspace of $K_0$ consisting of the vectors having self-inner product 0. The quotient space $K_0/N$ is a bonifide inner product space, and its completion, which we denote by $K$, is thus a Hilbert space. (It is our "new" Hilbert space.) We denote the inner product in $K$ by $\langle \cdot, \cdot \rangle$.

The space $H_1^+$ is identified in the obvious way with a subspace of $K_0$, and this identification is isometric, in other words, the inner product of two vectors in $H_1^+$ is the same when they are regarded as vectors in $K_0$ as when they play their original roles as vectors in $H_1^+$. Thus, we can identify $H_1^+$ with a subspace of the Hilbert space $K$. The same is true of $H_2^-$. The natural injections of $H_1^+$ and $H_2^-$ into $K$ will be denoted by $W_1$ and $W_2$, respectively.

The next step is to define a certain isometry $V_0$ in the space $K$. In preparation we verify that, if $x_1$ and $x_2$ are vectors in $H_1^+$ and $H_2^-$, respectively, then $\|x_1 \dotplus U_2^* x_2\|_0 = \|U_1 x_1 \dotplus x_2\|_0$. The equality depends on the intertwining relation $XT_1 = T_2 X$. We have

$$\|U_1 x_1 \dotplus x_2\|_0^2 = \|U_1 x_1\|_1^2 + \|x_2\|_2^2 + 2 \operatorname{Re} \langle X P_1 U_1 x_1, x_2 \rangle_2.$$

The inner product on the right-hand side does not change if we replace $x_2$ by $P_2 x_2$ (obviously) and $x_1$ by $P_1 x_1$ (since $U_1$ sends the subspace $H_1^+ \ominus H_1$ into itself). It therefore equals $\langle X T_1 P_1 x_1, P_2 x_2 \rangle_2$, and hence equals $\langle T_2 X P_1 x_1, P_2 x_2 \rangle_2$. The last inner product equals $\langle X P_1 x_1, P_2 U_2^* P_2 x_2 \rangle_2$. Here we can replace $P_2 x_2$ by $x_2$ (since $U_2^*$ sends the subspace $H_2^- \ominus H_2$ into itself), so we finally obtain

$$\|U_1 x_1 \; \dot{+} \; x_2\|_0^2 = \|U_1 x_1\|_1^2 + \|x_2\|_2^2 + 2\,\mathrm{Re}\,\langle XP_1 x_1, P_2 U_2^* x_2\rangle_2$$
$$= \|x_1\|_1^2 + \|U_2^* x_2\|_2^2 + 2\,\mathrm{Re}\,\langle XP_1 x_1, U_2^* x_2\rangle_2$$
$$= \|x_1 \; \dot{+} \; U_2^* x_2\|_0^2,$$

as desired.

We now define the operator $V_0$. Its domain, $D(V_0)$, is to be the subspace of $K$ that corresponds to the subspace $H_1^+ \; \dot{+} \; U_2^* H_2^-$ of $K_0$; in other words, $D(V_0) = W_1 H_1^+ + W_2 U_2^* H_2^-$. Its range, $R(V_0)$, is to be the subspace of $K$ that corresponds to the subspace $U_1 H_1^+ \; \dot{+} \; H_2^-$ of $K_0$; in other words, $R(V_0) = W_1 U_1 H_1^+ + W_2 H_2^-$. The definition is

$$V_0(W_1 x_1 + W_2 U_2^* x_2) = W_1 U_1 x_1 + W_2 x_2 \quad (x_1 \in H_1^+,\; x_2 \in H_2^-).$$

The calculation above shows that $V_0$ is an isometry. From the definition one sees that $W_j$ intertwines $U_j$ and $V_0$ to the extent possible: $W_1 U_1 x_1 = V_0 W_1 x_1$ for $x_1$ in $H_1^+$, and $W_2 U_2 x_2 = V_0 W_2 x_2$ for $x_2$ in $U_2^* H_2^-$.

Next, we take a unitary extension of $V_0$. This is a standard construction. First we extend $V_0$ by continuity to an isometry from the closure of $D(V_0)$ onto the closure of $R(V_0)$. Then we take a larger Hilbert space $K'$ containing $K$ as a subspace such that the orthogonal complements of $D(V_0)$ and $R(V_0)$ in $K'$ have the same dimension. [If the orthogonal complements of $D(V_0)$ and $R(V_0)$ in $K$ itself happen to have the same dimension, we can take $K' = K$ but are not obliged to do so. If those orthogonal complements have different dimensions, then $K' \ominus K$ must be infinite dimensional.] If we take any isometry of $K' \ominus D(V_0)$ onto $K' \ominus R(V_0)$ and use it to extend $V_0$, we get the desired unitary extension, which we shall denote by $V$.

The injections $W_1$ and $W_2$ extend naturally to isometries of $H_1'$ and $H_2'$ into $K'$ that intertwine $V$ with $U_1$ and $U_2$, respectively. Consider first $W_1$. Let $K_1$ be the closed invariant subspace of $V^*$ generated by $W_1 H_1^+$. Since $W_1 H_1^+$ is already invariant under $V$ (by the intertwining relation connecting $V_0$, $U_1$, and $W_1$), the subspace $K_1$ is $V$-invariant as well as $V^*$-invariant, so the restriction $V \mid K_1$ is a unitary operator. In $K_1$ we consider the subspace $W_1 H_1$, the natural image of $H_1$, and the transplantation to $W_1 H_1$ of the operator $T_1$, namely, the operator $W_1 T_1 W_1^*$. An examination of our construction shows that $V \mid K_1$ is a minimal unitary dilation of $W_1 T_1 W_1^*$. The uniqueness part of Sz.-Nagy's theorem thus yields the desired extension of $W_1$ to an isometry, which we also denote by $W_1$, of $H_1'$ onto $K_1$ that intertwines $U_1$ and $V: W_1 U_1 = V W_1$. To obtain a similar extension of $W_2$ we apply the same reasoning, but with $T_2^*$, $U_2^*$, and $V^*$ in place of $T_1$, $U_1$, and $V$. [The switch to adjoints is made because $D(V_0^{-1})$ contains $W_2 H_2$ while $D(V_0)$ possibly does not.] One gets an isometric extension of $W_2$, also called $W_2$, from $H_2'$ into $K'$ that intertwines $U_2^*$ and $V^*: W_2 U_2^* = V^* W_2$.

Now, believe it or not, the proof of the lifting theorem is all but finished, for the operator $Y = W_2^* W_1$ has all of the required properties. The inequality $\|Y\| \leq 1$ is obvious because $W_1$ and $W_2$ are isometries; that equality actually holds will follow once the relation between $Y$ and $X$ has been established. The equality $Y U_1 = U_2 Y$ follows immediately from the two equalities $W_1 U_1 = V W_1$ and $W_2^* V = U_2 W_2^*$. To establish the remaining properties we take a vector $x_1$ in $H_1^+$ and a vector $x_2$ in $H_2^-$ and note that

$$\langle Y x_1, x_2 \rangle_2 = \langle W_1 x_1, W_2 x_2 \rangle = \langle x_1 \overset{.}{+} 0, 0 \overset{.}{+} x_2 \rangle_0$$
$$= \langle X P_1 x_1, x_2 \rangle_2.$$

If $x_1$ is in $H_1$ and $x_2$ is in $H_2$, the left-hand side can be rewritten as $\langle P_2 Y x_1, x_2 \rangle_2$ and the right-hand side as $\langle X x_1, x_2 \rangle_2$, from which it follows that $X = P_2 Y \mid H_1$. If $x_2$ is in $H_{2,0}^-$ the right-hand side vanishes, from which it follows that $Y H_1^+ \subset H_2^+$. If finally $x_1$ is in $H_{1,0}^+$ the right-hand side again vanishes, from which it follows that $Y H_{1,0}^+ \subset H_{2,0}^+$. The lifting theorem is completely proved.

The reader may have noticed that one can recast Arocena's proof in the language of scattering theory, referring to $W_1$ and $W_2$ as wave operators and $W_2^* W_1$ $(= Y)$ as a scattering operator. The proof fits into the scattering-theoretic approach to unitary dilations developed by Adamyan and Arov [1].

## REFERENCES

1. V.M. Adamyan and D.Z. Arov, On unitary couplings of semi-unitary operators, *Am. Math. Soc. Transl. Ser.* 2, **95** (1970), 75–129.

2. V.M. Adamyan, D.Z. Arov, and M.G. Krein, Infinite Hankel matrices and generalized problems of Carathéodory–Fejér and I. Schur, *Funkcional. Anal. Prilozhen.* 2, **4** (1968), 1–17.

3. N.I. Akhiezer, *The Classical Moment Problem*, Hafner, New York, 1965.

4. R. Arocena, Unitary extensions of isometries and contractive intertwining dilations, *Operator Theory: Advances and Applications*, Birkhauser Verlag, Basel, 1989, Vol. 41, pp. 13–23.

5. P.R. Halmos, Normal dilations and extensions of operators, *Summa Brasil. Math.* **2** (1950), 125–134.

6. P.R. Halmos, Shifts on Hilbert spaces, *J. Reine Angew. Math.* **208** (1961), 102–112.

7. Z. Nehari, On bounded bilinear forms, *Ann. Math.* **65** (1957), 153–162.

8. M. Rosenblum and J. Rovnyak, *Hardy Classes and Operator Theory,* Oxford University Press, New York, 1985.

9. D. Sarason, Moment problems and operators in Hilbert space, Moments in Mathematics, *Proc. Symp. Appl. Math.,* Am. Math. Soc., Providence, R.I., 1987, Vol. 37, pp. 54–70.

10. J.J. Schäffer, On unitary dilations of contractions, *Proc. Am. Math. Soc.* **6** (1955), 322.

11. B. Sz.-Nagy, Sur les contractions de l'espace de Hilbert, *Acta Sci. Math.* **18** (1953), 87–92.

12. B. Sz.-Nagy and C. Foias, *Harmonic Analysis of Operators in Hilbert Space,* North-Holland, Amsterdam, 1970.

*Department of Mathematics*
*University of California*
*Berkeley, CA 94720*

# Some Universal Constraints for Discrete Möbius Groups

## F.W. Gehring and G.J. Martin

Fred Gehring, 1974

## 1. Introduction

When asked recently about how he does mathematics, Paul Halmos replied "Since a lot of my work has been in operator theory and infinite dimensional Hilbert spaces, and since the most easily accessible part of that is matrix theory and finite dimensional vector spaces, I start by looking at a 2 by 2 matrix."

This article is concerned with Möbius groups, another part of mathematics where 2 by 2 matrices play an important role. In particular, we derive here two universal constraints on elements of a discrete Möbius group which do not commute. The first involves the chordal distance one element moves the fixed points of the other while the second estimates the maximal chordal distance between images of a point under these transformations. Both results follow from a sharpened lower bound for the commutator parameter of Möbius transformations with equal trace.

Let $\mathbb{M}$ denote the group of all Möbius transformations of the extended complex plane $\overline{\mathbb{C}} = \mathbb{C} \cup \{\infty\}$. We associate with each

$$f = \frac{az + b}{cz + d} \in \mathbb{M}, \qquad ad - bc = 1, \tag{1.1}$$

the matrix

$$A = \begin{pmatrix} a & b \\ c & d \end{pmatrix} \in \mathrm{SL}(2, \mathbb{C}) \tag{1.2}$$

and set $\mathrm{tr}(f) = \mathrm{tr}(A)$ where $\mathrm{tr}(A)$ denotes the trace of $A$. For $f$ and $g$ in $\mathbb{M}$ we let $[f, g]$ denote the multiplicative commutator $fgf^{-1}g^{-1}$ and call the three complex numbers

$$\beta(f) = \mathrm{tr}^2(f) - 4, \quad \beta(g) = \mathrm{tr}^2(g) - 4, \quad \gamma(f, g) = \mathrm{tr}([f, g]) - 2 \tag{1.3}$$

This research was supported in part by grants from the U.S. National Science Foundation (FWG) and the Institut Mittag-Leffler (FWG and GJM).

the *parameters* of the two generator subgroup $\langle f, g \rangle$. They are independent of the choice of representative for $f$ and $g$ and they determine $\langle f, g \rangle$ up to conjugacy whenever $\gamma(f, g) \neq 0$ [1]. Then $\gamma(f, g) \neq 0$ if and only if $\mathrm{fix}(f) \cap \mathrm{fix}(g) = \emptyset$ where $\mathrm{fix}(f)$ denotes the fixed point set of $f$ in $\overline{\mathbb{C}}$.

The usual topology on $\mathbb{M}$ is induced by the metric

$$d(f, g) = \sup\big(q(f(z), g(z)): z \in \overline{\mathbb{C}}\big), \tag{1.4}$$

where $q$ denotes the chordal distance in $\overline{\mathbb{C}}$,

$$q(z_1, z_2) = \frac{2|z_1 - z_2|}{(|z_1|^2 + 1)^{1/2}(|z_2|^2 + 1)^{1/2}}. \tag{1.5}$$

A subgroup $G$ of $\mathbb{M}$ is *discrete* if there exists a constant $d = d(G) > 0$ such that $d(f, g) \geq d$ for each distinct pair $f, g$ in $G$, *nonelementary* if the limit set $L(G)$ contains at least three points and *Fuchsian* if each element in $G$ preserves some disk or half plane $D = D(G)$ in $\mathbb{C}$.

The following result follows from [3, Theorem 2.7 and Corollary 3.15].

**1.6. Lemma.** *If $f$ and $g$ are in $\mathbb{M}$, then*

$$\max\big(d(f, id), d(g, id)\big) \geq \left(\frac{4|\gamma(f, g)|^{1/2}}{|\gamma(f, g)|^{1/2} + 2}\right)^{1/2}. \tag{1.7}$$

Thus lower bounds for $|\gamma(f, g)|$ imply lower bounds for the maximum chordal displacement of $f$ and $g$ and hence geometric constraints on the group $\langle f, g \rangle$.

If $\langle f, g \rangle$ is a nonelementary discrete Fuchsian subgroup of $\mathbb{M}$, then

$$|\gamma(f, g)| \geq 2 - 2\cos(\pi/7) = 0.198\ldots. \tag{1.8}$$

See, for example, [12, Corollary]. On the other hand, Jørgensen has shown [7] that if $1 < a < \infty$, then the transformations

$$f_0(z) = -a^2 z, \qquad g_0(z) = \frac{(a^2 + a^{-2})z - 2}{2z - (a^2 + a^{-2})} \tag{1.9}$$

generate a nonelementary discrete subgroup of $\mathbb{M}$ with

$$\beta(f_0) = -(a + a^{-1})^2, \quad \beta(g_0) = -4, \quad \gamma(f_0, g_0) = 4(a - a^{-1})^{-2}. \tag{1.10}$$

Since $\gamma(f_0, g_0) \to 0$ as $a \to \infty$, (1.8) fails if $\langle f, g \rangle$ is not Fuchsian.

When $\langle f, g \rangle$ is not Fuchsian, one can obtain lower bounds for $|\gamma(f, g)|$ in a nonelementary discrete group $\langle f, g \rangle$ if one knows something more about the generators. For example,

$$|\gamma(f, g)| \geq 1 \tag{1.11}$$

if $f$ or $g$ is parabolic [9] and [13],

$$|\gamma(f,g)| \geq 2\cos(2\pi/7) - 1 = 0.246\ldots \tag{1.12}$$

if $f$ or $g$ is elliptic of order $n \geq 3$ [5, Theorem 3.1] and

$$|\gamma(f,g)| > 0.125 \tag{1.13}$$

if $\beta(f) = \beta(g)$ [8]. In addition,

$$|\gamma(f,g)| + |\beta(f)| \geq 1, \tag{1.14}$$

an important bound due to Jørgensen which holds in general [6].

We show in Section 3 that (1.13) can be sharpened to give

$$|\gamma(f,g)| > 0.193, \tag{1.15}$$

an estimate which is within 2.5% of the bound

$$|\gamma(f,g)| \geq 2 - 2\cos(\pi/7) = 0.198\ldots \tag{1.16}$$

which we conjecture holds in this case. Inequality (1.16) is true if, in addition, $\gamma(f,g)$ is real, a sharp bound which holds when $\langle f, g \rangle$ is Fuchsian.

Next we use (1.15) in Section 4 to obtain geometric information about noncommuting elements $f$ and $g$ of a discrete group $G$. For example, we prove that if $f$ is loxodromic, then $g$ moves a fixed point of $f$ at least chordal distance $\delta(|\beta(f)|)$ times the chordal diameter of fix$(f)$, where $\delta(t) = O(1/t)$ as $t \to \infty$ and $\delta(t) \to 1/2$ as $t \to 0$. We show also that if $f$ and $g$ are not both of order 2, 3, 4, or 6, then

$$\max\big(d(f,g), d(f^{-1}, g^{-1})\big) \geq d, \tag{1.17}$$

where $d$ is an absolute constant, $0.848 \leq d \leq 0.911\ldots$.

## 2. Preliminary Results

We collect here facts needed to establish the main results of this paper.

**2.1. Lemma.** *If $f$ and $g_j$ are elements of a discrete subgroup $G$ of* M *with $\gamma(f,g_j) \neq 0$ for $j = 1,2,\ldots$ and if $f$ is not of order 2, then*

$$\liminf_{j \to \infty} |\gamma(f,g_j)| > 0. \tag{2.2}$$

**Proof.** See [5, Lemma 2.19]. $\square$

We will use the following polynomial trace identities. See [4] for other identities and the abstract words in $f$ and $g$ which give rise to them.

**2.3. Lemma.** *If $f$ and $g$ are in $\mathbb{M}$ with $\gamma(f,g) = \gamma$ and $\beta(f) = \beta$, then*

$$\gamma(f, gfg^{-1}) = \gamma(\gamma - \beta), \tag{2.4}$$

$$\gamma(f, gf^{-1}g^{-1}fgfg^{-1}f^{-1}g) = \gamma[\gamma^2 - (\beta - 1)\gamma - (\beta - 1)]^2. \tag{2.5}$$

**Proof.** The identities

$$\mathrm{tr}(fh) + \mathrm{tr}(fh^{-1}) = \mathrm{tr}(f)\mathrm{tr}(h), \tag{2.6}$$

$$\mathrm{tr}([f,h]) = \mathrm{tr}^2(f) + \mathrm{tr}^2(h) + \mathrm{tr}^2(fh) - \mathrm{tr}(f)\mathrm{tr}(h)\mathrm{tr}(fh) - 2 \tag{2.7}$$

and (1.3) imply that

$$\mathrm{tr}(fgfg^{-1}) = \mathrm{tr}^2(f) - \mathrm{tr}([f,g]) = \beta - \gamma + 2$$

and hence that

$$\begin{aligned}
\gamma(f, gfg^{-1}) &= \mathrm{tr}([f, gfg^{-1}]) - 2 \\
&= 2\,\mathrm{tr}^2(f) + \mathrm{tr}^2(fgfg^{-1}) - \mathrm{tr}^2(f)\mathrm{tr}(fgfg^{-1}) - 4 \\
&= 2(\beta + 4) + (\beta - \gamma + 2)^2 - (\beta + 4)(\beta - \gamma + 2) - 4 \\
&= \gamma(\gamma - \beta).
\end{aligned}$$

This is (2.4). For (2.5) suppose first that $f$ and $g$ are represented by

$$A = \begin{pmatrix} u & 0 \\ 0 & u^{-1} \end{pmatrix}, \qquad B = \begin{pmatrix} a & b \\ c & d \end{pmatrix}.$$

Then

$$\gamma = -bc(u - u^{-1})^2, \qquad \beta = (u - u^{-1})^2,$$

and a long but elementary calculation shows that

$$BA^{-1}B^{-1}ABAB^{-1}A^{-1}B$$

$$= \begin{pmatrix} a[(\gamma + 1)^2 - \gamma u^{-2}] & b[\gamma^2 - (\beta - 1)\gamma - (\beta - 1)] \\ c[\gamma^2 - (\beta - 1)\gamma - (\beta - 1)] & d[(\gamma + 1)^2 - \gamma u^2] \end{pmatrix}.$$

Hence

$$\begin{aligned}
\gamma(f, gf^{-1}g^{-1}fgfg^{-1}f^{-1}g) &= -bc[\gamma^2 - (\beta - 1)\gamma - (\beta - 1)]^2(u - u^{-1})^2 \\
&= \gamma[\gamma^2 - (\beta - 1)\gamma - (\beta - 1)]^2.
\end{aligned}$$

A similar calculation yields this result when $f$ is parabolic. $\quad\square$

**2.8. Lemma.** *If $\langle f, g \rangle$ is an elementary discrete subgroup of $\mathbb{M}$ with $\gamma(f,g) \neq 0$ and $\beta(f) = \beta(g) \neq -4$, then*

$$|\gamma(f,g)| \geq (3 - \sqrt{5})/2 = 0.381\ldots. \tag{2.9}$$

**Proof.** Since $f$ and $g$ have no common fixed point, the classification of elementary discrete groups implies that $\langle f, g \rangle$ must be one of the groups $A_4$, $S_4$, or $A_5$. Then (2.9) follows from the fact that $[f, g]$ is elliptic of order 2, 3, 4, or 5. $\square$

**2.10. Lemma.** *If $\langle f, g \rangle$ is discrete with $\gamma(f, g) = \beta(f) \neq 0$, then either $f$ is elliptic of order 2, 3, 4, or 6 or $g$ is elliptic of order 2.*

**Proof.** By hypothesis, $f$ has two fixed points, $\langle f, gfg^{-1} \rangle$ is discrete and

$$\gamma(f, gfg^{-1}) = \gamma(f, g)[\gamma(f, g) - \beta(f)] = 0$$

by (2.4). Hence

$$g\big(\mathrm{fix}(f)\big) \cap \mathrm{fix}(f) = \mathrm{fix}(gfg^{-1}) \cap \mathrm{fix}(f) \neq \emptyset, \qquad \mathrm{fix}(f) \cap \mathrm{fix}(g) = \emptyset$$

and thus $g$ maps one point of $\mathrm{fix}(f)$ onto the other. The desired conclusion then follows from [11, Proposition 1]. $\square$

# 3. A General Lower Bound for the Commutator Parameter

We derive here the following pair of bounds for the commutator parameter of a two generator group.

**3.1. Theorem.** *Suppose that $\langle f, g \rangle$ is a discrete subgroup of $\mathbb{M}$ with $\gamma(f, g) \neq 0$ and $\beta(f) = \beta(g) \neq -4$. Then*

$$|\gamma(f, g)| > 0.193. \tag{3.2}$$

*If, in addition, $\gamma(f, g)$ is real, then*

$$|\gamma(f, g)| \geq 2 - 2\cos(\pi/7) = 0.198\ldots. \tag{3.3}$$

*Inequality (3.3) is sharp.*

**3.4. Theorem.** *Suppose that $\langle f, g \rangle$ is a discrete subgroup of $\mathbb{M}$ with $\gamma(f, g) \neq 0$, $\beta(f) \neq -4$, $\beta(g) \neq -4$, and $(\gamma(f, g), \beta(f)) \neq (-j, -j)$ for $j = 1, 2$ or $3$. Then*

$$|\gamma(f, g)| \, |\gamma(f, g) - \beta(f)| > 0.193. \tag{3.5}$$

*If, in addition, $\gamma(f, g)[\gamma(f, g) - \beta(f)]$ is real, then*

$$|\gamma(f, g)| \, |\gamma(f, g) - \beta(f)| \geq 2 - 2\cos(\pi/7) = 0.198\ldots. \tag{3.6}$$

*Inequality (3.6) is sharp.*

**Proof of Theorem 3.4.** The group $\langle f, gfg^{-1} \rangle$ is discrete with

$$\gamma(f, gfg^{-1}) = \gamma(f, g)[\gamma(f, g) - \beta(f)], \qquad \beta(f) = \beta(gfg^{-1}) \neq -4.$$

Next $\gamma(f, g) \neq \beta(f)$ since otherwise Lemma 2.10 would imply that $f$ is elliptic of order 3, 4, or 6 contradicting the hypothesis that $(\gamma(f, g), \beta(f)) \neq (-j, -j)$ for $j = 1, 2$, or 3. Hence $\langle f, gfg^{-1} \rangle$ satisfies the hypotheses of Theorem 3.1, and (3.5) and (3.6) follow from (3.2) and (3.3), respectively.

To show that (3.6) is sharp, let $\langle \phi, \psi \rangle$ denote the (2,3,7) triangle group with $\phi^2 = \psi^3 = (\phi\psi)^7 = id$ and set $f = [\phi, \psi]$ and $g = \phi\psi$. Then

$$\begin{cases} \gamma(f, g) = \gamma(\phi, \psi) = \mathrm{tr}([\phi, \psi]) - 2 = 2\cos(2\pi/7) - 1 = 0.246\ldots, \\ \beta(f) = \mathrm{tr}^2([\phi, \psi]) - 4 = 2[\cos(2\pi/7) + \cos(\pi/7) - 1] = 1.0489\ldots \end{cases} \tag{3.7}$$

by (2.7) and hence

$$\gamma(f, g)[\gamma(f, g) - \beta(f)] = 2\cos(\pi/7) - 2 = 0.198\ldots. \qquad \square \tag{3.8}$$

The proof of Theorem 3.1 depends on two preliminary results the first of which is of independent interest.

**3.9. Lemma.** *If $\langle f, g \rangle$ is a discrete subgroup of $\mathbb{M}$ with $\gamma(f, g) \neq 0$, then*

$$|\beta(f) - 1| \leq \frac{|\gamma(f, g)|^2}{1 - |\gamma(f, g)|} \quad or \quad |\beta(f) - 1| \geq 1 - |\gamma(f, g)|. \tag{3.10}$$

**Proof.** Suppose (3.10) does not hold and let $\gamma = \gamma(f, g)$ and $\beta = \beta(f)$. Then

$$\frac{|\gamma|^2}{1 - |\gamma|} < |\beta - 1| < 1 - |\gamma| \tag{3.11}$$

and hence

$$p(\gamma) = [\gamma^2 - (\beta - 1)\gamma - (\beta - 1)]^2 \neq 0$$

since otherwise we would have

$$|\beta - 1| = \left| \frac{\gamma^2}{\gamma + 1} \right| \leq \frac{|\gamma|^2}{1 - |\gamma|}$$

contradicting the first part of (3.11).

Let $\gamma_j = \gamma(f, g_j)$ where

$$g_{j+1} = g_j f^{-1} g_j^{-1} f g_j f g_j^{-1} f^{-1} g_j, \qquad g_1 = g$$

for $j = 1, 2, \ldots$. Then $\gamma_1 = \gamma$,

$$|\gamma_{j+1}| = |\gamma_j|\,|p(\gamma_j)| \tag{3.12}$$

by (2.5) and

$$|\gamma_j| + |\beta - 1| < 1, \qquad p(\gamma_j) \neq 0 \tag{3.13}$$

implies that

$$0 < |p(\gamma_j)| = |\gamma_j^2 - (\beta - 1)\gamma_j - (\beta - 1)|^2 < \left(|\gamma_j| + |\beta - 1|\right)^2 < 1. \tag{3.14}$$

Now (3.11) holds with $\gamma_1$ in place of $\gamma$ and thus implies (3.13) for $j = 1$. Hence $0 < |\gamma_2| < |\gamma_1|$ by (3.12) and (3.14), and (3.11) holds with $\gamma_2$ in place of $\gamma$. Thus we see by induction and (3.14) that

$$0 < |\gamma_{j+1}| \leq |\gamma_1|(|\gamma_1| + |\beta - 1|)^{2j} \to 0$$

as $j \to \infty$ contradicting Lemma 2.1. $\square$

**3.15. Lemma.** *If $w_0$, $w_1$, $w_2$ are complex numbers with*

$$0 < |w_0| < \min(|w_1|, |w_2|), \qquad 0 \leq \theta = |\arg(-w_2/w_1)| \leq \pi,$$

*then*

$$|w_0 - w_1| + |w_0 - w_2| \leq r_1 + r_2 + |w_0|\theta \tag{3.16}$$

*where*

$$r_j = \left(|w_j|^2 - |w_0|^2\right)^{1/2} + |w_0|\arcsin(|w_0|/|w_j|).$$

**Proof.** Let $B = \{z : |z| \leq |w_0|\}$ and let $E$ denote the union of the convex hulls of $B \cup \{w_1\}$ and $B \cup \{w_2\}$. Then

$$u(z) = |z - w_1| + |z - w_2|$$

is subharmonic in $D = \text{int}(E)$ and hence there exists a point $w$ in $\partial D$ such that

$$|w_0 - w_1| + |w_0 - w_2| = u(w_0) \leq u(w). \tag{3.17}$$

Inequality (3.16) then follows from (3.17) by estimating $u(w)$ from above by the length of the component of $\partial D \setminus \{w_1, w_2\}$ which contains $w$. $\square$

**Proof of Theorem 3.1.** We establish inequalities (3.2) and (3.3) by considering three different cases.

**3.18. Lemma.** *If $\langle f, g \rangle$ is discrete with $\gamma(f, g) \neq 0$ and $\beta(f) = \beta(g) \neq -4$ and if*

$$\min(|\beta(f)|, |\beta(fg)|, |\beta(fg^{-1})|) \geq 2[\cos(2\pi/7) + \cos(\pi/7) - 1] \\ = 1.0489\ldots, \tag{3.19}$$

*then*

$$|\gamma(f, g)| \geq 2 - 2\cos(\pi/7) = 0.198\ldots. \tag{3.20}$$

*Inequality (3.20) is sharp.*

**Proof.** If $A$ represents $f$, then by replacing $A$ by $-A$ if necessary, we may assume that $\text{tr}(f) = \text{tr}(g)$. Let

$$a = \text{tr}(fg) - 2, \quad b = \text{tr}(fg^{-1}) - 2, \quad c = 2[\cos(2\pi/7) + \cos(\pi/7) - 1].$$

Next by replacing $g$ by $g^{-1}$ we may also assume that $|a| \leq |b|$. Then

$$\beta(f) = \text{tr}^2(f) - 4 = a + b, \qquad \beta(fg) = \text{tr}^2(fg) - 4 = a^2 + 4a,$$

$$\gamma(f, g) = 2\,\text{tr}^2(f) + \text{tr}^2(fg) - \text{tr}^2(f)\text{tr}(fg) - 4 = -ab$$

by (2.6) and (2.7). Hence

$$(|a| + 2)^2 \geq |\beta(fg)| + 4 \geq c + 4 = [2\cos(2\pi/7) + 1]^2$$

by (3.19) and $|a| \geq d$ where $d = 2\cos(2\pi/7) - 1 = 0.246\ldots$.
If $|a| \geq 0.5$, then

$$|\gamma(f, g)| = |a|\,|b| \geq |a|^2 \geq 0.25$$

while if $|a| < 0.5$, then $|a| < c/2$ and

$$|\gamma(f, g)| = |a|\,|\beta(f) - a| \geq d(c - d) = 2 - 2\cos(\pi/7) = 0.198\ldots$$

by (3.19). This completes the proof of (3.20).

To show that (3.20) is sharp, let $\langle \phi, \psi \rangle$ be the (2,3,7) triangle group with $\phi^2 = \psi^3 = (\phi\psi)^7 = id$ and set $f = [\phi, \psi]$, $h = \phi\psi$, $g = hf^{-1}h^{-1}$. Then $\beta(f) = \beta(g) = c$,

$$\beta(fg) = \beta(fhf^{-1}h^{-1}) = \gamma(f, h)[\gamma(f, h) + 4] = c,$$
$$\beta(fg^{-1}) = \beta(fhfh^{-1}) = [\beta(f) - \gamma(f, h)][\beta(f) - \gamma(f, h) + 4] > c,$$
$$\gamma(f, g) = \gamma(f, h)[\gamma(f, h) - \beta(f)] = 2\cos(\pi/7) - 2 = 0.198\ldots$$

by (3.7) and (3.8). Hence $\langle f, g \rangle$ satisfies the hypotheses of Lemma 3.18 and (3.20) holds with equality. $\square$

**3.21. Lemma.** *If $\langle f, g \rangle$ is discrete with $\gamma(f, g) \neq 0$ and $\beta(g) \neq -4$, if*

$$|\beta(f)| \leq 2[\cos(2\pi/7) + \cos(\pi/7) - 1] = 1.0489\ldots \qquad (3.22)$$

*and if $\gamma(f, g)$ is real, then*

$$|\gamma(f, g)| \geq 2 - 2\cos(\pi/7) = 0.198\ldots \qquad (3.23)$$

*with equality only if $\gamma(f, g) < 0$.*

**Proof.** Let $\gamma = \gamma(f, g)$, $\beta = \beta(f)$ and set

$$c = 2[\cos(2\pi/7) + \cos(\pi/7) - 1], \qquad d = 2 - 2\cos(\pi/7).$$

We may assume that $|\gamma| \leq d$ since otherwise there is nothing to prove. Then $\gamma \neq \beta$ since otherwise Lemma 2.10 would imply that $f$ is elliptic of order 2, 3, 4, or 6 and hence that $|\gamma| = |\beta| \geq 1$. Thus $\langle [f,g], f \rangle = \langle f, gfg^{-1} \rangle$ is discrete with

$$\gamma' = \gamma([f,g], f) = \gamma(\gamma - \beta) \neq 0, \qquad \beta' = \beta([f,g]) = \gamma(\gamma + 4). \quad (3.24)$$

If $\gamma < 0$, then $-0.9 < \beta' < 0$, $[f,g]$ is an elliptic of order $n \geq 7$, $[f,g]$ and $f$ have no common fixed point by (3.24) and $f$ is not of order 2. Hence $\langle [f,g], f \rangle$ is nonelementary by the classification of elementary discrete groups,

$$|\gamma|(|\gamma| + c) \geq |\gamma'| \geq 2\cos(2\pi/7) - 1 = d(d + c)$$

by (1.12) and hence $|\gamma| \geq d$. If $\gamma > 0$, then $0 < \beta' < 0.9$, $|\gamma'| < 0.25$, and

$$|\beta' - 1| > 0.1 > \frac{|\gamma'|^2}{1 - |\gamma'|}.$$

Then (3.10) implies that

$$1 - \gamma(\gamma + 4) = |\beta' - 1| \geq 1 - |\gamma'| = 1 - \gamma|\gamma - \beta|$$

and hence that $|\beta| \geq 4$, a contradiction. Thus $\gamma > d$.   $\square$

**3.25. Lemma.** *If $\langle f, g \rangle$ is discrete with $\gamma(f,g) \neq 0$ and $\beta(g) \neq -4$ and if*

$$|\beta(f)| \leq 2[\cos(2\pi/7) + \cos(\pi/7) - 1] = 1.0489\ldots, \quad (3.26)$$

*then*

$$|\gamma(f,g)| > 0.193. \quad (3.27)$$

**Proof.** Let $\gamma = \gamma(f,g)$, $\beta = \beta(f)$ and set

$$c = 2[\cos(2\pi/7) + \cos(\pi/7) - 1], \qquad d = 0.193.$$

We may assume that $|\gamma| \leq d$ and hence that $\gamma \neq \beta$ as in the proof of Lemma 3.21. Next $\langle [f,g], f[f,g]f^{-1} \rangle$ is discrete with

$$0 < |\gamma([f,g], f[f,g]f^{-1})| = |-\gamma^2(\gamma - \beta)(\beta + 4)| \leq d^2(d + c)(c + 4) < 0.3$$

and

$$|\beta([f,g])| = |\beta(f[f,g]f^{-1})| = |\gamma(\gamma + 4)| \leq d(d + 4) < 1.$$

Thus Lemma 2.8 implies that $\langle [f,g], f[f,g]f^{-1} \rangle$ is nonelementary and we obtain

$$1 \leq |\gamma|^2|\gamma - \beta| \, |\beta + 4| + |\gamma| \, |\gamma + 4| \quad (3.28)$$

from (1.14). In particular, $|\beta + 4| > 4$; otherwise (3.28) would imply that

$$1 \leq 4|\gamma|^3 + (4c + 1)|\gamma|^2 + 4|\gamma| < 1.$$

Similarly $|\beta| > |\gamma|$ since otherwise we would have

$$1 \leq (2|\gamma|^3 + |\gamma|)(|\gamma| + 4) < 1.$$

Let $\theta = |\arg(\beta)| \in [0, \pi]$. By Lemma 3.15 with $w_0 = \gamma$, $w_1 = -4$, $w_2 = \beta$,

$$|\gamma + 4| + |\gamma - \beta| \leq r_1 + r_2 + |\gamma|\theta \leq r + |\gamma|\theta, \qquad (3.29)$$

where

$$r_1 = \left(4^2 - |\gamma|^2\right)^{1/2} + |\gamma|\arcsin(|\gamma|/4) < 4.0048,$$
$$r_2 = \left(|\beta|^2 - |\gamma|^2\right)^{1/2} + |\gamma|\arcsin(|\gamma|/|\beta|) < 1.0682$$

and $r = 5.073$. Next

$$|\beta + 4|^2 = 4^2 + |\beta|^2 + 8|\beta| \cos\theta < a + b\cos\theta, \qquad (3.30)$$

where $a = 17.2$ and $b = 8.4$, and we obtain

$$\begin{aligned}
1 &\leq |\gamma|^2|\beta + 4|(|\gamma + 4| + |\gamma - \beta|) + |\gamma|(1 - |\gamma| |\beta + 4|)(|\gamma| + 4) \\
&\leq |\beta + 4|(\theta - 1)|\gamma|^3 + [|\beta + 4|(r - 4) + 1]|\gamma|^2 + 4|\gamma| \\
&\leq |\beta + 4|(\theta - 1)d^3 + [|\beta + 4|(r - 4) + 1]d^2 + 4d \\
&\leq (a + b\cos\theta)^{1/2}[d(\theta - 1) + r - 4]d^2 + (d^2 + 4d) = f(\theta)
\end{aligned}$$

from (3.28), (3.29), and (3.30). Then it is not difficult to verify that $f(\theta) < 1$ for $0 \leq \theta \leq \pi$, a contradiction. Thus $|\gamma| > d = 0.193$. $\quad\square$

**3.31. Completion of the proof of Theorem 3.1.** If

$$\min\left(|\beta(f)|, |\beta(fg)|, |\beta(fg^{-1})|\right) \geq 2[\cos(2\pi/7) + \cos(\pi/7) - 1],$$

then (3.2) and (3.3) follow from Lemma 3.18. Otherwise since

$$\gamma(f, g) = \gamma(fg, g) = \gamma(fg^{-1}, g),$$

we may assume by relabeling that

$$|\beta(f)| < 2[\cos(2\pi/7) + \cos(\pi/7) - 1].$$

Then $\beta(g) \neq -4$ and we obtain (3.2) and (3.3) from Lemmas 3.25 and 3.21, respectively.

Finally the example for which (3.20) holds with equality in Lemma 3.18 shows that (3.3) is sharp. $\quad\square$

# 4. Applications to Discrete Groups

We derive here some geometric consequences of the inequalities in the previous section. The first concerns the chordal displacement of the fixed points of a nonparabolic element by other elements in a discrete group.

**4.1. Theorem.** *Suppose that $\langle f, g \rangle$ is a discrete subgroup of $\mathbb{M}$, that $f$ and $g$ have no common fixed point, and that $f$ is loxodromic or elliptic of order $n \geq 3$. If $\mathrm{fix}(f) = \{z_1, z_2\}$, then*

$$\max_{j=1,2} \left( \frac{q(g(z_j), z_j)}{q(z_1, z_2)} \right) \geq \delta(|\beta(f)|), \tag{4.2}$$

*where*

$$\frac{c}{t + 2c} \leq \delta(t) \leq \min\left( \frac{2}{t}, \frac{1}{2} \right) \tag{4.3}$$

*for $0 < t < \infty$ and $c$ is an absolute constant, $0.439 \leq c \leq 2$.*

**Proof.** Suppose that

$$g(\mathrm{fix}(f)) \cap \mathrm{fix}(f) \neq \emptyset. \tag{4.4}$$

Since $g$ does not fix $z_1$ and $z_2$, by relabeling we may assume that $g(z_2) = z_1$. Then $q(g(z_2), z_2) = q(z_1, z_2)$ and we obtain the first part of (4.3) since this bound does not exceed $1/2$ for $0 < t < \infty$.

Suppose next that (4.4) does not hold and let $h = gfg^{-1}$. Then $\langle f, h \rangle$ is discrete with $\gamma(f, h) \neq 0$ and $\beta(f) = \beta(h) \neq -4$ and thus

$$|\gamma(f, h)| > 0.193 > c^2 \tag{4.5}$$

by Theorem 3.1 where $c \geq 0.439$. Next by [3, (2.13) and (2.15)],

$$\left( \frac{c}{|\beta(f)|} \right)^2 \leq \left| \frac{\gamma(f, h)}{\beta(f)\beta(h)} \right| = \frac{q(w_1, z_1)q(w_2, z_2)q(w_1, z_2)q(w_2, z_1)}{q(z_1, z_2)^2 q(w_1, w_2)^2}, \tag{4.6}$$

where $w_j = g(z_j)$. Hence if

$$s = \max_{j=1,2} \left( \frac{q(w_j, z_j)}{q(z_1, z_2)} \right) < 1/2,$$

then

$$q(w_1, z_1)q(w_2, z_2) \leq s^2 q(z_1, z_2)^2, \quad q(w_1, w_2) \geq (1 - 2s)q(z_1, z_2). \tag{4.7}$$

Next by the Möbius invariant form of the triangle inequality,

$$|w_1 - z_2||w_2 - z_1| \leq |z_1 - z_2||w_1 - w_2| + |w_1 - z_1||w_2 - z_2|$$

whence

$$q(w_1, z_2)q(w_2, z_1) \leq q(z_1, z_2)q(w_1, w_2) + q(w_1, z_1)q(w_2, z_2) \tag{4.8}$$

by (1.5). Then (4.6), (4.7), and (4.8) imply that

$$\left(\frac{c}{|\beta(f)|}\right)^2 \le \frac{s^2 q(z_1, z_2)^3 q(w_1, w_2) + s^4 q(z_1, z_2)^4}{q(z_1, z_2)^2 q(w_1, w_2)^2} \le \left(\frac{s(1-s)}{1-2s}\right)^2$$

from which the first part of (4.3) follows with $c \ge 0.439$.

Suppose that $4 < t < \infty$, choose $1 < a < \infty$ so that $t = (a + a^{-1})^2$ and let

$$f = h f_0 h^{-1}, \qquad g = h g_0 h^{-1}, \qquad h = r\frac{z-1}{z+1}, \qquad (4.9)$$

where $f_0$ and $g_0$ are as in (1.9) and $0 < r < 1$. Then $\langle f, g \rangle$ is discrete, $f$ is loxodromic, and

$$\text{fix}(f) = \{-r, r\}, \qquad g(\pm r) = \pm r\frac{t-4}{t} = \pm s, \qquad |\beta(f)| = t.$$

Hence

$$\delta(t) \le \frac{q(g(\pm r), \pm r)}{q(-r, r)} = \frac{2}{t}\left(\frac{r^2+1}{s^2+1}\right)^{1/2} \qquad (4.10)$$

by (4.2). Since the left-hand side of (4.10) is independent of $r$, we can let $r \to 0$ to obtain the second part of (4.3) for the case where $4 < t < \infty$.

Suppose next that $0 < t \le 4$, choose $1 < b < a < \infty$ so that $t = (a-a^{-1})^2$ and let $f$, $g$, and $h$ be as in (4.9) where now

$$f_0(z) = a^2 z, \qquad g_0 = \frac{(b^2+1)z + 2b}{2bz + (b^2+1)}. \qquad (4.11)$$

Then $\langle f, g \rangle$ is discrete, $f$ is loxodromic, and

$$\text{fix}(f) = \{-r, r\}, \qquad g(\pm r) = \pm r\left(\frac{b-1}{b+1}\right)^2 = \pm s, \qquad \beta(f) = t.$$

Hence

$$\delta(t) \le \frac{q(g(\pm r), \pm r)}{q(-r, r)} < \frac{1}{2}\left(\frac{r^2+1}{s^2+1}\right)^{1/2},$$

and letting $r \to 0$ yields the second half of (4.3) for the case where $0 < t \le 4$.

Finally we obtain $c \le 2$ by letting $t \to \infty$ in (4.3)    $\square$

**4.12. Corollary.** *Suppose that $f$ is a loxodromic element of a discrete subgroup $G$ of $\mathbb{M}$. Then each element $g$ in $G$ either commutes with $f$ or moves one of its fixed points at least chordal distance $\delta(|\beta(f)|)$ times the chordal diameter of $\text{fix}(f)$ where $\delta(t)$ is as in (4.3).*

**4.13. Example.** For $1 < a < \infty$ and $1 < r < \infty$ let $f_0$ and $g_0$ be as in (1.9) and set

$$f = h f_0 h^{-1}, \qquad g = h g_0 h^{-1}, \qquad h = rz.$$

Then $\text{fix}(f) = \{0, \infty\}$, $|\beta(f)| > 4$, and

$$\frac{q(g(\infty), \infty)}{q(0, \infty)} = 2\left[r^2(|\beta(f)| - 2)^2 + 4\right]^{-1/2} \to 0$$

as $r \to \infty$. Hence there exists no universal lower bound for each of the ratios

$$\frac{q(g(z_1), z_1)}{q(z_1, z_2)}, \qquad \frac{q(g(z_2), z_2)}{q(z_1, z_2)}$$

in Theorem 4.1 but rather only for their maximum.

We derive next a lower bound for the distance in the metric of (1.4) between the generators of a nonelementary discrete group.

**4.14. Theorem.** *Suppose that $\langle f, g \rangle$ is a discrete subgroup of $\mathbb{M}$ and that $f$ and $g$ have no common fixed point and are not both of order 2. Then*

$$\max\left(d(f, g), d(f^{-1}, g^{-1})\right) \geq d, \tag{4.15}$$

*where $d$ is an absolute constant, $0.848 \leq d \leq 0.911\ldots$.*

**Proof.** Suppose that $g$ is not of order 2 and let $\gamma = \gamma(f, g) = \gamma(fg^{-1}, g^{-1})$ and $\beta = \beta(fg^{-1}) = \beta(g^{-1}f)$. If $fg^{-1}$ is of order $n$ where $n = 2, 3, 4$, or $6$, then

$$d(f, g) = d(fg^{-1}, id) \geq 2\sin(\pi/n) \geq 1 \tag{4.16}$$

by [3, Corollary 3.17]. Otherwise $\gamma \neq \beta$ by Lemma 2.10 applied to $\langle fg^{-1}, g^{-1} \rangle$,

$$\langle fg^{-1}, g^{-1}f \rangle = \langle fg^{-1}, g^{-1}(fg^{-1})g \rangle$$

is discrete with

$$\gamma(fg^{-1}, g^{-1}f) = \gamma(fg^{-1}, g^{-1})\left[\gamma(fg^{-1}, g^{-1}) - \beta(fg^{-1})\right] = \gamma(\gamma - \beta) \neq 0 \tag{4.17}$$

by (2.4) and Theorem 3.1 implies that

$$|\gamma(fg^{-1}, g^{-1}f)| > 0.193.$$

We then obtain

$$\max\left(d(f, g), d(f^{-1}, g^{-1})\right) = \max\left(d(fg^{-1}, id), d(g^{-1}f, id)\right) \geq 0.848\ldots$$

from Lemma 1.6 completing the proof of (4.15) with $d \geq 0.848$.

To obtain an upper bound for $d$, let $\langle \phi, \psi \rangle$ denote the $(2,3,7)$ triangle group acting on the upper half plane $H$ with $\phi^2 = \psi^3 = (\phi\psi)^7 = id$ and set $f = \phi\psi$ and $g = \psi\phi$. Then (2.7) and the fact $\langle f, g \rangle$ is Nielsen equivalent to $\langle \phi, \psi \rangle$ imply that

$$\begin{cases} \gamma = \text{tr}([\phi, \psi]) - 2 = \text{tr}^2(\psi) + \text{tr}^2(\phi\psi) - 4 = 2\cos(2\pi/7) - 1 \\ \beta = \text{tr}^2([\phi, \psi]) - 4 = 2[\cos(2\pi/7) + \cos(\pi/7) - 1] > 0 \end{cases} \tag{4.18}$$

since $fg^{-1} = [\phi, \psi]$. Hence $fg^{-1}$ and $g^{-1}f$ are hyperbolic while

$$\gamma(fg^{-1}, g^{-1}f) = \gamma(\gamma - \beta) < 0$$

by (4.17) and (4.18). Then as in the first part of [3, Lemma 2.27], there exists a Möbius self map $h$ of $H$ such that after a preliminary conjugation by $h$ we have fix$(fg^{-1}) = \{z_1, z_2\}$ and fix$(g^{-1}f) = \{w_1, w_2\}$ where $q(z_1, z_2) = q(w_1, w_2) = 2$. Then it follows from [3, Corollary 3.15] and from (4.18) that

$$\begin{cases} d(f,g) = d(fg^{-1}, id) = 2 \left( \dfrac{\cos(2\pi/7)+\cos(\pi/7)-1}{\cos(2\pi/7)+\cos(\pi/7)+1} \right)^{1/2} = 0.911\ldots \\[4mm] d(f^{-1},g^{-1}) = d(g^{-1}f, id) = 2 \left( \dfrac{\cos(2\pi/7)+\cos(\pi/7)-1}{\cos(2\pi/7)+\cos(\pi/7)+1} \right)^{1/2} = 0.911\ldots \end{cases}$$
$$(4.19)$$

and hence that $d \leq 0.911\ldots$.  □

We can reformulate Theorem 4.14 as follows.

**4.20. Theorem.** *Suppose that $f$ and $g$ are elements of a discrete subgroup $G$ of $\mathbb{M}$ which are not both of order 2, 3, 4, or 6. Then $f$ and $g$ commute or*

$$\max\big(d(f,g), d(f^{-1},g^{-1})\big) \geq d, \qquad (4.21)$$

*where $d$ is an absolute constant, $0.848 \leq d \leq 0.911\ldots$.*

**Proof.** It suffices to consider the case where $\gamma(f,g) = 0$ and $fg \neq gf$ whence

$$\text{fix}(f) \cap \text{fix}(g) \neq \emptyset, \qquad \text{fix}(f) \neq \text{fix}(g). \qquad (4.22)$$

Then $\langle f, g \rangle$ is elementary and $h = [f, g]$ is parabolic. We complete the proof by showing that $fg^{-1}$ is elliptic of order $n$ where $n = 2, 3, 4$, or 6 and then appealing to (4.16).

Suppose that fix$(h) = \{\infty\}$. If $f$ or $g$, say $f$, is parabolic, then (4.22) implies that $g$ is elliptic of order 2, 3, 4, or 6 and an elementary calculation shows that the same is true of $fg^{-1}$. If $f$ and $g$ are both elliptic, then

$$f = \lambda z + a, \qquad g = \mu z + b, \qquad fg^{-1} = (\lambda/\mu)z + c,$$

where $\lambda^p = \mu^q = (\lambda/\mu)^r = 1$, $p, q \in \{2,3,4,6\}$, and $r \in \{1,2,3,4,6\}$. By hypothesis $p \neq q$; hence $\lambda \neq \mu$ and $fg^{-1}$ is of order 2, 3, 4, or 6.  □

**4.23. Example.** If $f = \lambda z$ and $g = \lambda z - c$ where $\lambda^p = 1$, $0 < |c| \leq 2$ and $p = 2, 3, 4$, or 6, then $\langle f, g \rangle$ is discrete [10, Section V.D] while

$$d(f,g) = d(f^{-1},g^{-1}) = \frac{8|c|}{4 + |c|^2} \to 0$$

as $c \to 0$ by [3, Lemma 3.8]. Thus the hypothesis that $f$ and $g$ not both be of order 2, 3, 4, or 6 in Theorem 4.20 is necessary.

**4.24. Remark.** For each $1 < b < a < \infty$ let $f = f_0 g_0$ and $g = g_0$ where $f_0$ and $g_0$ are as in (4.11). Then $\langle f, g \rangle$ is nonelementary and discrete while

$$d(f, g) = d(f_0, id) = 2 \left( \frac{a^2 - 1}{a^2 + 1} \right) \to 0$$

as $a \to 1$ by [3, Corollary 3.15]. Hence there exists no universal lower bound for $d(f, g)$ and $d(f^{-1}, g^{-1})$ in Theorems 4.14 and 4.20 but only for their maximum.

Theorem 4.14 gives a geometric estimate of how different two Möbius transformations must be in order to generate a nonelementary discrete group. We conclude this paper with a similar geometric measure of how far from commuting the generators of such a group must be.

**4.25. Theorem.** *Suppose that $\langle f, g \rangle$ is a discrete subgroup of $\mathbb{M}$ and that $f$ and $g$ have no common fixed point and are not both of order 2. If $fg$ is also not of order 2, then*

$$\max\left( d(fg, gf), d((fg)^{-1}, (gf)^{-1}) \right) \geq d, \tag{4.26}$$

*where $d$ is an absolute constant, $0.848 \leq d \leq 0.911$.*

**Proof.** Suppose that $g$ is not of order 2 and let $\gamma = \gamma(f, g)$ and $\beta = \beta(fg)$. If $\gamma = \beta$, then Lemma 2.10 applied to $\langle fg, g \rangle = \langle f, g \rangle$ implies $\beta([f, g]) = \gamma(\gamma + 4) = -k$ where $k = 3$ or $4$ and thus that $[f, g]$ is elliptic of order 2 or 3. Hence

$$d(fg, gf) = d([f, g], id) \geq \sqrt{3}$$

by (4.16) with $fg$ and $gf$ in place of $f$ and $g$. Otherwise $\langle fg, gf \rangle = \langle fg, g(fg)g^{-1} \rangle$ is discrete with

$$\gamma(fg, gf) = \gamma(fg, g)[\gamma(fg, g) - \beta(fg)] = \gamma(\gamma - \beta) \neq 0$$

and (4.26) follows with $d \geq 0.848$ from Theorem 4.14.

Next let $\langle f, g \rangle$ be the group for which (4.15) holds with equality in Theorem 4.14. Then $f = \phi\psi$ and $g = \psi\phi$ where $\langle \phi, \psi \rangle$ is the (2,3,7) triangle group with $\phi^2 = \psi^3 = (\phi\psi)^7 = id$ and we obtain

$$d(\phi\psi, \psi\phi) = d\left( (\phi\psi)^{-1}, (\psi\phi)^{-1} \right) = 0.911\ldots$$

from (4.19). Hence the group $\langle \phi, \psi \rangle$ shows that $d \leq 0.911\ldots$ in (4.26).  $\square$

## REFERENCES

1. F.W. Gehring and G.J. Martin, Stability and extremality in Jørgensen's inequality, *Complex Variables* **12** (1989), 277–282.

2. F.W. Gehring and G.J. Martin, Iteration theory and inequalities for Kleinian groups, *Bull. Am. Math. Soc.* **21** (1989), 57–63.

3. F.W. Gehring and G.J. Martin, Inequalities for Möbius transformations and discrete groups, *J. Reine Angew. Math.* (to appear).

4. F.W. Gehring and G.J. Martin, Discreteness in Kleinian groups and iteration theory (to appear).

5. F.W. Gehring and G.J. Martin, Commutators, collars and the geometry of Möbius groups (to appear).

6. T. Jørgensen, On discrete groups of Möbius transformations, *Am. J. Math.* **98** (1976), 739–749.

7. T. Jørgensen, Comments on a discreteness condition for subgroups of SL(2, ℂ), *Can. J. Math.* ₁**31** (1979), 87–92.

8. T. Jørgensen, Commutators in SL(2, ℂ), *Riemann Surfaces and Related Topics: Proceedings of the 1978 Stony Brook Conference,* Princeton University Press, Princeton, 1980, 301–303.

9. A. Leutbecher, Über Spitzen diskontinuierlicher Gruppen von lineargebrochenen Transformationen, *Math. Z.* **100** (1967), 183–200.

10. B. Maskit, *Kleinian Groups,* Springer-Verlag, New York, 1988.

11. B. Maskit, Some special 2-generator Kleinian groups, *Proc. Am. Math. Soc.* **106** (1989), 175–186.

12. G. Rosenberger, All generating pairs of all two-generator Fuchsian groups, *Arch. Math.* **46** (1986), 198–204.

13. H. Shimizu, On discontinuous groups operating on the product of the upper half planes, *Ann. Math.* **77** (1963), 33–71.

*Department of Mathematics*
*University of Michigan*
*Ann Arbor, MI 48109*

*Department of Mathematics*
*University of Auckland*
*Private Bag*
*New Zealand*

# Noncommutative Variations on Theorems of Marcel Riesz and Others†

## T.A. Gillespie

Alastair Gillespie,
1976

Paul Halmos spent part of a sabbatical year at Edinburgh University in 1973 at a time when operator theory was seeing some spectacular developments. The Brown–Douglas–Fillmore theory, bringing ideas from algebraic topology into operator theory, was rapidly coming to fruition; the Romanian school seemed to be solving important problems almost by the month; Lomonosov proved his celebrated invariant subspace theorem. The list could go on and on, but one common feature of many of these new results was that they were inspired by earlier work of Halmos and, in particular, were answering questions raised by him in his celebrated "Ten Problems" paper [13]. All this meant that Halmos was very much at the information centre of things and we were caught up in the excitement of the times.

Although the 1973 visit is, I think, the longest continuous time spent by Halmos in Scotland, his connections with the country go back a lot earlier. He declares himself a Scotophile in his automathography [15] and we, in turn, would call him an honorary Scot. Indeed, that is more precise than mere rhetoric. He is an Honorary Fellow of the Royal Society of Edinburgh and was given an Honorary D.Sc. by the University of St. Andrews to mark his long association with the four-yearly Colloquium held there under the auspices of the Edinburgh Mathematical Society. These Colloquia are just the sort of things that Halmos relishes in — a happy mixture of expository mathematics and recreation — a mathematical holiday, in fact.

To return to Halmos's 1973 visit to Edinburgh, this was when he was caught up in what he called noncommutative approximation theory: how best can a given operator be approximated by an operator from some prescribed class? The problem he was concerned with at the time was when the prescribed class consists of normal operators with spectrum contained

---

†For Paul Halmos with affectionate respect.

in a given set and the given operator is also normal. This is close to commutative approximation theory, that is, the approximation of functions by functions, since every normal operator can be thought of as multiplication by a bounded function on an $L^2$ space [12]. The noncommutative twist comes with the realization that noncommuting normal operators cannot be represented this way simultaneously as multiplications. As with so much of Halmos's work, the problem was novel, simply stated, and had an elegant solution given in [14]. That particular paper, apart from its mathematical interest, has another claim to fame. The final draft of the manuscript was completed and, being on leave, Halmos asked the Secretary of the Edinburgh Mathematical Society whether he could have the manuscript typed locally before submission. On seeing the immaculate writing, the Secretary immediately said "no need" and the hand-written manuscript went through the whole process to the printer as it was. Would the same thing happen in today's world of word processors and the electronic transfer of ASCII files?

Noncommutative perspectives of commutative (that is, function-theoretic in some form) analysis appear in many places in abstract analysis. The aim of the present article is to focus on a few such results arising mainly from the work of Marcel Riesz and, in particular, from his celebrated convexity theorem.

## Commutative Interpolation

Roughly speaking, interpolation theorems give information about an operator $T$ on a space $X$ from knowledge of $T$ on two other spaces $X_1$ and $X_2$, where $X$ is in some sense an intermediate space between $X_1$ and $X_2$. The properties of $T$ in both the hypotheses and the conclusion usually involve some form of boundedness and there is a vast literature on the subject. However, the whole story started with the Convexity Theorem of Marcel Riesz [18].

In essence, Riesz's theorem is a result about the norms of matrices. To state it precisely, we need some notation. Let $\mathbb{F}$ denote either the reals $\mathbb{R}$ or the complexes $\mathbb{C}$, and let $\| \cdot \|_p$ be the usual $p$ norms

$$\|x\|_p = \left\{ \sum_{j=1}^{n} |x_j|^p \right\}^{1/p} \quad \text{for } 1 \le p < \infty, \qquad \|x\|_\infty = \sup_j |x_j|$$

on $\mathbb{F}^n$, where $x = (x_1, \ldots, x_n)$. Given an $m \times n$ matrix $A$, we can consider the norm $\|A\|_{p,q}$ of $A$ as a mapping from $\mathbb{F}^n$ with $\| \cdot \|_p$ to $\mathbb{F}^m$ with $\| \cdot \|_q$; thus

$$\|A\|_{p,q} = \sup\{\|Ax\|_q : x \in \mathbb{F}^n \text{ and } \|x\|_p \le 1\}.$$

**The Riesz Convexity Theorem for Matrices.** *Let $(p_1, q_1)$ and $(p_2, q_2)$*

*be two pairs of indices with* $1 \leq p_1 \leq q_1 \leq \infty$ *and* $1 \leq p_2 \leq q_2 \leq \infty$, *and
let* $(p, q)$ *be such that*

$$\left(\frac{1}{p}, \frac{1}{q}\right) = (1 - t)\left(\frac{1}{p_1}, \frac{1}{q_1}\right) + t\left(\frac{1}{p_2}, \frac{1}{q_2}\right),$$

*where* $0 < t < 1$. *Then*

$$\|A\|_{p,q} \leq \|A\|_{p_1,q_1}^{(1-t)} \|A\|_{p_2,q_2}^{t}. \tag{1}$$

This result is best remembered as follows. Associate with each pair of indices $(p, q)$, where $1 \leq p, q \leq \infty$, the point $(p^{-1}, q^{-1})$ in the unit square $S = [0, 1] \times [0, 1]$, with the usual convention that $\infty^{-1} = 0$, and think of $\|A\|_{p,q}$ as a function of $(p^{-1}, q^{-1}) \in S$. Then Eq. (1) says that $\log \|A\|_{p,q}$ is convex on the lower triangle

$$S_- = \{(x, y) \in S : 0 \leq y \leq x \leq 1\}.$$

Anyone seeing the result for the first time might fleetingly entertain the thought that it should also be valid on the upper triangle

$$S_+ = \{(x, y) \in S : 0 \leq x \leq y \leq 1\};$$

just consider the transpose of $A$ and replace each index $r$ by its conjugate $r'$ $(r^{-1} + r'^{-1} = 1)$. However, this strategy leaves you inside $S_-$ and nothing new emerges. In fact, as Riesz noted in [18], the result fails in general on $S_+$ when $\mathbb{F} = \mathbb{R}$. The question of what happens when $\mathbb{F} = \mathbb{C}$ was resolved by G. Thorin some years later ([24]) when, using complex variable methods, he showed that the Riesz convexity theorem extends to the whole unit square $S$ when $\mathbb{F} = \mathbb{C}$.

**The Riesz–Thorin Convexity Theorem for Matrices.** *Let $A$ be an $m \times n$ matrix over $\mathbb{C}$. Then $\log \|A\|_{p,q}$ is convex as a function of $(p^{-1}, q^{-1})$ on the unit square $S = [0, 1] \times [0, 1]$.*

In practice, the above convexity theorems are usually applied to operators acting between infinite-dimensional $L^p$ spaces, but the infinite-dimensional versions follow easily from the matrix case. To be a bit more precise, let $(\mathcal{M}, \mu)$ and $(\mathcal{N}, \nu)$ be two measure spaces and let $\mathcal{D}$ be a linear space of measurable functions defined on $\mathcal{M}$ such that $\mathcal{D}$ is suitably dense in $L^p(\mu)$ for all $p$ (for instance, $\mathcal{D}$ might be the space of simple integrable functions on $\mathcal{M}$). Suppose that $T$ is a linear mapping defined on $\mathcal{D}$ such that $Tf$ is a measurable function on $\mathcal{N}$ for all $f$ in $\mathcal{D}$. As for matrices, define

$$\|T\|_{p,q} = \sup\{\|Tf\|_q : f \in \mathcal{D} \text{ and } \|f\|_p \leq 1\}.$$

Then $\log \|T\|_{p,q}$ is convex as a function of $(p^{-1}, q^{-1})$ on

(a) the lower triangle $S_-$ in the case of real $L^p$ spaces;

(b) the unit square $S$ in the case of complex $L^p$ spaces.

Here $\|T\|_{p,q}$ is allowed to be infinite (when $T$ is not $\|\cdot\|_p$ to $\|\cdot\|_q$ bounded), but such pairs of indices are of no particular interest. The interesting situation is when $T$ is both $(p_1, q_1)$ and $(p_2, q_2)$ bounded, then $T$ is also $(p, q)$ bounded for any $(p^{-1}, q^{-1})$ on the line segment joining $(p_1^{-1}, q_1^{-1})$ and $(p_2^{-1}, q_2^{-1})$, with

$$\|T\|_{p,q} \leq \|T\|_{p_1,q_1}^{(1-t)} \|T\|_{p_2,q_2}^t,$$

where

$$(p^{-1}, q^{-1}) = (1-t)(p_1^{-1}, q_1^{-1}) + t(p_2^{-1}, q_2^{-1}).$$

As Riesz showed in his original paper, interpolation provides a beautifully simple proof of the following classical result.

**The Hausdorff–Young Theorem.** *Let* $1 \leq p \leq 2$ *and let* $f \colon [0, 1] \to \mathbb{C}$ *be $p$-integrable, with Fourier series* $f(t) \sim \sum_{n=-\infty}^{\infty} \hat{f}(n) e^{2\pi i n t}$. *Then*

$$\left\{ \sum_{n=-\infty}^{\infty} |\hat{f}(n)|^{p'} \right\}^{1/p'} \leq \left\{ \int_0^1 |f(t)|^p dt \right\}^{1/p} \qquad (\equiv \|f\|_p < \infty). \qquad (2)$$

To prove this by interpolation, take $(\mathcal{M}, \mu)$ to be $[0, 1]$ with Lebesgue measure and $(\mathcal{N}, \nu)$ to be $\mathbb{Z}$ with counting measure (that is, assign unit mass to each point), let $\mathcal{D}$ be the space of trigonometric polynomials, and let $Tf = \{\hat{f}(n)\}_{n \in \mathbb{Z}}$. Then $\|T\|_{2,2} = 1$ by Parseval's equality, whilst $\|T\|_{1,\infty} = 1$ from elementary properties of integrals. Since $p^{-1} + p'^{-1} = 1$, we can interpolate between $(2^{-1}, 2^{-1})$ and $(1^{-1}, \infty^{-1}) = (1, 0)$ in $S$ to obtain Eq. (2) for trigonometric polynomials. Finally, a simple approximation argument extends Eq. (2) to all of $L^p[0, 1]$.

# Noncommutative Interpolation

What would constitute a noncommutative version of the Riesz or Riesz–Thorin convexity theorem in finite dimensions? For simplicity, let's restrict to the case $\mathbb{F} = \mathbb{C}$ and ask what might be a reasonable noncommutative analogue of $\mathbb{C}^n$ and its associated norms $\|\cdot\|_p$. It is natural, in the Halmos spirit of noncommutative approximation theory, to replace $\mathbb{C}^n$ by $M_n(\mathbb{C})$, the space of $n \times n$ complex matrices. One candidate for a $p$-norm on $M_n(\mathbb{C})$ would be the operator norm $\|\cdot\|_{p,p}$. However, if we realize $\mathbb{C}^n$ inside $M_n(\mathbb{C})$ as the subspace of diagonal matrices, then $\|\cdot\|_{p,p}$ always reduce to $\|\cdot\|_\infty$ and $p$ has been lost.

A better way to proceed is as follows. For $A \in M_n(\mathbb{C})$, the matrix $A^*A$ is positive and so has a unique positive square root $(A^*A)^{1/2}$. Let $\lambda_1, \ldots, \lambda_n$ be the eigenvalues of this square root, repeated according to multiplicity, and define the $p$-norm of $A$ to be

$$|A|_p = \|(\lambda_1, \ldots, \lambda_n)\|_p.$$

For a diagonal matrix, $|\cdot|_p$ is indeed the $p$-norm in $\mathbb{C}^n$ of its diagonal so, from that point of view at least, we are in business. Of course, we still have to check that $|\cdot|_p$ is a norm. To do this, the only nontrivial thing to verify is the triangle inequality, and this follows easily from the formula

$$|A|_p = \sup\{\|\mathrm{diag}(UAV)\|_p : U, V \in \mathcal{U}_n\} \tag{3}$$

and the triangle (Minkowski) inequality for $\|\cdot\|_p$ in $\mathbb{C}^n$. In Eq. (3), $\mathcal{U}_n$ denotes the group of $n \times n$ unitary matrices and $\mathrm{diag}\, B$ is the diagonal sequence $(b_{11}, \ldots, b_{nn})$ of an $n \times n$ matrix $B = (b_{ij})$. Equation (3) says that $|\cdot|_p$ is the supremum of the $p$-norms of the diagonals of all matrices representing the mapping $x \to Ax$ of $\mathbb{C}^n$ into itself relative to possibly different orthonormal bases in the domain and range spaces $\mathbb{C}^n$. The proof of Eq. (3) is not difficult (see the proof of [20], Lemma 2.3.4).

Equation (3) is useful in that it involves $A$ directly; one of the difficulties when working with the $p$-norm $|\cdot|_p$ is that it is not easy to compute with, apart, that is, from the case $p = 2$, when we have the Hilbert–Schmidt norm

$$|A|_2 = \left\{ \sum_{j,k=1}^{n} |a_{jk}|^2 \right\}^{1/2}. \tag{4}$$

Even $|\cdot|_1$ and $|\cdot|_\infty$ are not computable in general, which contrasts with the norms $\|\cdot\|_{p,p}$, which are easy to calculate for $p = 1$ and $\infty$ but difficult for $p = 2$ (because $|\cdot|_\infty = \|\cdot\|_{2,2}$).

To illustrate the use of Eq. (3), we give an economical proof of the noncommutative analogue of Hölder's inequality

$$|AB|_1 \le |A|_p |B|_{p'}. \tag{5}$$

Fix $U, V \in \mathcal{U}_n$ and let $U_1 \in \mathcal{U}_n$ be such that

$$\|\mathrm{diag}(UABV)\|_1 = \mathrm{tr}(U_1 UABV),$$

where tr denotes trace. (Just take $U_1$ to be a diagonal matrix with appropriate unimodular diagonal entries.) Writing $A = W(A^*A)^{1/2}$ and $(A^*A)^{1/2} = R^*DR$ with $W, R \in \mathcal{U}_n$ and

$$D = \begin{bmatrix} \lambda_1 & & \text{O} \\ & \ddots & \\ \text{O} & & \lambda_n \end{bmatrix},$$

and using the familiar commutativity property of the trace, we see that

$$\text{tr}(U_1 U A B V) = \text{tr}(D R B V_1)$$

for some unitary $V_1$. Hence

$$\|\text{diag}(U A B V)\|_1 = \text{tr}(D R B V_1)$$
$$\leq \|(\lambda_1, \ldots, \lambda_n)\|_p \, \|\text{diag}(R B V_1)\|_{p'}$$
$$\leq |A|_p |B|_{p'}$$

by the commutative version of Hölder's inequality and Eq. (3) for $B$. Now a further application of Eq. (3), this time for $AB$, gives Eq. (5).

We can also obtain a noncommutative version of the Riesz–Thorin convexity theorem from Eq. (3). Given a linear mapping $\Phi$ of $M_n(\mathbb{C})$ into $M_m(\mathbb{C})$, denote by $|\Phi|_{p,q}$ the norm of $\Phi$ with respect to $|\cdot|_p$ on $M_n(\mathbb{C})$ and $|\cdot|_q$ on $M_m(\mathbb{C})$.

**A Noncommutative Riesz–Thorin Convexity Theorem for Matrix Spaces.** *Let* $\Phi : M_n(\mathbb{C}) \to M_m(\mathbb{C})$ *be a linear mapping. Then* $\log |\Phi|_{p,q}$ *is convex as a function of* $(p^{-1}, q^{-1})$ *on the unit square* $S = [0,1] \times [0,1]$.

As before, this follows easily from the commutative version of the theorem and Eq. (3). To see this, consider for all pairs $U_1, V_1 \in \mathcal{U}_n$ and $U_2, V_2 \in \mathcal{U}_m$ the mappings $T(U_1, V_1, U_2, V_2)$ from $\mathbb{C}^n$ into $\mathbb{C}^m$ given by

$$T(U_1, V_1, U_2, V_2) : \lambda \to \text{diag}[U_2 \Phi(U_1 D_\lambda V_1) V_2],$$

where $D_\lambda$ is the diagonal matrix with diagonal $\lambda = (\lambda_1, \ldots, \lambda_n) \in \mathbb{C}^n$. Equation (3), together with the fact that each $A \in M_n(\mathbb{C})$ has a factorization of the form $A = U_1 D_\lambda V_1$ for some $U_1, V_1 \in \mathcal{U}_n$, shows that

$$|\Phi|_{p,q} = \sup\{\|T(U_1, V_1, U_2, V_2)\|_{p,q} : U_1, V_1 \in \mathcal{U}_n \text{ and } U_2, V_2 \in \mathcal{U}_m\}.$$

Since each of the functions $\|T(U_1, V_1, U_2, V_2)\|_{p,q}$ has the required convexity property on $S$ by the commutative Riesz–Thorin theorem, so has $|\Phi|_{p,q}$ and the theorem is proved.

To formulate an infinite-dimensional version of the above noncommutative Riesz–Thorin theorem, we need the appropriate infinite-dimensional versions of the matrix spaces $M_n(\mathbb{C})$ with their $|\cdot|_p$ norms. These are provided by the von Neumann–Schatten spaces $\mathcal{C}_p$, introduced originally in [17] and studied by many authors. The setting is an infinite-dimensional complex Hilbert space $\mathcal{H}$, which we shall take to be separable for simplicity. The inner product and norm on $\mathcal{H}$ will be denoted by $(\cdot, \cdot)$ and $\|\cdot\|$, respectively. Given a compact operator $A$ on $\mathcal{H}$, the operator $(A^*A)^{1/2}$ is compact and self-adjoint, and so can be represented as

$$(A^*A)^{1/2} x = \sum_{n=1}^{\infty} \lambda_n (x, u_n) u_n,$$

where $\lambda_1, \lambda_2, \ldots$ are the (necessarily nonnegative) eigenvalues of $(A^*A)^{1/2}$ repeated with multiplicity and $u_1, u_2, \ldots$ form an orthonormal basis of $\mathcal{H}$. For $1 \leq p < \infty$, define

$$|A|_p = \left\{ \sum_{n=1}^{\infty} \lambda_n^p \right\}^{1/p} \leq \infty$$

if $A$ is compact; $|A|_p = \infty$ otherwise. The space $\mathcal{C}_p$ is then defined to be the set of all (compact) $A$ for which $|A|_p$ is finite; it can be shown to be a Banach space under the norm $|\cdot|_p$. There is no universal agreement about the definition of $\mathcal{C}_\infty$. Some define it as the space $\mathcal{B}(\mathcal{H})$ of all bounded linear operators on $\mathcal{H}$ and others take it to be the space $\mathcal{K}(\mathcal{H})$ of all compact operators on $\mathcal{H}$. Sometimes the context dictates the appropriate definition and sometimes it doesn't really matter but, whichever definition is taken, there is unanimity about the norm $|\cdot|_\infty$. It is always taken to be the operator norm

$$|A|_\infty = \sup\{\|Ax\| : x \in \mathcal{H} \text{ with } \|x\| \leq 1\} \quad (= |(A^*A)^{1/2}|_\infty).$$

For the sake of being specific, take $\mathcal{C}_\infty = \mathcal{B}(\mathcal{H})$ in what follows.

It is customary to think of the $\mathcal{C}_p$ spaces as noncommutative analogues of the classical $\ell^p$ spaces of $p$-summable scalar sequences $x = (x_1, x_2, \ldots)$ with their norms

$$\|x\|_p = \left\{ \sum_{n=1}^{\infty} |x_n|^p \right\}^{1/p} \quad (p < \infty) \quad \text{and} \quad \|x\|_\infty = \sup_n |x_n|.$$

Under this analogy, our interpretation $\mathcal{C}_\infty = \mathcal{B}(\mathcal{H})$ corresponds to the space $\ell^\infty$ of bounded sequences, whilst $\mathcal{C}_\infty = \mathcal{K}(\mathcal{H})$ corresponds to the space $c_0$ of sequences converging to 0. The analogy can be made a little more precise in terms of the formula for $|\cdot|_p$ corresponding to Eq. (3). Fix some distinguished orthonormal basis $\{e_n : n \in \mathbb{N}\}$ for $\mathcal{H}$ and identify each $A \in \mathcal{B}(\mathcal{H})$ with its matrix

$$\begin{bmatrix} a_{11} & a_{12} & \cdots \\ a_{21} & a_{22} & \cdots \\ \vdots & \vdots & \end{bmatrix},$$

where $a_{jk} = (Ae_k, e_j)$ and, following our earlier notation, put

$$\text{diag}(A) = (a_{11}, a_{22}, \ldots).$$

Precisely which matrices occur in this way is something of a mystery, but nevertheless we do have the formula

$$|A|_p = \sup\{\|\text{diag}(UAV)\|_p : U, V \in \mathcal{U}(\mathcal{H})\}, \tag{6}$$

where $\mathcal{U}(\mathcal{H})$ is the set of unitary operators on $\mathcal{H}$. The validity of Eq. (6) when $p = \infty$ is a triviality. For $1 \leq p < \infty$, first show that the finiteness of the right-hand side of Eq. (6) implies that $A$ is compact and then the proof is much the same as in finite dimensions.

The duality between $\ell^p$ and $\ell^{p'}$ has its counterpart for $\mathcal{C}_p$ spaces as follows. Given $A \in \mathcal{C}_p$ and $B \in \mathcal{C}_{p'}$, $AB$ belongs to $\mathcal{C}_1$ and has a well-defined trace given by

$$\text{tr}(AB) = \sum_{n=1}^{\infty} (ABu_n, u_n),$$

where $\{u_n\}$ is any orthonormal basis of $\mathcal{H}$, the series here converging absolutely. Under the duality pairing $\langle A, B \rangle \to \text{tr}(AB)$, the dual space of $\mathcal{C}_p$ is identified with $\mathcal{C}_{p'}$ for $1 \leq p < \infty$.

Having set the stage, we can now formulate the analogue of the Riesz–Thorin theorem for the $\mathcal{C}_p$ spaces. Let $\mathcal{D}$ be a dense subspace of $\mathcal{C}_1$ and let $\Phi$ be a linear mapping from $\mathcal{D}$ into $\mathcal{B}(\mathcal{H})$. (For instance, $\mathcal{D}$ might consist of the finite rank operators or the operators whose matrices with respect to some preferred orthonormal basis have only finitely many nonzero entries.) Because the spaces $\mathcal{C}_p$ increase and the norms $\| \cdot \|_p$ decrease as $p$ increases from 1 to $\infty$, there is no need to have any other conditions on $\mathcal{D}$; the condition at $p = 1$ ensures that $\mathcal{D}$ is large enough in $\mathcal{C}_p$ for all $p$ to be able to interpolate. Denote by $|\Phi|_{p,q}$ the norm of $\Phi$ with respect to $|\cdot|_p$ and $|\cdot|_q$, that is,

$$|\Phi|_{p,q} = \sup\{|\Phi(A)|_q : A \in \mathcal{D} \text{ with } |A|_p \leq 1\}.$$

**The Riesz–Thorin Convexity Theorem for $\mathcal{C}_p$ Spaces.** *As a function of $(p^{-1}, q^{-1})$, $\log |\Phi|_{p,q}$ is convex on the unit square $S = [0, 1] \times [0, 1]$.*

As with the matrix space case, this result follows easily from its commutative version (just adapt what we did before, using the commutative Riesz–Thorin theorem for the $\ell^p$ spaces) or, equally well, deduce it from its noncommutative finite-dimensional counterpart. By way of history, the result appears in the work of Gohberg and Krein ([10], p. 113), where the approach is to adapt the usual complex analysis proof of the commutative version. On the other hand, the account sketched here is closer to that given by Dunford and Schwartz ([9], p. 1137). It is worth noting that, since the method of handling properties of norms on $\mathcal{C}_p$ is to use Eqs. (3) or (6) to reduce considerations to sequence spaces, we could equally well prove mixed interpolation theorems for mappings from spaces of measurable functions into $\mathcal{B}(\mathcal{H})$ and for mappings from spaces of operators on $\mathcal{H}$ to function spaces. Here is a simple sample of the latter type of result; variants are easily formulated and proved.

**Theorem.** *Let $\mathcal{D}$ be a subspace of $\mathcal{B}(\mathcal{H})$ such that $\mathcal{D}$ is dense in $\mathcal{C}_1$ and let $\Phi$ be a linear mapping defined on $\mathcal{D}$ such that $\Phi(A)$ is a complex sequence for all $A \in \mathcal{D}$. Then the norm of $\Phi$ with respect to $|\cdot|_p$ on $\mathcal{D}$ and $\| \cdot \|_q$ on* *sequence spaces is logarithmically convex as a function of $(p^{-1}, q^{-1})$ on the unit square $S = [0,1] \times [0,1]$.*

# The Hausdorff–Young Theorem Revisited

The proof Riesz gave of the Hausdorff–Young inequality (2) is an archetypal application of interpolation methods. The tactic is to ask the right question, in this case "how does the $p$ integrability of a function control the size of its Fourier coefficients (measured in terms of a sequence $q$ norm)?," find two cases where there is an easy answer (here $p = q = 2$ and $p = 1$, $q = \infty$), and then interpolate to get something less obvious.

With this in mind, can we find a noncommutative analogue of the Hausdorff–Young inequality? A first attempt might go as follows. Fix a distinguished orthonormal basis $\{e_n\}_{n \in \mathbb{N}}$ of our Hilbert space $\mathcal{H}$ as in the previous section. Then each $A \in \mathcal{B}(\mathcal{H})$ has a corresponding matrix $(a_{jk})$ and we can think of the entries $a_{jk} = (Ae_k, e_j)$ as the Fourier coefficients of $A$. There are two easy inequalities (actually one is an equality)

$$|A|_2 = \left\{ \sum_{j,k=1}^{\infty} |a_{jk}|^2 \right\}^{1/2} \quad \text{and} \quad |A|_1 \le \sum_{j,k=1}^{\infty} |a_{jk}|. \tag{7}$$

We saw the finite-dimensional version of the first of these earlier [Eq. (4)] and the second comes easily from the triangle inequality for $|\cdot|_1$. Interpolation of the $\ell^p$ to $\mathcal{C}_q$ variety between $(2^{-1}, 2^{-1})$ and $(1,1)$, this time for the sequence spaces $\ell^p(\mathbb{N} \times \mathbb{N})$, gives

$$|A|_p \le \left\{ \sum_{j,k=1}^{\infty} |a_{jk}|^p \right\}^{1/p} \quad \text{for } 1 \le p \le 2. \tag{8}$$

That is, the map $(a_{jk}) \to A$ is norm-reducing from $\ell^p(\mathbb{N} \times \mathbb{N})$ to $\mathcal{C}_p$ for $p$ in the above range. Taking adjoints, we do get something new on this occasion, namely the reverse inequality in the complementary range

$$\left\{ \sum_{j,k=1}^{\infty} |a_{jk}|^p \right\}^{1/p} \le |A|_p \quad \text{for } 2 \le p \le \infty, \tag{9}$$

where the left-hand side is interpreted as $\sup |a_{jk}|$ in the case $p = \infty$. Notice that Eq. (8) gives a sufficient condition for a matrix to belong to $\mathcal{C}^p$ for $p \le 2$.

Although Eqs. (8) and (9) are, in a certain sense, versions of the Hausdorff–Young theorem for operators, it is possible to do better if we consider what might be called the "Fourier vectors" of $A$. The action of $A$ on $\mathcal{H}$ can be expressed as $Ax = \sum_{n=1}^{\infty}(x, e_n)Ae_n$, which becomes

$$A = \sum_{n=1}^{\infty}(\cdot, e_n)Ae_n$$

when $x$ is suppressed. This leads us to consider the vectors $Ae_n$ as the "Fourier vectors" of $A$ and to ask, "can we relate $|A|_p$ to the $q$ norm of the sequence $\{\|Ae_n\|\}$?" From this perspective, the inequalities (8) and (9) would suggest that perhaps

$$|A|_p \leq \left\{ \sum_{n=1}^{\infty} \|Ae_n\|^p \right\}^{1/p} \quad \text{for } 1 \leq p \leq 2 \tag{10}$$

and

$$\left\{ \sum_{n=1}^{\infty} \|Ae_n\|^p \right\}^{1/p} \leq |A|_p \text{ for } 2 \leq p \leq \infty, \tag{11}$$

with a similar supremum interpretation of the left-hand side of Eq. (11) when $p = \infty$.

These inequalities, which do indeed strengthen Eqs. (8) and (9), were proved by Gohberg and Markus [11]. The cases $p = 1$ (noted earlier by Stinespring [23]) and $p = \infty$ are straightforward, whilst the equality in Eq. (7) is Eqs. (10) and (11) rephrased when $p = 2$. Gohberg and Markus did not use interpolation to prove their inequalities, but these easy cases suggest that it ought to give the results for general $p$. To do this as things stand, we would have to prove appropriate interpolation theorems between $\mathcal{C}_p$ spaces and the spaces $\ell^p(\mathcal{H})$ of sequences $\{x_n\}_{n\in\mathbb{N}}$ such that $\{\|x_n\|\}_{n\in\mathbb{N}}$ belongs to scalar $\ell^p$. Whilst this is perfectly possible, we can in fact proceed with the tools at our disposal.

Express $\mathcal{H}$ as the direct sum of countably many copies of itself; i.e., write

$$\mathcal{H} = \mathcal{K} \oplus \mathcal{K} \oplus \dots,$$

where each $\mathcal{K}$ is a copy of $\mathcal{H}$. For $A \in \mathcal{B}(\mathcal{H})$, let $\tilde{A}$ denote the operator on $\mathcal{H}$ given by

$$\tilde{A} = (\cdot, e_1)Ae_1 \oplus (\cdot, e_2)Ae_2 \oplus \dots,$$

where, as before, $(\cdot, e_j)Ae_j$ denotes the mapping $x \to (x, e_j)Ae_j$. A simple calculation gives

$$|\tilde{A}|_p = \left\{ \sum_{n=1}^{\infty} \|Ae_n\|^p \right\}^{1/p}$$

for $1 \leq p \leq \infty$; thus $|\tilde{A}|_2 = |A|_2$ and $|\tilde{A}|_\infty \leq |A|_\infty$ from the known easy cases of Eqs. (10) and (11). Applying our $\mathcal{C}_p$ interpolation theorem to the map $A \to \tilde{A}$, we obtain Eq. (11) for the intermediate exponents $2 < p < \infty$. The reverse inequalities (10) now follow by dualizing the map $A \to \{Ae_n\}$ of $\mathcal{C}_p$ into $\ell^p(\mathcal{H})$.

# Convolutions and Schur Products

The convolution product $f * g$ of two integrable functions on the unit circle $\mathbb{T}$ (thought of for notational convenience as periodic functions on $\mathbb{R}$ with period one) is defined as

$$(f * g)(t) = \int_0^1 f(t - s)g(s)ds.$$

A classical inequality of W.H. Young relates appropriate norms of a convolution product to norms of its factors as follows:

$$\|f * g\|_r \leq \|f\|_p \|g\|_q, \text{ where } r^{-1} = p^{-1} + q^{-1} - 1. \tag{12}$$

(As always, the implied assumption $1 \leq p, q, r \leq \infty$ is present.) Interpolation gives a quick proof of Eq. (12). Fix $f$ and consider the mapping $T \colon g \to f * g$ (defined initially on, say, the simple functions). Minkowski's integral inequality $\|f * g\|_p \leq \|f\|_p \|g\|_1$, which is really just a continuous analogue of the triangle inequality for $\| \cdot \|_p$ once we think of $f * g$ as a continuous weighted sum of the translates $f_s(t) = f(t - s)$ of $f$, shows that $\|T\|_{1,p} \leq \|f\|_p$, whilst Hölder's inequality gives $\|T\|_{p',\infty} \leq \|f\|_p$. Now interpolate to get Eq. (12).

Is there a noncommutative analogue of Eq. (12)? Multiplying two functions together by convolution corresponds to multiplying their sequences of Fourier coefficients pointwise on $\mathbb{Z}$, i.e., $(f * g)\hat{}(n) = \hat{f}(n)\hat{g}(n)$. With the analogy between the entries of the matrix of an operator and Fourier coefficients, this leads us to consider the Schur (or Hadamard) product $A * B = (a_{ij}b_{ij})$ of two matrices $A = (a_{ij})$ and $B = (b_{ij})$. Assuming that $A$ and $B$ are bounded as mappings of the sequence space $\ell^2$ into itself, i.e., in our previous terminology, that they are the matrices of operators in $\mathcal{B}(\mathcal{H})$ with respect to some given orthonormal basis, then what can be said about $|A * B|_p$? The term Schur product here is in recognition of the celebrated inequality

$$|A * B|_\infty \leq |A|_\infty |B|_\infty$$

of Schur [22].

Consider the map $\Phi : A \to A * B$, defined initially for matrices with only finitely many nonzero entries, and let $1 \leq p \leq \infty$. Duality and Schur's inequality easily imply that

$$|A * B|_1 \leq |A|_1 |B|_\infty. \tag{13}$$

(The important thing to note here is that $\mathrm{tr}[(A*B)C] = \mathrm{tr}[B(A^t*C)]$.) Thus $|\Phi|_{\infty,\infty} \leq |B|_\infty$ and $|\Phi|_{1,1} \leq |B|_\infty$. Interpolation gives the intermediate inequality $|\Phi|_{p,p} \leq |B|_\infty$, or

$$|A * B|_p \leq |A|_p |B|_\infty. \tag{14}$$

Predualizing Eq. (14), but thinking now of the map $B \to A * B$ with $A$ fixed, we arrive at

$$|A * B|_1 \leq |A|_p |B|_{p'}.$$

Finally, interpolate $B \to A * B$ between $(p', 1)$ and $(\infty, p)$ to obtain

$$|A * B|_r \leq |A|_p |B|_q,$$

where $r^{-1} = p^{-1} + q^{-1} \leq 1$.

It should be remarked that Bennett [2] has obtained definitive (and much deeper) results concerning inequalities for the norms of Schur products of matrices when considered as operators acting on $\ell^p$ spaces.

## Analytic Projections

Another famous result proved by Marcel Riesz in the 1920's established the $L^p$ boundedness of the mapping which sends a Fourier series to its analytic half for $p$ in the range $1 < p < \infty$ (see [19]). More precisely, if $f \in L^p[0, 1]$ and has Fourier series $\sum_{n=-\infty}^{\infty} a_n e^{2\pi i n t}$, then the series $\sum_{n=0}^{\infty} a_n e^{2\pi i n t}$ is the Fourier series of an $L^p$ function, say $Qf$, and $\|Qf\|_p \leq \alpha_p \|f\|_p$ for some constant $\alpha_p$ independent of $f$. In more functional analytic language, $H^p(\mathbb{T})$ is a complemented subspace of $L^p(\mathbb{T})$ when $1 < p < \infty$. Actually, Riesz expressed the result in terms of the conjugacy map

$$\sum_{n=-\infty}^{\infty} a_n e^{2\pi i n t} \to i \sum_{n=-\infty}^{-1} a_n e^{2\pi i n t} - i \sum_{n=1}^{\infty} a_n e^{2\pi i n t}$$

(conjugacy here refers to harmonic conjugation), but the two formulations are equivalent. There are many more proofs of the result and a number of them use interpolation, but more work has to be done here to arrive at a position where it is possible to interpolate. The problem is that, although the theorem is easy when $p = 2$ [$Q$ is just the orthogonal projection of $L^2(\mathbb{T})$ onto $H^2(\mathbb{T})$], it is actually false for the extreme cases $p = 1$ and $\infty$.

An elegant approach is to use the "bootstrap" method of Cotlar [8], which involves the following strategy. Show that, if the desired boundedness holds for a particular value of $p$, then it also holds for $2p$. Since we have the result for $p = 2$, we thus get it for $p = 4, 8, \ldots$. We can now interpolate to cover the range $2 \leq p < \infty$ and finally dualize to handle $1 < p \leq 2$. The passage from $p$ to $2p$ comes from the ingenious identity

$$g^2 = 2(2Q - I)(fg) - f^2, \tag{15}$$

where $g = (2Q - I)f$. All that is behind Eq. (15) is the fact that the ranges of $Q$ and $I - Q$ are both closed under pointwise multiplication. (As usual, we have to restrict all our functions to some good class initially; trigonometric polynomials will do here.) Assuming that $Q$ is $L^p$ bounded with norm $\alpha_p$, Eq. (15) implies that

$$\|g\|_{2p}^2 = \|g^2\|_p \leq 2\|(2Q - I)(fg)\|_p + \|f^2\|_p$$
$$\leq 2(2\alpha_p + 1)\|(fg)\|_p + \|f\|_{2p}^2$$
$$\leq 2(2\alpha_p + 1)\|f\|_{2p}\|g\|_{2p} + \|f\|_{2p}^2.$$

Hence $\|g\|_{2p} \leq \beta_p\|f\|_{2p}$, where $\beta_p$ is the positive root of the quadratic equation $t^2 - 2(2\alpha_p + 1)t - 1 = 0$.

What about a noncommutative analogue here? Replacing Fourier coefficients by matrix entries, an obvious candidate for $Q$ is the map $\psi$ which replaces the entries below the diagonal in a matrix by 0 but leaves the upper-triangular entries alone; that is,

$$\psi: \begin{bmatrix} a_{11} & a_{12} & \cdots \\ a_{21} & a_{22} & \cdots \\ a_{31} & a_{32} & \cdots \\ \vdots & & \ddots \ \vdots \end{bmatrix} \rightarrow \begin{bmatrix} a_{11} & a_{12} & \cdots \\ & a_{22} & \cdots \\ & \mathbf{O} & \ddots \ \vdots \end{bmatrix}.$$

The ranges of $\psi$ and $1 - \psi$ are certainly closed under matrix multiplication if, say, we restrict to matrices with only finitely many nonzero entries. The analogue of Eq. (15) is

$$B^2 = (2\psi - 1)(AB + BA) - A^2,$$

where $B = (2\psi - 1)(A)$. To make the bootstrap proof go through, all we need are the inequalities $|AB|_p \leq |A|_{2p}|B|_{2p}$ and $|BA|_p \leq |B|_{2p}|A|_{2p}$, together with the equalities

$$|A|_{2p}^2 = |A^2|_p \quad \text{and} \quad |B|_{2p}^2 = |B^2|_p.$$

The inequalities here are true, but the equalities are not in general. They are true, however, if $A$ self-adjoint with zero diagonal (for then $iB$ is also self-adjoint and $|C|_{2p}^2 = |C^2|_p$ is valid for self-adjoint $C$). Finally, it is a simple matter to check that the boundedness of $\psi$ on all of $\mathcal{C}_p$ follows from its boundedness on the subspace of self-adjoint $\mathcal{C}_p$ operators with zero diagonals. Thus the bootstrap method works and $\psi$ is indeed bounded as a mapping of $\mathcal{C}_p$ into itself whenever $1 < p < \infty$, a result originally proved by Macaev [16].

As in the commutative case, the result fails for $p = 1$ and $p = \infty$. For unboundedness when $p = \infty$, consider Hilbert's matrix $(a_{jk})$, where $a_{jk} = (j - k)^{-1}$ for $j \neq k$ and $a_{jj} = 0$; then observe that unboundedness when $p = \infty$ implies unboundedness when $p = 1$.

The $L^p$ boundedness of the conjugacy map for Fourier series has been the motivation for other developments in abstract analysis. One direction is to consider (Bochner) integrable functions $f: \mathbb{T} \to X$, where $X$ is a Banach space. Such functions will still have Fourier series, the only difference being that the coefficients will be elements of $X$, and there are natural $p$ norms

$$\|f\|_p = \left\{ \int_0^1 \|f(t)\|^p dt \right\}^{1/p}.$$

The analogue of the Riesz conjugacy theorem does not hold for arbitrary Banach spaces, but those for which it does hold have recently been identified, both geometrically and in terms of vector-valued martingales [6], [7]. They are called UMD spaces, shorthand for spaces with the unconditionality property for martingale differences, and often provide a good setting in which to extend results from scalar-valued harmonic analysis to Banach spaces (see [5] and [21] for two rather different instances of this). UMD spaces also provide a good framework in which to develop certain aspects of operator theory which, in the Hilbert space case, are classical [4].

# Coda

In conclusion, it must again be said that the subject of interpolation theory in Banach spaces is a vast one. The aim of this article has been merely to focus on the original theorem of Marcel Riesz and to describe some consequences both of it and its noncommutative $\mathcal{C}_p$ analogue. Comprehensive accounts of interpolation theory in general can be found in [1] and [3].

REFERENCES

1. C. Bennett and R. Sharpley, *Interpolation of Operators*, Academic Press, New York, 1988.

2. G. Bennett, Schur multipliers, *Duke Math. J.* **44** (1977), 603–639.

3. J. Bergh and J. Löfström, *Interpolation Spaces: An Introduction*, Springer-Verlag, Berlin, 1976.

4. E. Berkson, T.A. Gillespie, and P.S. Muhly, Abstract spectral decompositions guaranteed by the Hilbert transform, *Proc. London Math. Soc.* (3) **53** (1986), 489–517.

5. E. Berkson, T.A. Gillespie and P.S. Muhly, Generalized analyticity in UMD spaces, *Ark. för Mat.* **27** (1989), 1–14.

6. J. Bourgain, Some remarks on Banach spaces in which martingale difference sequences are unconditional, *Ark. för Mat.* **21** (1983), 163–168.

7. D.L. Burkholder, A geometric condition that implies the existence of certain singular integrals of Banach-space-valued functions, *Proc. of Conf. on Harmonic Analysis in Honor of A. Zygmund (Chicago, 1981)*, edited by W. Beckner *et al.*, Wadsworth, Belmont, California, 1983, pp. 270–286.

8. M. Cotlar, A unified theory of Hilbert transforms and ergodic theorems, *Rev. Mat. Cuyana* **1** (1955), 105–167.

9. N. Dunford and J.T. Schwartz, *Linear Operators,* Part II, Wiley Interscience, New York, 1963.

10. I.C. Gohberg and M.G. Kreĭn, *Theory and Applications of Volterra Operators in Hilbert Space,* Translations Math. Monographs 24, Amer. Math. Soc., Providence, 1970.

11. I.C. Gohberg and A.S. Markus, Some relations between eigenvalues and matrix elements of linear operators, *Mat. Sb.* **64** (1964), 481–496; *Amer. Math. Soc. Transl.* (2) **52** (1966), 201–216.

12. P.R. Halmos, What does the spectral theorem say?, *Am. Math. Monthly* **70** (1963), 241–247.

13. P.R. Halmos, Ten problems in Hilbert space, *Bull. Am. Math. Soc.* **76** (1970), 887–933.

14. P.R. Halmos, Spectral approximants of normal operators, *Proc. Edinburgh Math. Soc.* (2) **19** (1974), 51–58.

15. P.R. Halmos, *I Want to Be a Mathematician,* Springer-Verlag, New York, 1985.

16. V.I. Macaev, Volterra operators produced by perturbation of self-adjoint operators, *Dokl. Akad. Nauk SSSR* **139** (1961), 810–813; *Soviet Math. Dokl.* **2** (1961), 1013–1016.

17. J. von Neumann and R. Schatten, The cross space of linear transformations III, *Ann. Math.* **49** (1948), 557–582.

18. M. Riesz, Sur les maxima des formes bilinéaires et sur les fonctionnelles linéaires, *Acta Math.* **49** (1926), 465–497.

19. M. Riesz, Sur les fonctions conjuguées, *Math. Z.* **27** (1928), 218–244.

20. J.R. Ringrose, Compact non-self-adjoint operators, *Mathematical Studies,* van Nostrand Reinhold, Princeton, 1971, Vol. 35.

21. J.L. Rubio de Francia, Fourier series and Hilbert transforms with values in UMD Banach spaces, *Studia Math.* **81** (1985), 95–105.

22. I. Schur, Bemerkungen zur Theorie der beschränkten Bilinearformen mit unendlich vielen Varänderlichen, *J. Reine Angew. Math.* **140** (1911), 1–28.

23. W.F. Stinespring, A sufficient condition for an integral operator to have a trace, *J. Reine Angew. Math.* **200** (1958), 200–207.

24. G. Thorin, An extension of a convexity theorem due to M. Riesz, *Kungl. Fys. Säll. Lund Förh.* **8** (1938), 166–170.

*Department of Mathematics*
*University of Edinburgh*
*Mayfield Road*
*Edinburgh EH9 35Z*
*Scotland, UK*

# On Products of Involutions

## William H. Gustafson

Bill Gustafson, 1970

## Involutions, Paul Halmos, and Me

In 1974, A. Sampson [16] gave a rather technical matrix-theoretic proof that any complex matrix of determinant $\pm 1$ can be written as a product of finitely many involutions. A brash, young (at the time) algebraist, I quickly saw a group-theoretic proof that gives the same result for matrices over any field, and involves very little computation. Indeed, my proof occupied one and a half typed pages, most of it devoted to a couple of special cases. I sent a copy off to Hans Schneider for publication in *Linear Algebra and its Applications,* and distributed copies in the mailboxes of a few selected colleagues. One of those colleagues was Paul Halmos (that was a different time and place for both of us). A few hours later, Paul passed me in the hall.

"Four," he intoned.

"Huh?" I replied in a burst of insight, "What on earth are you talking about?"

"Shut up," he explained. "In 1958, Kakutani and I proved (see [11]) that every unitary operator on an infinite-dimensional Hilbert space is a product of four involutions. I'll bet that all that's needed in your case."

"I can't imagine how to do that," I said.

He said, "Let's try." We did, and we succeeded. We worked on the proof over a period of months as a casual, after-dinner sort of thing. I'm not sure we ever discussed it anywhere but in the study of Paul's house. If you've never visited Paul at home, you should know that the study is the room that says "Department of Mathematics" on the door. It also doubles as the library for mystery novels.

I wrote to Hans Schneider to withdraw my short paper; it turned out that he had never received it. When our joint paper was done, we duly sent it off to Hans. We soon received it back from him, along with a paper by Heydar Radjavi, in which the same result was proved. Heydar's paper, which had arrived in Hans's mail on the same day as ours, also contained some further results and examples. It was decided that we would write a three-author paper [9] on the main result, and Heydar could publish his supplements separately. As far as I know, they were never published, so I will note some of them later.

For me, the greatest benefit of the effort has been a continuing friendship with Paul. We subsequently collaborated on two expository papers (with a number of other coauthors). We keep in touch by electronic mail and visit one another from time to time. We have discussed some further research problems in linear algebra, but they have not (yet) come to fruition. Working with Paul is very enjoyable. He's prefectly organized and has remarkable insight. He does not pressure collaborators, and makes no assumptions about what one ought to know. He is eager to learn new things, especially from outside his own area.

Away from mathematics, Paul is the same — organized, low-pressured, and anxious to learn. He loves languages (he is fluent in English, Hungarian, German, and Spanish) and for years always attended a nonmathematical university course each semester. I can recall his taking courses in such diverse subjects as law and semiotics. Those who have read his "automathography" [10] will know that his undergraduate and early graduate studies were concerned as much with philosophy as with mathematics. Like me, he enjoys the mystery novels and humor of the past. We share an enthusiasm for the likes of Robert Benchely, Sir Arthur Conan Doyle, and Rex Stout. Indeed, at times I feel as though I am Watson to his Holmes (intriguingly, the anglicization of the family name adopted by one of Paul's brothers) or Archie Goodwin to his Nero Wolfe (note however that Paul is happily married, does not grow orchids, and does not weight anywhere near a seventh of a ton!). He is an organized repository of traditional knowledge who is sometimes bewildered by practicalities. I recall a time when he decided that he wanted a small tabletop lectern, of the sort often found in classrooms, to put in his bathroom so that he could read while brushing his teeth. He seemed genuinely astounded when John Ewing and I told him that any competent amateur carpenter could easily make one for him. Like Wolfe, Paul adheres to a strict daily schedule that includes several periods of work, interspersed with a four-mile walk (to be completed in less than one hour), a cocktail hour before dinner (two bottles of imported beer, no more), and breaks spent chatting with friends and colleagues. Paul and his charming wife Ginger adhere strictly to the Pritikin diet, which Larry Wallen once described as "the foodless diet." He quit smoking years ago when he gave away the fact that he was holding a good poker hand by having two cigarettes going at the same time.

I hope I have conveyed some of the satisfaction to be found in my personal and professional relationships with Paul Halmos. Now it's time to congratulate him on the occasion of his 75th birthday, get down to work and see what these involutions are all about.

# Involutions in Some Groups

An *involution* in a group $G$ is just an element $x$ of order two; i.e., $x^2 = e \neq x$, where $e$ is the identity element of $G$. The involution $x$ is its own inverse, and for any $y \in G$, the conjugate $yxy^{-1}$ is also an involution. It follows that the set of all products of involutions in $G$ is a normal subgroup $H$ of $G$. Therefore, if $G$ is simple and contains at least one involution, then $H = G$, and hence all elements of $G$ can be written as products of involutions. By the way, this works "for all values of 2," as Paul would put it. That is to say, for any positive integer $k$, the set of products of elements of order $k$ is a normal subgroup of $G$, and is everything if $G$ is simple and contains at least one such element. I know nothing further about the case $k \neq 2$, so I won't mention it again. The famous theorem of Thompson and Feit tells us that every finite nonabelian simple group has even order, where Sylow's theorem guarantees the presence of involutions. The recently finished classification of finite simple groups makes much use of involutions, especially of the structure of their centralizers.

Another case where involutions are present in a simple group is the projective special linear group $\mathrm{PSL}(n, F)$ of degree $n$ over the field $F$. This group is obtained by starting with the group $\mathrm{SL}(n, F)$ of $n \times n$ matrices of determinant one over $F$ and factoring out its center, which consists of all scalar matrices $\alpha I_n$, where $\alpha^n = 1$. $\mathrm{PSL}(n, F)$ is a simple group, unless $n = 2$ and $F$ has fewer than four elements ( see [17, p. 294]). The image in $\mathrm{PSL}(2, F)$ of the element

$$x = \begin{pmatrix} 0 & 1 \\ -1 & 0 \end{pmatrix}$$

is an involution, since $x^2 = -I_2$, and $(-1)^2 = 1$. For $n \geq 3$, the matrix

$$\begin{pmatrix} 0 & 1 & 0 & \\ 1 & 0 & 0 & \\ 0 & 0 & -1 & \\ & & & I_{n-3} \end{pmatrix}$$

is an involution in $\mathrm{SL}(n, F)$ and its image in $\mathrm{PSL}(n, F)$ is also an involution.

The simplicity of $\mathrm{PSL}(n, F)$ (for $|F| > 3$) implies that all proper normal subgroups of $\mathrm{SL}(n, F)$ are contained in the center. When $n \geq 3$, $\mathrm{SL}(n, F)$ contains the noncentral involution above, and hence all its elements are products of involutions. For $n = 2$, things are a bit more difficult. If $F$ has characteristic two, the matrix

$$\begin{pmatrix} 1 & 0 \\ 1 & 1 \end{pmatrix}$$

is a noncentral involution in $\mathrm{SL}(2, F)$ and all is well. However, when $F$ is not of characteristic two, $\mathrm{SL}(2, F)$ contains only one involution, $-I_2$. For, if $A^2 = I$, where

$$A = \begin{pmatrix} a & b \\ c & d \end{pmatrix},$$

then $a^2 + bc = 1 = d^2 + bc$ and $c(a + d) = 0 = b(a + d)$. Thus, $a = \pm d$, and if $b$ or $c$ is nonzero, then the minus sign holds. It follows that if $A \neq \pm I_2$, then

$$A = \begin{pmatrix} a & b \\ c & -a \end{pmatrix},$$

with $a^2 + bc = 1 = -\det A$, so $A \notin \mathrm{SL}(2, F)$. Therefore, $\mathrm{SL}(2, F)$ is *not* generated by involutions in this case. The cure for this problem is to consider instead the group $\pm\mathrm{SL}(n, F)$, of matrices whose determinant is $\pm1$. Its proper normal subgroups are $\mathrm{SL}(n, F)$ and the subgroups of its center. It always contains the noncentral involution $\mathrm{diag}(-1, 1, \ldots, 1)$. It follows that every element of $\pm\mathrm{SL}(n, F)$ is a product of involutions if $|F| > 3$ or $n > 2$. The two exceptional cases are easily handled by finite-group methods, and we have now seen my proof of Sampson's result.

Let us look at the situation in the finite symmetric groups. Let $S_n$ denote the group of all permutations of $\{1, \ldots, n\}$. The *$k$-cycle* $(i_1, i_2, \ldots, i_k)$ is the permutation that sends $i_j$ to $i_{j+1}$ for $j = 1, \ldots, k - 1$, sends $i_k$ to $i_1$, and leaves all other symbols fixed. Any element $\sigma$ of $S_n$ is a product (= composite) $\sigma_1 \sigma_2 \cdots \sigma_t$ of *disjoint* cycles; i.e., any symbol that occurs in $\sigma_i$ is left fixed by the other $\sigma_j$'s. The order of $\sigma$ as a group element is then the least common multiple of the lengths of the $\sigma_i$'s. Hence, the involutions in $S_n$ are exactly the products of disjoint *transpositions* (2-cycles). Note that every element of $S_n$ is a product of transpositions, since each $k$-cycle can be so expressed. For instance, $(1, 2, \ldots, k) = (1, k)(1, k - 1) \cdots (1, 2)$. These transpositions are, of course, not disjoint. This actually reveals that every element of $S_n$ can be written as a product of at most $n - 1$ involutions.

Let us see if we can do much better. Consider first the $k$-cycle $\gamma = (1, 2, \ldots, k)$. If $\sigma \in S_n$, then $\sigma\gamma\sigma^{-1} = (\sigma(1), \ldots, \sigma(k))$. Since $\gamma^{-1} = (1, k, k - 1, \ldots, 2)$, we have $\sigma\gamma\sigma^{-1} = \gamma^{-1}$, where $\sigma$ is the involution that interchanges 2 and $k$, 3 and $k - 1$, and so on. That is, $\sigma = (2, k)(3, k - 1) \cdots (u, v)$, where $(u, v) = (l + 1, l + 2)$ if $k = 2l + 1$ is odd, and $(u, v) = (l, l + 2)$ if $k = 2l$ is even. In both cases, 1 is a fixed point of $\sigma$; in the even case, $l + 1$ is also fixed. We call $\sigma$ the *standard conjugator* of $\gamma$, following Berggren [4]. If we agree to let $k$ represent the zero class modulo $k$, we can view $\gamma$ and $\sigma$ as the functions $i \mapsto i + 1 \pmod{k}$ and $i \mapsto 2 - i \pmod{k}$, respectively. From this, we easily see that $\gamma = \sigma\tau$, where $\tau = \sigma\gamma$ is the involution given by $i \mapsto 1 - i \pmod{k}$. Any other $k$-cycle has the form $\pi\gamma\pi^{-1}$, and hence is a product of two involutions conjugate to $\sigma$ and $\tau$. We call this conjugate of $\sigma$ a standard conjugator also. Hence, each cycle is a product of two involutions that leave fixed all symbols outside of the cycle. Now let $\varphi$ be an element of $S_n$, and write it as a product of disjoint cycles $\gamma_1 \cdots \gamma_m$. Express each $\gamma_i$ as a product $\sigma_i\tau_i$ of involutions, as above. Noting that disjoint cycles commute with one another, we have $\varphi = ST$, where $S = \sigma_1 \cdots \sigma_m$ is called the standard conjugator of $\varphi$, and $T = \tau_1 \cdots \tau_m$. Since the $\sigma_i$'s are pairwise disjoint, $S^2 = \sigma_1^2 \cdots \sigma_m^2 = e$. Similarly, $T^2 = e$. Thus, we have:

**Proposition.** *Every element of $S_n$ is a product of two involutions.*

If an element $g$ of a group can be written as a product $xy$ of two involutions, then $g^{-1} = y^{-1}x^{-1} = yx$ is conjugate to $g$, since $yx = x(xy)x = xgx^{-1}$. Groups in which every element is conjugate to its inverse are called *ambivalent;* we have just seen that the symmetric groups have this property. It is a useful property, since it implies that all complex characters of the group are real-valued (see [4], [5], and [14] for more about these groups). We should remark that in groups in general, elements that are conjugate to their inverses need not be products of involutions. Indeed, let $G$ be the set of ordered pairs of integers, with the operation $(a, b) * (c, d) = (a + (-1)^b c, b + d)$. Then $G$ is a group with identity element $(0, 0)$, and one has $(a, b)^{-1} = ((-1)^{b+1}a, -b)$. We see that $y * x * y^{-1} = x^{-1}$, where $x = (1, 0)$ and $y = (0, 1)$. However, $G$ contains no involutions, or indeed any elements of finite order. On a grander scale, Higman, Neumann, and Neumann [12] have shown that every torsion-free group can be embedded in a torsion-free group in which any two nonidentity elements are conjugate. Such a group is clearly ambivalent, but contains no involutions. By Cayley's Theorem, every finite group can be embedded in some $S_n$, which is ambivalent. Let us show how to embed every abelian group $G$ in a (generally nonabelian) ambivalent group that is just twice as large as $G$. Let $\text{Dih}(G) = G \times \{\pm 1\}$, as a set. We define multiplication by $(a, x)(b, y) = (ab^x, xy)$. This yields a group structure in which the identity element is $(e, 1)$, and $(a, x)^{-1} = (a^{-x}, x)$. We call $\text{Dih}(G)$ the *dihedralization* of $G$. The usual dihedral groups are the ones where $G$ is cyclic. The set of all $(a, 1)$ forms a normal subgroup of $\text{Dih}(G)$ that is isomorphic to $G$ and has index two. The elements $(a, -1)$ are themselves involutions, and one has $(a, 1) = (a, -1)(e, -1)$, a product of involutions. The automorphism $a \mapsto a^{-1}$ of $G$ is made into conjugation by $(e, -1)$ in $\text{Dih}(G)$. This construction doesn't work for nonabelian $G$, since inversion is not an automorphism in that case.

What happens in the alternating group $A_n$? It is the subgroup of $S_n$ consisting of *even* permutations, i.e., those that can be written as a product of an even number of transpositions. As a curiosity, note that the elements of odd order in $S_n$ are products of cycles of odd length, which, in turn, are products of even numbers of transpositions. Hence, the elements of *odd* order in $S_n$ are *even* permutations. We just have to live with that. We might hope that, as in $S_n$, every element of $A_n$ be a product of two involutions. That is certainly not true for $A_2 = \{e\}$, nor for $A_3$, which is cyclic of order three, and hence contains no involutions. In $A_4$, which has order 12, the involutions, together with $e$, form a normal subgroup of order 4. For $n \geq 5$, $A_n$ is simple, hence every element is a product of some number of involutions. Since most of these groups have characters that are not real-valued, there must usually exist elements that are not products of two involutions. Let us show that three involutions suffice, and that two suffice precisely when $n$ is 5, 6, 10, or 14 (Berggren [4] proved that these are the values of $n$ for which $A_n$ is ambivalent and our approach follows his closely).

Let $\gamma \in A_n$, and let $\sigma$ be its standard conjugator. If $\sigma$ is also in $A_n$, then so is the involution $\tau = \sigma\gamma$, and $\gamma = \sigma\tau$ is a product of two involutions in $A_n$. If $\sigma$ is odd, so is $\tau$. If $\gamma$ has two fixed points $u$ and $v$, then $\sigma(u,v)$ and $(u,v)\tau$ are involutions in $A_n$, and $\gamma$ is their product. Conversely, if $\gamma$ is a product $\lambda\psi$ of two involutions in $A_n$, then $\gamma^{-1} = \lambda\gamma\lambda^{-1} = \sigma\gamma\sigma^{-1}$, and $\sigma\lambda$ lies in the *centralizer* $C = C_{s_n}(\gamma) = \{\xi \in S_n \mid \xi\gamma = \gamma\xi\}$. Now, $C$ is a subgroup of $S_n$ and either is or is not a subgroup of $A_n$. Suppose that $C \subseteq A_n$. Then the number of conjugates of $\gamma$ in $S_n$ is the index $[S_n : C] = n!/|C|$, while the number of its conjugates in $A_n$ is $[A_n : C]$, which is just half as many. Since all $S_n$ conjugates of $\gamma$ are in $A_n$, the $S_n$ class of $\gamma$ splits into two equal-size classes in $A_n$. One of these contains $\gamma$, the other consists of all $\xi\gamma\xi^{-1}$ where $\xi$ is odd. In particular, if $\sigma$ is odd, then $\gamma^{-1}$ is in the other class, and so $\gamma$ is not conjugate to $\gamma^{-1}$ in $A_n$. When, if ever, does all this happen? It can be shown that the class of $\gamma$ splits in $A_n$ if and only if $\gamma$ is a product of cycles of distinct odd lengths (see Scott [17, p. 300]). Here, each fixed point counts as a cycle of length one, so, as we saw before, there can be at most one of them. We saw earlier that the standard conjugator of a $k$-cycle is a product of $l$ transpositions when $k = 2l + 1$ and of $l - 1$ transpositions when $k = 2l$. Thus, if $\gamma$ is a product of disjoint cycles of odd length, its standard conjugator is odd just when an odd number of those cycles have length congruent to three modulo four. Thus, we can construct an element $\gamma$ of $A_n$ that is not a product of two involutions in $A_n$ whenever $n = n_1 + \cdots + n_k$, where the $n_i$'s are distinct odd numbers, an odd number of which are congruent to three modulo four. If $n = 4k$, we have $n = 3 + (4k - 3)$, while for $n = 4k + 3$, we need only $n$ itself. When $k > 1$, we write $4k + 1 = 1 + 3 + (4k - 3)$, and for $k > 3$, $4k + 2 = 1 + 3 + 5 + (4k - 7)$. The only values of $n$ larger than 4 not covered here are 5, 6, 10, and 14, and experimentation shows that they are authentic exceptions.

Let $\gamma \in A_n$, with $n \geq 5$. We now show that $\gamma$ is the product of at most three involutions. Write $\gamma = \sigma\tau$, where $\sigma$ is the standard conjugator of $\gamma$. As we have noted, this exhibits $\gamma$ as a product of two involutions in $A_n$, if $\sigma \in A_n$. Otherwise, we borrow one transposition $\alpha$ from $\sigma$ and one transposition $\beta$ from $\tau$. It is not difficult to see that we can make $\alpha$ and $\beta$ commute with one another, so that $\xi = \alpha\beta$ is an involution. Then we have $\gamma = (\sigma\alpha)\xi(\beta\tau)$, a product of three involutions.

Now let us try to see exactly which elements of $A_n$ are products of two involutions. Since we have dealt with the case where the standard conjugator is even, our previous discussion shows that there are two unresolved situations. First is the case where the expression of $\gamma$ as a product of disjoint cycles involves a cycle of even length. Let $\gamma = \lambda\pi$, where $\lambda$ is a cycle of even length and $\pi$ is some odd permutation disjoint from $\pi$. We write $\lambda = \mu\varphi$, where $\mu$ and $\varphi$ are involutions of opposite parity. Since $\mu\varphi = (\mu\varphi\mu^{-1})\mu$, we may assume that $\mu$ is even. In the same way, we can write $\pi = \xi\psi$, with $\xi$ and $\psi$ involutions, where $\xi$ is even and $\psi$ is odd. We then have $\gamma = (\mu\xi)(\varphi\psi)$, a product of two involutions in $A_n$.

Second, consider the case where the decomposition of $\gamma$ as a product of disjoint cycles involves two cycles of the same odd length $k$. Thus, we write $\gamma$ as a disjoint product $\lambda_1\lambda_2\pi$, where $\lambda_1 = (i_1, \ldots, i_k)$, $\lambda_2 = (j_1, \ldots, j_k)$, and $\pi$ is an even permutation. We may assume that the standard conjugator $\sigma$ of $\gamma$ is odd. The permutation $c = (i_1, j_1) \cdots (i_k, j_k)$ is an odd involution. Conjugation by $c$ interchanges $\lambda_1$ and $\lambda_2$, and does not affect $\pi$, so $c$ commutes with $\gamma$. Also, we see that $\sigma c\sigma = c = c^{-1}$ and $c\sigma c = \sigma$. It follows that $(\sigma c)^2 = e = (\sigma c\gamma)^2$. Since $\sigma$ and $c$ are odd, we have the factorization $\gamma = (\sigma c)(\sigma c\gamma)$ as a product of two involutions in $A_n$. We have now proved the following:

**Proposition.** *Let $n \geq 5$. Every element $\gamma$ of $A_n$ is a product of three involutions. Two involutions suffice unless $\gamma$ is the product of cycles of distinct odd lengths, an odd number of which are congruent to three modulo four.*

**Corollary (Berggren).** *$A_n$ is ambivalent if and only if $n$ is 5, 6, 10, or 14.*

Finally, we remark that each element of $A_n$ is a product of two involutions when viewed as an element of $A_{n+2}$. Hence, in the groups $A_\infty = \bigcup A_n$ and $S_\infty = \bigcup S_n$, every element is a product of two involutions, and so these groups are ambivalent.

## Matrices

If $A$ is a square matrix over a commutative field and $A^2 = I$, then it follows that $\det A = \pm 1$. Hence, any matrix that is a product of involutions has the same property. Thus, we confine our attention to matrices belonging to the groups $\pm SL(n, F)$. Here is a closer look at involutory matrices. They satisfy the polynomial $x^2 - 1 = (x - 1)(x + 1)$. If the characteristic of $F$ is not two, it follows that such a matrix is similar to a block matrix

$$\begin{pmatrix} I_k & 0 \\ 0 & -I_{n-k} \end{pmatrix},$$

with $k \leq n$. If $F$ has characteristic two, an involution is similar to the direct sum of an identity matrix and some blocks of the form

$$\begin{pmatrix} 1 & 1 \\ 0 & 1 \end{pmatrix}.$$

This is because $x^2 - 1 = (x-1)^2 = (x+1)^2$ in characteristic two. Involutory matrices are often encountered in the forms

$$\pm \begin{pmatrix} I_k & 0 \\ B & -I_l \end{pmatrix}, \quad \begin{pmatrix} 0 & C \\ C^{-1} & 0 \end{pmatrix},$$

and $\pm J_n$, where

$$J_n = \begin{pmatrix} 0 & \cdots & 1 \\ \vdots & \reflectbox{$\ddots$} & \vdots \\ 1 & \cdots & 0 \end{pmatrix}_{n \times n}.$$

Here, $B$ is any $l \times k$ matrix, and $C$ is a nonsingular square matrix. If $A$ is an involution, so is $-A$.

Let us now discuss the matrices that are products of two involutions. As we noted earlier, such a matrix is conjugate, i.e., similar to its inverse. We first show the remarkable fact that the converse holds: any matrix that is similar to its inverse is a product of two involutions. This was first proved in the case where the characteristic of the underlying field is not two by Wonnenburger [20] and later in the general case by Djoković [6]. Our proof is close to his and to that of Hoffman and Paige [13].

We will write $A \sim B$ to indicate that the matrices $A$ and $B$ are similar. Suppose that $A \sim A^{-1}$, say $YAY^{-1} = A^{-1}$. From the theory of canonical forms, $A$ is also similar to a block diagonal matrix $A' = \mathrm{diag}(A_1, \ldots, A_k)$, where each $A_i$ is the companion matrix of a polynomial and is *indecomposable*, in the sense that it is not similar to any nontrivial $\mathrm{diag}(B, C)$. $A'$ is unique up to the order in which the blocks occur. If $XAX^{-1} = A'$, then we have $(XYX^{-1})A'(XYX^{-1})^{-1} = (A')^{-1}$. It suffices to show that $A'$ is a product of two involutions. Now, $(A')^{-1} = \mathrm{diag}(A_1^{-1}, \ldots, A_k^{-1})$ is similar to $A'$, and each block is indecomposable. It follows that some of the $A_i$'s, say the first $l$ of them, are similar to their own inverses, while the rest come in pairs, within which each element is similar to the inverse of the other. Hence, $A'$ is similar to a direct sum of the $A_i$, with $1 \leq i \leq l$, and blocks of the form $T_j = \mathrm{diag}(A_j, A_j^{-1})$. It suffices to show that the $A_j$'s for $1 \leq j \leq l$ and the $T_j$'s are each products of two involutions. Since

$$T_j = \begin{pmatrix} 0 & I \\ I & 0 \end{pmatrix} \begin{pmatrix} 0 & A_j^{-1} \\ A_j & 0 \end{pmatrix}$$

is a product of two involutions, it suffices to deal with $A_1, \ldots, A_l$. Each is the companion matrix

$$
C(f) = \begin{pmatrix}
0 & \cdots & 0 & -a_0 \\
1 & \cdots & 0 & -a_1 \\
\vdots & \ddots & \vdots & \vdots \\
0 & \cdots & 1 & -a_{n-1}
\end{pmatrix}
$$

of a monic polynomial $f(x) = x^n + a_{n-1}x^{n-1} + \cdots + a_1 x + a_0$, for suitable $n$. It is not hard to check that $C(f)^{-1}$ is similar to $C(\tilde{f})$, where $\tilde{f}(x) = a_0^{-1}x^n f(1/x) = x^n + a_0^{-1}a_1 x^{n-1} + \cdots + a_0^{-1}a_{n-1}x + a_0^{-1}$. From the uniqueness of the rational canonical form, it follows that $a_0 = a_0^{-1}$, and $a_0 a_j = a_{n-j}$, for $j = 1, \ldots, n-1$. Recalling the involution $J_n$ described earlier, we see that

$$
A_i J_n = \begin{pmatrix}
-a_0 & 0 & \cdots & 0 \\
-a_1 & 0 & \cdots & -1 \\
\vdots & \vdots & \iddots & \vdots \\
-a_{n-1} & -1 & \cdots & 0
\end{pmatrix}
$$

is an involution, so $A_i = (A_i J_n)J_n$ is a product of two involutions. We therefore have the following:

**Two-Involution Theorem.** *A square matrix over a field is a product of two involutions if and only if it is similar to its inverse.*

Here is a situation where this theorem applies. Suppose that $AA^t = I$, where $A^t$ is the transpose of $A$. Then $A^{-1} = A^t$, and since every matrix over a field is similar to its transpose, $A$ is a product of two involutions. More generally, we get the same conclusion if $A^t X A = X$, for some nonsingular $X$, for then, $A^t = X A^{-1} X^{-1}$. Let $A$ belong to a finite subgroup $G$ of $\mathrm{GL}(n, F)$, where $F$ is a subfield of the field $\mathbb{R}$ of real numbers. The matrix $X = \sum_{B \in G} B^t B$ is positive definite and symmetric, and we have $A^t X A = X$, since $BA$ varies through $G$ as $B$ does. Thus, if a matrix over such a field has a power equal to the identity, it is a product of two involutions. In case $F = \mathbb{R}$, it suffices to assume that $A$ belongs to a compact topological subgroup $G$ of $\mathrm{GL}(n, \mathbb{R})$, since we can then define $X = \int_G B^t B \, d\mu$, where $\mu$ is Haar measure on $G$. These remarks do not apply to complex matrices. If $\omega$ is a complex root of unity and $\omega^n \neq 1$, then $\omega I_n$ has finite multiplicative order, but is not a product of involutions because its determinant is not $\pm 1$. Later, we prove a proposition that serves as partial compensation for this misfortune.

We should remark that this characterization of products of two involutory matrices requires commutativity of the base field, as Ellers [7] has pointed out. For example, the quaternionic matrix

$$
\begin{pmatrix}
i & 0 \\
0 & 1
\end{pmatrix}
$$

is similar to its inverse, but is not a product of two involutions.

Now, let's try to factor the elements of $\pm\mathrm{SL}(n, F)$ as products of as few involutions as possible. As a simple but illuminating first case, we will consider a cyclic matrix $A$. That is, we assume that $A$ is similar to a companion matrix. The case with $n = 3$ is typical, so let us suppose that $A$ is similar to

$$A' = \begin{pmatrix} 0 & 0 & a \\ 1 & 0 & b \\ 0 & 1 & c \end{pmatrix}.$$

We will show that $A'$ (and hence $A$) can be written as a product of three involutions. Observe that $\det A' = a = \pm 1$. We see that $A' = XR$, where

$$X = \begin{pmatrix} -1 & 0 & 0 \\ -b/a & 1 & 0 \\ -c/a & 0 & 1 \end{pmatrix}, \quad \text{and} \quad R = \begin{pmatrix} 0 & 0 & -a \\ 1 & 0 & 0 \\ 0 & 1 & 0 \end{pmatrix}.$$

As we noted earlier, any $n \times n$ matrix, such as $X$, that has the block form

$$\begin{pmatrix} -I_k & 0 \\ B & I_{n-k} \end{pmatrix}$$

is an involution. Since $a = \pm 1$, $RR^t = I$, and hence $R$ is a product $YZ$ of two involutions. Thus, $A' = XYZ$ is a product of three involutions, as claimed. This, in the $n \times n$ case, was one of Radjavi's unpublished remarks. He showed further that any direct sum of two companion matrices that lies in $\pm\mathrm{SL}(n, F)$ is a product of three involutions. This result was rediscovered and published by Ballantine [1].

Now, let us sketch the proof that a matrix of determinant $\pm 1$ is a product of 4 involutions. For an arbitrary $A \in \pm\mathrm{SL}(n, F)$, there is a direct sum $A'$ of companion matrices similar to $A$, and it is harmless to assume that $A' = A$. Some of those companion matrices may be $1 \times 1$, and it is these that force us to use 4 involutions, instead of just 3. Let us take as an example that illustrates all possible difficulties the following matrix:

$$A = \begin{pmatrix} 0 & a & & & & & \\ 1 & b & & & & & \\ & & 0 & c & & & \\ & & 1 & d & & & \\ & & & & e & & \\ & & & & & f & \\ & & & & & & g \end{pmatrix}.$$

We write $A = XR$, where

$$X = \begin{pmatrix} -1 & 0 & & & & & \\ -b/a & 1 & & & & & \\ & & -1 & 0 & & & \\ & & -d/c & 1 & & & \\ & & & & 0 & 1 & \\ & & & & 1 & 0 & \\ & & & & & & 1 \end{pmatrix}$$

and

$$R = \begin{pmatrix} 0 & -a & & & & & \\ 1 & 0 & & & & & \\ & & 0 & -c & & & \\ & & 1 & 0 & & & \\ & & & & 0 & e & \\ & & & & f & 0 & \\ & & & & & & g \end{pmatrix}.$$

$X$ is clearly an involution. $R$ is a *monomial* matrix (called a *weighted permutation matrix* in [9]). That is, each row and each column of $R$ contains exactly one nonzero entry. Were we to replace all its nonzero entries with ones, we would obtain a matrix corresponding to a transformation that permutes the underlying ordered basis $\{v_1, \ldots, v_7\}$. In terms of the subscripts, that is the permutation $\sigma = (1, 2)(3, 4)(5, 6)(7)$, where we show the fixed point 7 as a 1-cycle. In general, we pair up the $1 \times 1$ blocks as far as possible, so that the resulting permutation has at most one fixed point. Let $\tau = (2, 3)(4, 5)(6, 7)$, an involution in $S_7$ chosen to tie together the cycles in $\sigma$. We have $\tau\sigma = (1, 3, 5, 7, 6, 4, 2)$, a cycle on all 7 symbols. If $Y$ is the permutation matrix corresponding to $\tau$ (obtained from the identity matrix by permuting its columns according to $\tau$), then $Y$ is an involution, and

$$YR = \begin{pmatrix} 0 & -a & 0 & 0 & 0 & 0 & 0 \\ 0 & 0 & 0 & -c & 0 & 0 & 0 \\ 1 & 0 & 0 & 0 & 0 & 0 & 0 \\ 0 & 0 & 0 & 0 & 0 & f & 0 \\ 0 & 0 & 1 & 0 & 0 & 0 & 0 \\ 0 & 0 & 0 & 0 & 0 & 0 & g \\ 0 & 0 & 0 & 0 & e & 0 & 0 \end{pmatrix}.$$

Changing to the ordered basis $\{v_1, v_3, v_5, ev_7, gev_6, fgev_4, -cfgev_2\}$, we see that $YR$ is similar to

$$S = \begin{pmatrix} 0_{1 \times 6} & \delta \\ I_{6 \times 6} & 0_{6 \times 1} \end{pmatrix},$$

where $\delta = acfge = \det A = \pm 1$. Now, $SS^t = I$, so $S$ is a product of two involutions. Hence, $YR$ is also a product $ZW$ of two involutions, and so $A = XR = XYYR = XYZW$ is a product of four involutions. This completes the sketch of the proof of the following:

**Four-Involutions Theorem.** *A matrix over a field is a product of four involutions if and only if its determinant is $\pm 1$.*

At this point, I would like to mention an alternate proof given by Sourour [18]. It only works when $|F|$ contains at least $n + 2$ elements, but that is a small restriction. Sourour gave the proof as one of several corollaries of a beautiful theorem, whose proof is rather technical:

**Theorem (Sourour).** *Let $A$ be a nonscalar, nonsingular $n \times n$ matrix over an arbitrary field $F$. Let $\beta_1, \ldots, \beta_n$, $\gamma_1, \ldots, \gamma_n$ be elements of $F$ such that $\prod_{j=1}^n \beta_j \gamma_j = 1$. Then there exist matrices $B$ and $C$ over $F$ such that $A = BC$, the eigenvalues of $B$ are $\beta_1, \ldots, \beta_n$, and those of $C$ are $\gamma_1, \ldots, \gamma_n$. Further, one may assume that $B$ is upper triangularizable and simultaneously, $C$ is lower triangularizable.*

To obtain the theorem on involutions, let us first consider the case where $\det A = 1$. We write $A = BC$, with the spectrum of each of $B$ and $C$ of the form $\{\beta_1, \beta_1^{-1}, \ldots, \beta_m, \beta_m^{-1}\}$ or $\{1, \beta_1, \beta_1^{-1}, \ldots, \beta_m, \beta_m^{-1}\}$, according to the parity of $n$. By the assumption that $F \geq n + 2$, we can arrange that each of $B$ and $C$ has distinct eigenvalues, so is diagonalizable. Since

$$\begin{pmatrix} \beta & 0 \\ 0 & \beta^{-1} \end{pmatrix} = \begin{pmatrix} 0 & 1 \\ 1 & 0 \end{pmatrix} \begin{pmatrix} 0 & \beta^{-1} \\ \beta & 0 \end{pmatrix}$$

and the factors are involutions, the proof is complete in this case. If $\det A = -1$, just apply the first part to $-A$ when $n$ is odd. For even $n$, replace the pair $\beta_1$, $\beta_1^{-1}$ by $1$, $-1$ in the spectrum of $B$, but not in that of $C$. If the characteristic is not two, this gives the diagonal form of $B$ a block $\operatorname{diag}(1, -1)$, which is itself an involution. In characteristic two, $B$ becomes the direct sum of a diagonal matrix with an involution of the form

$$\begin{pmatrix} 1 & a \\ 0 & 1 \end{pmatrix}.$$

While we have Sourour's theorem in mind, let us obtain a result that is similar in spirit to the results on involutions, but a bit different. If a matrix has finite order $k$ in $\operatorname{GL}(n, F)$, then $(\det A)^k = 1$ in $F$; i.e., $\det A$ is a root of unity. The determinant of a product of such matrices is therefore also a root of unity. We now prove a special case of the converse of this assertion.

**Proposition.** *A complex matrix whose determinant is a root of unity is the product of two matrices of finite order.*

**Proof.** Let $A$ be $n \times n$, and let $\det A = \zeta$ be an $m$th root of unity. If $A = \xi I_n$ is scalar, then $\xi^n = \zeta$, where $\xi$ is a root of unity and $A$ itself has finite order.

Now assume that $A$ is not scalar. Let $\omega = e^{2\pi i/n}$, a primitive $n$th root of unity. Observe that $\prod_{k=0}^{n-1} \omega^k = (-1)^{n+1}$. If $(-1)^{n+1}\zeta \notin \{\omega, \omega^2, \ldots, \omega^{n-1}\}$, let $S = \{(-1)^{n+1}\zeta, \omega, \omega^2, \ldots, \omega^{n-1}\}$. Otherwise, choose a root of unity $\gamma \neq \pm 1$ whose order is relatively prime to $mn$ and take

$$S = \{(-1)^{n+1}\zeta\gamma, \omega, \omega^2, \ldots, \omega^k\gamma^{-1}, \ldots, \omega^{n-1}\},$$

where $(-1)^{n+1}\zeta = \omega^k$. By Sourour's theorem, $A = BC$, where the eigenvalues of $B$ are $(-1)^{n+1}, \omega, \ldots, \omega^{n-1}$, while those of $C$ are the elements of $S$. $B$ and $C$ are diagonalizable with roots of unity as eigenvalues, and so are of finite order.

It would be nice to know whether a result like this holds in other fields that do not have an abundant supply of roots of unity. Were it true for subfields of the reals, it would give another proof of the four-involution theorem in that case. In a similar vein, it would be interesting to know whether each element of $\pm\mathrm{SL}(n, \mathbb{R})$ is a product of two matrices, each of which lies in a compact group.

Thus far, we have seen that the matrices that are products of 2 involutions are precisely those that are similar to their inverses, while those that are products of four involutions are those of determinant $\pm 1$. What about products of other numbers of involutions? There is little point in talking about products of more than 4 involutions, since such matrices have determinant $\pm 1$ and so can be rewritten as products of four involutions. That leaves the case of three involutory factors, and here, a complete characterization is not known. Before we state the known fragments, let us see that this really is a separate case. We will present Radjavi's example of a matrix that is a product of four involutions, but not of three.

We must assume that $|F|$ is not 2, 3, or 5; these are genuine exceptions, as we will see later. We consider the matrix

$$A = \begin{pmatrix} \alpha I_{2p \times 2p} & 0 \\ 0 & \alpha^{-2p} \end{pmatrix},$$

where $p \geq 2$ and $\alpha^{4(2p+1)} \neq 1$. If $|F| = 11$, we choose $p = 3$ and $\alpha = 2$. Otherwise, take $p = 2$, and note that the identity $x^{20} = 1$ holds only in fields with 2, 3, 5, or 11 elements, so that a suitable $\alpha$ can be chosen. Suppose that $A = K_1 K_2 K_3$, where the $K_i$'s are involutions. If necessary, multiply $K_2$ and $K_3$ by $-I$ to ensure that the dimension $q$ of the fixed-point space $U$ of $K_3$ is at least $p+1$. Let $V = \ker(A - \alpha I)$, $W = \ker(AK_3 - \alpha I)$ and $W' = \ker(AK_3 - \alpha^{-1}I)$. If $r = \dim W$, then $r = \dim W'$, since $AK_3 = K_1 K_2$ is similar to its inverse. We have $2p + 1 \geq \dim(V + U) = 2p + q - \dim(V \cap U)$, whence $r = \dim W \geq \dim(V \cap U) \geq q - 1 \geq p$. By the choice of $\alpha$, $\alpha \neq \alpha^{-1}$, whence $W \cap W' = (0)$. Thus, $2r \leq 2p + 1$, and it follows that $r = p$ and $q = p + 1$. Now we form a new basis. Start with a basis of $V \cap U$, extend it to one of $V$, and finally, tack on the last coordinate vector. The matrix

$A$ stays the same, while $K_3$ takes on the form

$$\begin{pmatrix} I_{p\times p} & X \\ 0 & K_0 \end{pmatrix},$$

where $K_0$ is an involution whose $(-1)$ eigenspace has dimension $q' \geq p$ (with equality except perhaps in characteristic two). Let $A_0$ be the lower, right-hand $(p+1) \times (p+1)$ block of $A$. Then $\dim \ker(A_0 - \alpha I) = p$, whence $\dim \ker(A_0 K_0 + \alpha I) \geq q' \geq p-1$. It follows that $AK_3$ has a triangular form whose diagonal contains $p$ $\alpha$'s and at least $p-1-\alpha$'s. Since $AK_3$ is similar to its inverse, the diagonal must also have at least $2p-1$ elements of the form $(\pm\alpha)^{-1}$. By the choice of $\alpha$, the diagonal must therefore accomodate at least $4p-2$ elements, but with $p \geq 2$, they just can't fit into the $2p+1$ diagonal positions.

Here's a simpler but more special example, due to Halmos and Kakutani. Let $g$ belong to the center of a group $G$, and suppose that $g = xyz$ is a product of three involutions. Then $g^4 = gxgygz = g(xg)y(gz) = g(yz)y(xy) = y(gz)y(xy) = yxyyxy = e$. Therefore, if $\omega$ is a primitive $n$th root of unity in $F$ and $n$ is neither 2 or 4, then $A = \omega I_n$, which lies in the center of $\mathrm{GL}(n, F)$, is not a product of 3 involutions, but $\det A = 1$. For the record, when $n = 3$,

$$A = \begin{pmatrix} 0 & 0 & 1 \\ 0 & 1 & 0 \\ 1 & 0 & 0 \end{pmatrix} \begin{pmatrix} 0 & 1 & 0 \\ 1 & 0 & 0 \\ 0 & 0 & 1 \end{pmatrix} \begin{pmatrix} 0 & \omega & 0 \\ \omega^2 & 0 & 0 \\ 0 & 0 & 1 \end{pmatrix} \begin{pmatrix} 0 & 0 & \omega^2 \\ 0 & 1 & 0 \\ \omega & 0 & 0 \end{pmatrix}.$$

Now let's see a few results with matrices that *are* products of three involutions. Our discussion of the four-involutions theorem showed that if the element $A$ of $\pm\mathrm{SL}(n, F)$ is cyclic (similar to a companion matrix), then it is a product of three involutions. Radjavi showed that the same is true if $A$ is similar to a direct product of two companion matrices; again, Ballantine rediscovered and published the result. Hence, if $|F|$ is 2, 3, or 5, then every element of $\pm\mathrm{SL}(n, F)$ is a product of 3 involutions. It also follows that $A$ is a product of three involutions if its characteristic polynomial is the product of two irreducible polynomials. Liu [15] showed that if a complex matrix of determinant $\pm1$ is similar to a direct sum of companion matrices, none of which is $1 \times 1$, then it is a product of 3 involutions.

Ballantine [2] showed that if the $n \times n$ matrix $A$ is a product of 3 involutions, then $\dim \ker(A - \beta I) \leq 3n/4$, for any $\beta \in F$ such that $\beta^4 \neq 1$. Liu [15] showed that for $A \in \pm\mathrm{SL}(n, \mathbb{C})$, $A$ is a product of 3 involutions if $\dim \ker(A - \alpha I) \leq [n/2]$, for all $\alpha \in \mathbb{C}$. On the other hand, if $A$ is such a product, then $m \leq [3n/4]$ and $m \leq (2n+r)/3$, where for any nonzero $\beta \in \mathbb{C}$, $m = \dim \ker(A - \beta I)$ and $r = \dim \ker(A - \beta^{-3}I)$. He also gave a simple characterization of the $5 \times 5$ complex matrices that are products of 3 involutions.

# Some Further Results and Speculation

Our discussion suggests some areas for further investigation. First, we have seen that there are many simple groups in which every element is a product of involutions, and one could ask how many factors are required. With the classification of finite simple groups now available, we expect that a qualified group theorist could resolve the issue without difficulty. It would also be nice to have a conceptual proof (independent of the classification) that some small number, say four or five, suffices in every finite nonabelian simple group.

Another place to look is in the linear groups over general rings. For non-commutative rings, the structure of these groups is not fully known. Results on normal subgroups are usually stated in the stable case, which is to say, for groups such as $GL(R)$, which is obtained by taking the direct limit of the $GL(n, R)$ with respect to the maps $GL(n, R) \to GL(n+1, R)$ given by $A \mapsto \mathrm{diag}(A, 1)$. The results often apply to $GL(n, R)$ only for sufficiently large $n$. The lack of suitable theories of determinants and canonical forms is also a difficulty. As the example of Ellers cited earlier shows, one doesn't even know what to conjecture. On the positive side, let $E(n, R)$ denote the group generated by the matrices $I_n + ae_{ij}$, where $i \neq j$, $e_{ij}$ is the usual matrix unit with one in the $(i, j)$ position and zeroes elsewhere, and $a \in R$. Waterhouse [19] noted that one has

$$I_n + ae_{ij} = (I_n + e_{ij} - 2e_{jj})(I_n + (a+1)e_{ij} - 2e_{jj})$$

and

$$(I_n + be_{ij} - 2e_{jj})^2 = I_n.$$

Thus, each element of $E(n, R)$ is a product of involutions (which are generally *not* in $E(n, R)$). We might add that one has $(I + ae_{ij})(I + be_{kl}) = [(I + ae_{ij})D][D(I + be_{kl})]$, where $D = \mathrm{diag}(d_1, \dots, d_n)$ is chosen so that $d_m^2 = 1$ for each $m$ and $d_i + d_j = 0 = d_k + d_l$. The factors in square brackets are involutions. We point out that when $R$ is commutative, $E(n, R) \subseteq SL(n, R)$, with equality if $R$ is a field, a semilocal ring, or a Euclidean domain. Equality also holds for some other classes of rings if $n$ is large enough, which usually means larger than two. There are principal ideal domains for which the inclusion is proper when $n = 2$ (see Bass [3]).

Now, let us assume that $R$ is commutative. Let $G_n$ denote the group of products of involutions in $GL(n, R)$. In general, the determinants of elements of $G_n$ needn't be $\pm 1$; they are simply involutions in $R$, of which there may be many (think of a direct product of rings). The right place to study products of involutions in this setting is therefore the group $IL(n, R) = \{A \in GL(n, R) \mid (\det A)^2 = 1\} = \{A \in GL(n, R) \mid A^2 \in SL(n, R)\}$. $IL(n, R) = \pm SL(n, R)$ if $1$ and $-1$ are the only involutions in $R$. If $A \in IL(n, R)$, then $DA \in SL(n, R)$, where $D = \mathrm{diag}(\det A, I_{n-1})$. If $E(n, R) = SL(n, R)$, then $DA \in G_n$, and so $A = D(DA)$ is a product of involutions. Hence, in this situation, every matrix that could be a product of involutions is one. Generally, we have no idea how many factors are needed.

The proofs of the two- and four-involution theorems for matrices over fields depend heavily on the rational canonical form, except for Sourour's proof, which depends on the presence of inverses. For matrices over commutative rings, the theory of the rational canonical form does not hold in general. Given an $n \times n$ matrix $A$ over a commutative ring $R$, define an endomorphism $T_A$ of the $R$ module $R_{n \times n}$ of all $n \times n$ matrices by $T_A(X) = AX - XA$. Guralnick [8] has shown that if $R$ is a local ring with maximal ideal $P$, then $A$ is $R$-similar to a matrix in rational canonical form if and only if the image of $T_A$ is a direct summand of $R_{n \times n}$, or equivalently, $T_A(R_{n \times n}) \cap PR_{n \times n} = PT_A(R_{n \times n})$. If this condition holds for $A$ and $\det A = \pm 1$, then the determinants of the companion matrices in the canonical form are all units of $R$, and the proof for fields that $A$ is a product of four involutions goes through virtually as is. On the other hand, consider a local ring $R$ in which two is a nonzero nonunit. One such ring is $\mathbb{Z}_2$, which consists of all rational numbers that have odd denominators when written in lowest terms; the unique maximal ideal is the set of rationals with odd denominators and even numerators. Let

$$A = \begin{pmatrix} 1 & 2 \\ 0 & 1 \end{pmatrix}.$$

One finds that every $T_A(X)$ has all its entries in $P$. In particular, $T_A(e_{21}) = \text{diag}(2, -2) = 2\,\text{diag}(1, -1)$, but $\text{diag}(1, -1)$ is not in the image of $T_A$. Thus, $A$ is not similar to a matrix in rational canonical form. On the other hand, $A$ is clearly a product of two involutions with entries in $R$.

For commutative rings that are not local, Guralnick's result gives a necessary, but not sufficient condition for similarity to a matrix in rational canonical form. For instance, consider the matrix of integers

$$A = \begin{pmatrix} 11 & -7 \\ -36 & 23 \end{pmatrix}.$$

A tedious calculation shows that the image of $T_A$ is a direct summand of $\mathbb{Z}_{2 \times 2}$. However, if we try to find a matrix

$$X = \begin{pmatrix} \alpha & \beta \\ \gamma & \delta \end{pmatrix}$$

such that $XAX^{-1}$ is the companion matrix of the minimal polynomial $x^2 - 34x + 1$, we find that $\alpha = -23\gamma - 36\delta$ and $\beta = -7\gamma - 11\delta$, whence $\det X = 7\gamma^2 - 12\gamma\delta - 36\delta^2$. For integers $\gamma$ and $\delta$, this cannot be $\pm 1$, as we see by reducing modulo 12. On the other hand, we can make $\det X$ a unit in each localization $\mathbb{Z}_p = \{m/n \in \mathbb{Q} \mid (p, n) = 1\}$: take $\gamma = 1$ and $\delta = 0$ if $p \neq 7$ and take $\gamma = 0$ and $\delta = 1$ when $p = 7$. Thus, $A$ is locally similar to a companion matrix, but not globally. By the way, for the nonscalar integral matrix

$$\begin{pmatrix} a & b \\ c & d \end{pmatrix}$$

to be integrally similar to a companion matrix, it is necessary and sufficient that the quadratic form $Q(x, y) = cx^2 + (d - a)xy - by^2$ represent $\pm 1$. This implies the necessary but not sufficient condition that $c$, $d - a$, and $b$ have greatest common divisor one. This condition *is* sufficient locally, since for each prime $p$, at least one of $Q(1, 0)$, $Q(0, 1)$, and $Q(1, 1)$ is not a multiple of $p$.

In casual experiments with elements of $\pm\mathrm{SL}(2, \mathbb{Z})$, I have found that when one factors them as products of small numbers of rational involutions, the factors are themselves often integral. For instance, we have the factorization

$$B = \begin{pmatrix} 1 & 2 \\ 3 & 7 \end{pmatrix} = \begin{pmatrix} -7 & 2 \\ -24 & 7 \end{pmatrix} \begin{pmatrix} -1 & 0 \\ -3 & 1 \end{pmatrix},$$

even though $B$ is not integrally similar to a companion matrix (the quadratic form $3x^2 + 6xy - 2y^2 = 3(x - y)^2 - 5y^2$ does not represent $\pm 1$ modulo 5). The only case where I have not been able to factor an element of $\pm\mathrm{SL}(2, \mathbb{Z})$ as a product of 3 or fewer integral involutions is that of the matrix $A$ in the previous paragraph. Over the rationals, it factors nicely:

$$\begin{pmatrix} 11 & -7 \\ -36 & 23 \end{pmatrix} = \begin{pmatrix} 781 & 238 \\ -17940/7 & -781 \end{pmatrix} \begin{pmatrix} 23 & 7 \\ -528/7 & -23 \end{pmatrix}.$$

On the other hand, we can express $A$ as a product of elementary matrices by reducing it to $-I_2$ through a sequence of elementary row and column operations, each achieved by multiplying one side or the other by an elementary matrix. Using our earlier remark about products of two elementary matrices, we find that

$$A = \begin{pmatrix} -1 & 0 \\ 3 & 1 \end{pmatrix} \begin{pmatrix} -1 & 4 \\ 0 & 1 \end{pmatrix} \begin{pmatrix} -1 & 0 \\ -3 & 1 \end{pmatrix} \begin{pmatrix} 1 & -1 \\ 0 & -1 \end{pmatrix}.$$

The same procedure expresses any element of $\pm\mathrm{SL}(n, \mathbb{Z})$ as a product of involutions, but more factors will usually be needed, since the reduction to diagonal form will usually take more steps, especially for large $n$. It is not difficult to show that if $A \in \pm\mathrm{SL}(2, \mathbb{Z})$ has top row $(a, b)$, then $A$ is a product of no more than $k + 3$ involutions, where $k$ is the number of steps in the Euclidean algorithm applied to $a$ and $b$.

## REFERENCES

1. C. Ballantine, Some involutory similarities, *Linear Multilinear Algebra* **3** (1975), 19–23; addendum, ibid. **4** (1976), 69.

2. C. Ballantine, Products of involutory matrices I, *Linear Multilinear Algebra* **5** (1977), 53–62.

3. H. Bass, *Some Problems in "Classical" Algebraic K-Theory,* Lecture Notes in Mathematics, Springer-Verlag, New York, 1973, Vol. 342. pp. 3–73.

4. J. Berggren, Finite groups in which every element is conjugate to its inverse, *Pacific J. Math.* **28** (1969), 289–293.

5. J. Berggren, Solvable and supersolvable groups in which every element is conjugate to its inverse, *Pacific J. Math.* **37** (1971), 21–27.

6. D. Djoković, Products of two involutions, *Arch. Math.* **18** (1967), 582–584.

7. E. Ellers, Products of two involutory matrices over skewfields, *Linear Algebra Appl.* **26** (1979), 59–63.

8. R. Guralnick, Similarity of matrices over local rings, *Linear Algebra Appl.* **41** (1981), 161–174.

9. W. Gustafson, P. Halmos, and H. Radjavi, Products of involutions, *Linear Algebra Appl.* **13** (1976), 157–162.

10. P. Halmos, *I Want to be a Mathematician,* Springer-Verlag, New York, 1985.

11. P. Halmos and S. Kakutani, Products of symmetries, *Bull. Am. Math. Soc.* **64** (1958), 77–78.

12. G. Higman, B. Neumann, and H. Neumann, Embedding theorems for groups, *J. London Math. Soc.* **24** (1949), 247–254.

13. F. Hoffman and E. Paige, Products of involutions in the general linear group, *Indiana Univ. Math. J.* **20** (1971), 1017–1020.

14. A. Kerber, *Representations of Permutation Groups I,* Lecture Notes in Mathematics, Springer-Verlag, New York, 1971, Vol. 240.

15. K. Liu, Decompositions of matrices into three involutions, *Linear Algebra Appl.* **111** (1988), 1–24.

16. A. Sampson, A note on a new matrix decomposition, *Linear Algebra Appl.* **8** (1974), 459–463.

17. W. Scott, *Group Theory,* Prentice-Hall, Englewood Cliffs, New Jersey, 1964.

18. A. Sourour, A factorization theorem for matrices, *Linear Multilinear Algebra* **19** (1986), 141–147.

19. W. Waterhouse, Solution of advanced problem 5876, *Am. Math. Monthly* **81** (1974), p. 1035.

20. M. Wonenburger, Transformations which are products of two involutions, *J. Math. Mech.* **16** (1966), 327–338.

*Department of Mathematics*
*Texas Technical University*
*Lubbock, TX 79409*

# Paul Halmos and Toeplitz Operators

## Sheldon Axler

Don Sarason and
Sheldon Axler, 1975

Paul Halmos has written two papers and several snippets about Toeplitz operators. Another of his papers was motivated by a major result in Toeplitz operator theory. This article is the story of Halmos's work on Toeplitz operators and its influence upon the field.

A *Toeplitz matrix* is a matrix that is constant on each line parallel to the main diagonal, so that it looks like this:

$$
\begin{pmatrix}
a_0 & a_{-1} & a_{-2} & a_{-3} & \\
 & & & & \ddots \\
a_1 & a_0 & a_{-1} & a_{-2} & \\
 & & & & \ddots \\
a_2 & a_1 & a_0 & a_{-1} & \\
 & & & & \ddots \\
a_3 & a_2 & a_1 & a_0 & \\
 & \ddots & \ddots & \ddots & \ddots
\end{pmatrix}
\quad (1)
$$

For us, Toeplitz matrices have infinitely many rows and columns, indexed by the nonnegative integers; the entries of the matrix are complex numbers.

When does the Toeplitz matrix (1) represent a bounded linear operator on the usual Hilbert space $\ell^2$ of square-summable sequences? The answer to this question points toward the fascinating connection between Toeplitz operators and function theory. It turns out that the Toeplitz matrix (1) is the matrix of a bounded operator on $\ell^2$ if and only if there is a bounded measurable function on the unit circle whose sequence of Fourier coefficients equals $(a_n)_{n=-\infty}^{\infty}$. Recall that for $f \in L^\infty(\mathbf{T})$ [here $\mathbf{T}$ denotes the unit circle in the complex plane and $L^\infty(\mathbf{T})$ denotes the set of bounded, complex-valued, measurable functions on $\mathbf{T}$], the $n$th *Fourier coefficient of $f$* is defined by

$$a_n = \int_0^{2\pi} f(e^{i\theta})e^{-in\theta}\, d\theta/(2\pi). \tag{2}$$

Thus the Toeplitz matrix (1) is the matrix of a bounded operator on $\ell^2$ if and only if there exists $f \in L^\infty(\mathbf{T})$ such that (2) holds for every integer $n$.

For $f \in L^\infty(\mathbf{T})$, the *Toeplitz operator with symbol* $f$, denoted $T_f$, is the operator whose matrix is (1), where $a_n$ is the $n$th Fourier coefficient of $f$. The connection between Toeplitz operators and analytic function theory is now close at hand. A function $f \in L^\infty(\mathbf{T})$ with Fourier coefficients $(a_n)_{n=-\infty}^\infty$ is called *analytic* if $a_n = 0$ for all $n < 0$. The reason for this terminology is that the Fourier series of $f$, which is the formal sum

$$\sum_{n=-\infty}^\infty a_n z^n,$$

is the Taylor series expansion of a function analytic on the unit disk if $a_n = 0$ for all $n < 0$:

$$\sum_{n=0}^\infty a_n z^n. \tag{3}$$

Note that a function $f \in L^\infty(\mathbf{T})$ is analytic if and only if the Toeplitz matrix corresponding to $T_f$ is a lower triangular matrix.

Halmos's first paper on Toeplitz operators was a joint effort with Arlen Brown that was published in 1964 [4]. Brown and Halmos emphasized the difficulties flowing from the observation that the linear map $f \mapsto T_f$ is not multiplicative ($T_f T_g$ is not necessarily equal to $T_{fg}$). They proved that for $f, g \in L^\infty(\mathbf{T})$, the product $T_f T_g$ equals $T_{fg}$ if and only if either $\bar{f}$ or $g$ is analytic.

Among many other results, Brown and Halmos showed that the spectrum of a Toeplitz operator cannot consist of exactly two points (the *spectrum* of a linear operator $S$ is the set of complex numbers $\lambda$ such that $S - \lambda I$ is not invertible; here $I$ denotes the identity operator). In the best Halmosian tradition, the Brown–Halmos paper contains an open problem stated as a yes–no question: Can the spectrum of a Toeplitz operator consist of exactly three points? A bit later, in [7] (which was written after the Brown–Halmos paper, although published slightly earlier) Halmos asked a bolder question: Is the spectrum of every Toeplitz operator connected? Considering what was known at the time, this has always struck me as an audacious question. The answer was known to be yes when the symbols are required to be analytic or real valued, but these are extremely special and unrepresentative cases. For the general complex-valued function, even the possibility that the spectrum could consist of exactly three points had not been eliminated. Nevertheless, Harold Widom [15] soon proved that every Toeplitz operator has connected spectrum. Ron Douglas has written that Widom's proof is unsatisfactory because "the proof gives us no hint as to why the result is true" [6, page 196], but no alternate proof has been found.

The Brown–Halmos paper set the tone for much of the later work on Toeplitz operators. Most of the results in the paper now seem easy, perhaps because in 1967 Halmos incorporated them into the chapter on Toeplitz operators in his marvelous and unique *Hilbert Space Problem Book* [8], from which several generations of operator theorists have learned the tools of the trade. Papers have been published in the 1960s, 1970s, 1980s, and 1990s that extend and generalize results that first appeared in the Brown–Halmos paper. Although it is probably the most cited paper ever written on Toeplitz operators, Halmos records in his automathography [12, pages 319–321] that it was rejected by the first journal to which it was submitted.

In 1970 Halmos gave a series of lectures [9] in which he posed ten open problems in operator theory. One of them dealt with Toeplitz operators. A Toeplitz operator is called *analytic* if its symbol is analytic. Halmos asked whether every subnormal Toeplitz operator is normal or analytic. Recall that a linear operator $S$ is called *normal* if it commutes with its adjoint ($S^*S = SS^*$). An operator $S$ on a Hilbert space $H$ is called *subnormal* if there is a Hilbert space $K$ containing $H$ and a normal operator $N$ on $K$ such that $N|_H = S$. In [4], Brown and Halmos had completely described the normal Toeplitz operators. Clearly every normal operator is subnormal (take $K = H$), and it is also easy to see that every analytic Toeplitz operator is subnormal. Halmos's question asks whether these are the only subnormal Toeplitz operators.

In 1979 Halmos described [10] what had happened to the ten problems in the intervening years. The problem about Toeplitz operators was still unsolved, but good work had been done on the problem, and strong evidence indicated that the question had an affirmative answer. In the spring of 1983 I believed that the time was right for a breakthrough on this problem, and so I organized a seminar about it at Michigan State University. We went through every paper on the topic, including the first draft of a manuscript by the Chinese mathematician Sun Shun-Hua. Sun claimed to have proved that no nonanalytic Toeplitz operator can lie in a certain important subclass of the subnormal operators. There was an uncorrectable error in the proof (and the result is false), but Sun introduced clever new ideas to the subject. His proof worked for all but a single family of operators, and so this particular family was an excellent candidate for a counter-example that no one expected to exist.

The spring quarter ended and I sent a copy of my seminar notes to Carl Cowen at Purdue University. When I returned from a long trip abroad, I found a letter from Cowen, who had amazingly answered Halmos's question (negatively!) by proving that each operator in the suspicious family is a subnormal Toeplitz operator that is neither normal nor analytic. Cowen's result is pretty and can be easily described. Let $b$ be a real number with $0 < b < 1$. Let $f$ be a one-to-one analytic mapping of the unit disk onto the ellipse with vertices $1/(1 + b)$, $-1/(1 + b)$, $i/(1 - b)$, and $-i/(1 - b)$. Cowen proved that the Toeplitz operator with symbol $f + b\overline{f}$ is subnormal, but that it is neither normal nor analytic.

I told my Ph.D. student John Long about Cowen's wonderful result, although I did not show Long the proof. Within a week, Long came back with a beautiful and deep proof that was shorter and more natural than Cowen's. Since there was no reason to publish Cowen's original proof, Cowen and Long decided to publish Long's proof in a joint paper [5]. So the contributions to that paper are as follows: Cowen first proved the result and provided the crucial knowledge of the correct answer (including the idea of using ellipses); the proof in the paper is due to Long. At no time did the two co-authors actually work together.

Halmos's second paper on Toeplitz operators was a joint effort with José Barría that was published in 1982 [3]. The main object of investigation in this paper is $\mathcal{T}$, which is the norm-closed algebra of operators generated by $\{T_f : f \in L^\infty(\mathbf{T})\}$. Perhaps the most important tool in the study of $\mathcal{T}$ is the existence of a homomorphism:

$$\varphi : \mathcal{T} \to L^\infty(\mathbf{T}),$$

such that $\varphi(T_f) = f$ for every $f \in L^\infty(\mathbf{T})$. The key point here is that $\varphi$ is multiplicative, so that $\varphi(T_f T_g) = \varphi(T_f)\varphi(T_g) = fg$. The homomorphism $\varphi$ was discovered and exploited by Douglas [6, Chapter 7].

The map $\varphi$ was a magical and mysterious homomorphism to me until I read the Barría–Halmos paper, where the authors actually construct $\varphi$. Here's how they do it: If $S$ happens to be a Toeplitz operator, then the matrix of $S$ is constant along each line parallel to the main diagonal, as in the Toeplitz matrix (1). Barría and Halmos prove that if $S \in \mathcal{T}$, then $S$ is an asymptotic Toeplitz operator in the sense that in the matrix of $S$, the limit (moving to the south-east) along each line parallel to the main diagonal exists. In the matrix of $S$, for each line parallel to the main diagonal, replace each entry in that line with the limit along that line. The matrix thus constructed is constant along each line parallel to the main diagonal, and thus is a Toeplitz matrix. The nature of the construction insures that this Toeplitz matrix represents a bounded operator, and so is equal to $T_f$ for some $f \in L^\infty(\mathbf{T})$. Starting with $S \in \mathcal{T}$, we have now obtained a function $f \in L^\infty(\mathbf{T})$; the process is completed by defining $\varphi(S)$ to be $f$.

The Barría–Halmos construction of $\varphi$ is completely different in spirit and technique from Douglas's proof that $\varphi$ exists. I knew Douglas's proof well (an idea that I got from reading it was a key ingredient in my first published paper [1]), but until the Barría–Halmos paper came along, I never guessed that $\varphi$ could be explicitly constructed or that so much additional insight could be squeezed from a new approach.

As mentioned earlier, in their classic paper Brown and Halmos [4] gave a necessary and sufficient condition on functions $f, g \in L^\infty(\mathbf{T})$ for $T_f T_g - T_{fg}$ to equal 0. One of the fruitful strands of generalization stemming from this result involves asking for $T_f T_g - T_{fg}$ to be small in some sense. In this context, the most useful way an operator can be small is to be compact.

[Recall that a linear operator on a Hilbert space is called *compact* if the operator maps the closed unit ball onto a compact set. The set of compact operators equals the norm closure of the bounded operators with finite dimensional range.] In 1978 Sun-Yung Alice Chang, Don Sarason, and I published a paper [2] in which we gave a sufficient condition on functions $f, g \in L^\infty(\mathbf{T})$ for $T_f T_g - T_{fg}$ to be compact. This condition included all previously known sufficient conditions. To describe this condition, let $H^\infty$ denote the set of all analytic functions in $L^\infty(\mathbf{T})$ and for $f \in L^\infty(\mathbf{T})$ let $H^\infty[f]$ denote the smallest closed subalgebra of $L^\infty(\mathbf{T})$ containing $H^\infty$ and $f$. Let $f, g \in L^\infty(\mathbf{T})$. What we proved was that if

$$H^\infty[\overline{f}] \cap H^\infty[g] \subset H^\infty + C(\mathbf{T}), \tag{4}$$

then $T_f T_g - T_{fg}$ is compact [here $C(\mathbf{T})$ denotes the set of continuous, complex-valued functions on $\mathbf{T}$].

We could prove that our condition (4) was necessary as well as sufficient if we put some additional hypotheses on $f$ and $g$. We conjectured that our condition was necessary without the additional hypotheses, but we were unable to prove so. A brilliant proof verifying the conjecture was published by Alexander Volberg [14] in 1982. A key step in Volberg's proof uses the following specific case of a theorem about interpolation of operators that had been proved 26 years earlier by Elias Stein [13, Theorem 2]: Let $d\mu$ be a positive measure on some space $X$ and let $v$ and $w$ be positive measurable functions on $X$. Suppose that $S$ is a linear operator on both $L^2(v\, d\mu)$ and $L^2(w\, d\mu)$, with norms $\|S\|_v$ and $\|S\|_w$, respectively. Then the norm $\|S\|_{\sqrt{vw}}$ of $S$ on the space $L^2(\sqrt{vw}\, d\mu)$ satisfies the inequality

$$\|S\|_{\sqrt{vw}} \leq \sqrt{\|S\|_v \, \|S\|_w}.$$

I had a preprint of Volberg's paper, and in the spring of 1981 I told Halmos about the special interpolation result that it used. Within a few days Halmos surprised me by producing a clean Hilbert space proof of the interpolation result that Volberg had needed. With his typical efficiency, Halmos put his inspiration into publishable form quickly and submitted the paper to the journal to which Volberg had sent his article. I was a referee for both papers; it was an unusual pleasure to see how a tool used in one paper had led to an improved proof of the tool. Halmos's short and delightful paper [11] containing his proof of the interpolation result appeared in the same issue of the *Journal of Operator Theory* as Volberg's paper.

I would like to close with a few words about my personal debt to Paul Halmos. Paul is my mathematical grandfather. His articles and books have been an important part of my mathematical heritage. I first met Paul for a few seconds when I was a graduate student, and then for a few minutes when I gave my first conference talk right after receiving my Ph.D. degree. Four years later I got to know Paul well when I spent a year's leave at

Indiana University. Paul is one of the three people who showed me how to be a mathematician (the other two are my thesis advisor Don Sarason and the late Allen Shields). Watching Paul, I saw how an expert proved a theorem, gave a talk, wrote a paper, composed a referee's report, and edited a journal. I'm lucky to have had such an extraordinary model.

## REFERENCES

1. Sheldon Axler, Factorization of $L^\infty$ functions, *Ann. Math.* **106** (1977), 567–572.

2. Sheldon Axler, Sun-Yung A. Chang, and Donald Sarason, Products of Toeplitz operators, *Integral Equations Operator Theory* **1** (1978), 285–309.

3. José Barría and P.R. Halmos, Asymptotic Toeplitz operators, *Trans. Am. Math. Soc.* **273** (1982), 621–630.

4. Arlen Brown and P.R. Halmos, Algebraic properties of Toeplitz operators, *J. Reine Angew. Math.* **213** (1964), 89–102.

5. Carl C. Cowen and John J. Long, Some subnormal Toeplitz operators, *J. Reine Angew. Math.* **351** (1984), 216–220.

6. Ronald G. Douglas, *Banach Algebra Techniques in Operator Theory*, Academic Press, New York, 1972.

7. P.R. Halmos, A glimpse into Hilbert space, *Lectures on Modern Mathematics*, Vol. I, edited by T.L. Saaty, John Wiley and Sons, New York, 1963, 1–22.

8. P.R. Halmos, *A Hilbert Space Problem Book*, Van Nostrand Company, Princeton, 1967.

9. P.R. Halmos, Ten problems in Hilbert space, *Bull. Am. Math. Soc.* **76** (1970), 887–993.

10. P.R. Halmos, Ten years in Hilbert space, *Integral Equations Operator Theory* **2** (1979), 529–564.

11. P.R. Halmos, Quadratic interpolation, *J. Operator Theory* **7** (1982), 303–305.

12. P.R. Halmos, *I Want to Be a Mathematician*, Springer-Verlag, New York, 1985.

13. Elias M. Stein, Interpolation of linear operators, *Trans. Am. Math. Soc.* **83** (1956), 482–492.

14. A.L. Volberg, Two remarks concerning the theorem of S. Axler, S.-Y.A. Chang and D. Sarason, *J. Operator Theory* **7** (1982), 209–218.

15. Harold Widom, On the spectrum of a Toeplitz operator, *Pacific J. Math.* **14** (1964), 365–375.

*Department of Mathematics*
*Michigan State University*
*East Lansing, MI 48824*

# Lebesgue's "Fundamental Theorem of Calculus" Revisited

## S.K. Berberian[1]

S.K. Berberian,
1961

## 0. Introduction

As the title attempts to suggest, this article is written for readers who (like the author) have been there, liked what they saw, but feel there ought to be a better way to get there. Concisely, the theorem in question states that

$$\int_a^b f \, d\lambda = F(b) - F(a), \qquad (*)$$

where $\lambda$ is Lebesgue measure on the closed interval $[a, b]$, $f$ is Lebesgue-integrable,[2] $F$ is absolutely continuous, and $F'(x) = f(x)$ for almost every $x$ in the open interval $(a, b)$. Apart from the definition of the integral, the key concepts here are "almost everywhere" and "absolute continuity":

(a.e.) A statement $P(x)$, defined for $x$ in a set S of real numbers, is said to be true *almost everywhere* (briefly, a.e.) in S if the set $N = \{x \in S : P(x)$ false $\}$ is negligible (that is, for every $\varepsilon > 0$ there exists a covering of N by a sequence of intervals whose total length is $< \varepsilon$).

(AC) A function $F : [a, b] \to \mathbb{R}$ (where $\mathbb{R}$ is the field of real numbers and $[a, b]$ is a closed interval of $\mathbb{R}$, is said to be *absolutely continuous* (briefly, AC) if, for every $\varepsilon > 0$, there exists a $\delta > 0$ such that, for finite lists $[a_1, b_1], \ldots, [a_n, b_n]$ of nonoverlapping closed subintervals of $[a, b]$,

$$\sum (b_k - a_k) < \delta \Rightarrow \sum |F(b_k) - F(a_k)| < \varepsilon.$$

---

[1]Presented in outline at the MAA Texas sectional meeting at the University of North Texas in Denton, April 6, 1990.

[2]Defined via simple functions (as in [H] or [B]); to simplify questions of algebraic operations, we require that integrable functions be finite-valued (as in [B]).

Suppose $f:[a,b] \to \mathbb{R}$ and $F:[a,b] \to \mathbb{R}$. We say that $F$ is a *primitive* for $f$ (in the sense of Lebesgue)[3] if (i) $F$ is absolutely continuous, and (ii) for almost every $x$ in $(a,b)$, $F$ is differentiable at $x$ and $F'(x) = f(x)$. Lebesgue's "Fundamental theorem of calculus" can be expressed succinctly (in four bites) as follows, arranged in the proposed order of proof:

**(I)** *$F$ is absolutely continuous if and only if there exists a Lebesgue-integrable function $f$ such that*

$$F(x) = F(a) + \int_a^x f \, d\lambda \quad \text{for all} \quad x \in [a,b].$$

**(II)** *If $F$ is absolutely continuous and if, for almost every $x \in (a,b)$, $F$ is differentiable with $F'(x) = 0$, then $F$ is a constant function* (briefly, the primitives for the zero function are the constant functions).

**(III)** *If $f$ is Lebesgue-integrable then the indefinite integral*

$$F(x) = \int_a^x f \, d\lambda \quad (a \le x \le b)$$

*is a primitive for $f$.*

**(IV)** *$f$ is Lebesgue-integrable if and only if it has a primitive.*

Item (IV) characterizes the class of Lebesgue-integrable functions independently of the concept of measure; all one needs are the concepts of negligible set, absolutely continuous function, and derivative.

Item (I) says, in effect, that the indefinite integrals in the Lebesgue theory are precisely the absolutely continuous functions that vanish at the left endpoint.

Item (III) is an existence theorem (a primitive for $f$ is exhibited) and (II) is a uniqueness theorem (any two primitives for $f$ differ only by a constant); together, they imply at once the formula ($*$) mentioned at the beginning.

Previewing the proofs, after noting that an AC function maps negligible sets to negligible sets, item (I) follows easily from the Radon–Nikodym theorem for totally finite measures; the proof is given in the next section. Item (II) is the most accessible of the four; a proof (patterned on an argument of E.J. McShane) is given in Section 2, using nothing fancier than the Jordan decomposition and the mapping theorem for negligible sets just mentioned. Item (III) is genuinely difficult; a strategy for a reasonably intuitive proof (again leaning on McShane's exposition) is sketched in Section 3. Item (IV) follows from (I) and (III); here's the easy proof:

---

[3]Lebesgue thought the concept important enough to be included in the title of his book [L].

**(I) & (III)** $\Rightarrow$ **(IV):** If $f$ is Lebesgue-integrable then it has a primitive by (III). Conversely, assuming $F$ is a primitive for a function $f\colon [a, b] \to \mathbb{R}$, we have to show that $f$ is Lebesgue-integrable. By (I), the function $x \mapsto F(x) - F(a)$ is the indefinite integral of a Lebesgue-integrable function $g$, and $F' = g$ a.e. by (III), thus $f = g$ a.e.; since $g$ is Lebesgue-integrable, so is $f$.   $\Diamond$

Implicit in the preceding argument is that *every absolutely continuous function is differentiable almost everywhere.* More generally, a continuous function of bounded variation can be shown to be differentiable a.e. by a recursive application of F. Riesz's "Rising sun lemma" [SzN, §3.1.3, pp. 107–111]; in fact, *every* function of bounded variation is differentiable a.e. (proved using the Vitali covering theorem in [H&S] and [R], and by other means in [McS, p. 205, 34.3] and [SzN, p. 114]). Neither of these generalizations is needed for Lebesgue's fundamental theorem of calculus.

# 1. Absolute Continuity; Proof of (I)

By the Jordan decomposition, a function $F\colon [a, b] \to \mathbb{R}$ of bounded variation has a canonical decomposition $F = P - N$, where $P$ and $N$ are increasing, and when $F$ is continuous (or absolutely continuous) the same is true of $P$ and $N$ ([SzN, p. 97], [H&S, p. 283, 18.13]).

**Lemma 1.** *If $F = G - H$, where $G$ and $H$ are increasing continuous functions on $[a, b]$, then, for every closed subinterval I of $[a, b]$,*

$$\lambda(F(\mathrm{I})) \leq \lambda(G(\mathrm{I})) + \lambda(H(\mathrm{I})).$$

**Proof.** Since $F$ is continuous, $F(\mathrm{I})$ is a closed interval, say $F(\mathrm{I}) = [F(c), F(d)]$, where $c, d \in \mathrm{I}$. Let J be the closed subinterval of I with endpoints $c$ and $d$ (whose order we do not know). Since $G$ is continuous and monotone, $G(\mathrm{J})$ is the closed interval with endpoints $G(c)$ and $G(d)$, therefore $\lambda(G(\mathrm{J})) = |G(d) - G(c)|$; similarly, $\lambda(H(\mathrm{J})) = |H(d) - H(c)|$, therefore

$$\begin{aligned}
\lambda(F(\mathrm{I})) = F(d) - F(c) &= [G(d) - G(c)] - [H(d) - H(c)] \\
&\leq |G(d) - G(c)| + |H(d) - H(c)| \\
&= \lambda(G(\mathrm{J})) + \lambda(H(\mathrm{J})) \leq \lambda(G(\mathrm{I})) + \lambda(H(\mathrm{I})).   \quad \Diamond
\end{aligned}$$

**Lemma 2.** *If $F\colon [a, b] \to \mathbb{R}$ is absolutely continuous and N is a negligible subset of $[a, b]$, then $F(\mathrm{N})$ is negligible.*

**Proof.** Write $F = G - H$ with $G, H$ increasing and absolutely continuous. Given any $\varepsilon > 0$, choose $\delta > 0$ in the definition of absolute continuity so as to "work" for both $G$ and $H$. Since N is negligible, there is a sequence of intervals $\mathrm{I}_k = [a_k, b_k]$ with $\mathrm{N} \subset \bigcup \mathrm{I}_k$ and $\sum(b_k - a_k) < \delta$. The intervals

$I_k$ can be taken to be nonoverlapping ("disjointify" in the usual way, then restore any missing endpoints). Replacing $I_k$ by $I_k \cap [a, b]$, we can suppose further that $I_k \subset [a, b]$; then

$$G(N) \subset \bigcup G(I_k) = \bigcup [G(a_k), G(b_k)].$$

For each positive integer $n$, the nonoverlapping intervals $I_1, \ldots, I_n$ have total length $< \delta$, therefore $\sum_1^n \lambda(G(I_k)) < \varepsilon$ by the choice of $\delta$; since $n$ is arbitrary, $\sum \lambda(G(I_k)) \leq \varepsilon$. Similarly $\sum \lambda(H(I_k)) \leq \varepsilon$; since

$$\lambda(F(I_k)) \leq \lambda(G(I_k)) + \lambda(H(I_k))$$

by Lemma 1, we conclude that $\sum \lambda(F(I_k)) \leq 2\varepsilon$. Thus, for every $\varepsilon > 0$, we have found a covering $F(N) \subset \bigcup F(I_k)$ of $F(N)$ by intervals $F(I_k)$ of total length $\leq 2\varepsilon$, therefore $F(N)$ is negligible.   $\diamond$

**Theorem 1.** *A function $F: [a, b] \to \mathbb{R}$ is absolutely continuous if and only if there exists a Lebesgue-integrable function $f: [a, b] \to \mathbb{R}$ such that*

$$F(x) = F(a) + \int_a^x f \, d\lambda \quad \text{for all } x \in [a, b].$$

**Proof.** "If": Suppose $F$ has a representation of the indicated sort. If $a \leq x < y \leq b$ then

$$F(y) - F(x) = \int_x^y f \, d\lambda = \int_{[x,y]} f \, d\lambda.$$

If $f$ is bounded, say $|f| \leq K$, then $|F(y) - F(x)| \leq K|x - y|$, therefore $F$ is absolutely continuous (even Lipschitz).

For general $f$, writing $f = f^+ - f^-$ we can suppose that $f \geq 0$ (so that $F$ is increasing). Given any $\varepsilon > 0$, choose a simple function $g$ such that $0 \leq g \leq f$ and

$$\int_a^b (f - g) d\lambda \leq \varepsilon.$$

Since $g$ is bounded (and $\geq 0$), its indefinite integral

$$G(x) = \int_a^x g \, d\lambda$$

is absolutely continuous (and increasing) by the preceding discussion; choose $\delta > 0$ to go along with $\varepsilon$ and $G$ in the definition of absolute continuity. If $a \leq x < y \leq b$ then

$$|F(y) - F(x)| = F(y) - F(x) = \int_x^y f \, d\lambda$$

$$= \int_x^y (f - g) d\lambda + \int_x^y g \, d\lambda$$

$$= \int_{[x,y]} (f - g) d\lambda + |G(y) - G(x)|.$$

Let $I_k = [a_k, b_k]$ $(k = 1, \ldots, n)$ be nonoverlapping subintervals of $[a, b]$ with $\sum \lambda(I_k) < \delta$ and let $A = \bigcup I_k$. By the above equations,

$$|F(b_k) - F(a_k)| = \int_{[a_k, b_k]} (f - g)d\lambda + |G(b_k) - G(a_k)|;$$

summing over $k$ yields

$$\sum |F(b_k) - F(a_k)| = \int_A (f - g)d\lambda + \sum |G(b_k) - G(a_k)|$$

(because integration is an additive set function and singletons are negligible), therefore

$$\sum |F(b_k) - F(a_k)| \leq \int_a^b (f - g)d\lambda + \sum |G(b_k) - G(a_k)| \leq \varepsilon + \varepsilon$$

by the choice of $g$ and $\delta$. This shows that $F$ is absolutely continuous.

"Only if": Assuming $F$ is absolutely continuous, we have to show that the function $x \mapsto F(x) - F(a)$ is an indefinite integral. By the Jordan decomposition, we can suppose that $F$ is increasing; replacing $F$ by the function $x \mapsto F(x) + x$, we can suppose that $F$ is strictly increasing. Then $F: [a, b] \to [F(a), F(b)]$ is a homeomorphism, therefore preserves Borel sets. Define a (finite) measure $\mu$ on the Borel sets of $[a, b]$ by the formula $\mu(B) = \lambda(F(B))$. By Lemma 2, we know that

$$\lambda(B) = 0 \Rightarrow \mu(B) = 0$$

(that is, $\mu$ 'absolutely continuous' with respect to $\lambda$ in the sense of measures); by the Radon–Nikodym theorem [H, p. 128, 31.B], there exists a Lebesgue-integrable Borel function $f$ such that

$$\mu(B) = \int_B f \, d\lambda$$

for all Borel sets B, and application of the formula to the sets B $= [a, x]$ yields the desired representation of $F$.   ◊

# 2. Criteria for Monotonicity: Proof of (II)

The conclusion in (II) is that certain functions are constant. The constant functions are the functions that are both increasing and decreasing; this is how monotonicity enters the present discussion. In particular, (II) will be obtained as a corollary of the following theorem, inspired by an argument in [McS, p. 200, 34.1]:

**Theorem 2.** *If $F: [a, b] \to \mathbb{R}$ is continuous and if there exists a set N $\subset$ $[a, b]$ such that*

(i) $F(N)$ *has empty interior, and*

(ii) *for every* $x \in (a, b) - N$, $F$ *is differentiable at* $x$ *with* $F'(x) > 0$, *then* $F$ *is increasing. If, in addition,*

(iii) $N$ *has empty interior,*

*then* $F$ *is strictly increasing.*

**Proof.** Assuming (i) and (ii), suppose to the contrary that there exist points $c$, $d$ such that $a \leq c < d \leq b$ and $F(c) > F(d)$. Since $F(N)$ has empty interior $(F(d), F(c)) \not\subset F(N)$; choose $k \in (F(d), F(c))$ such that $k \notin F(N)$, thus

$$F(d) < k < F(c) \quad \text{and} \quad (\forall x \in N) \quad F(x) \neq k.$$

The set $S = \{x \in [c, d]: F(x) \geq k\}$ is nonempty ($c \in S$) and bounded; let $s = \sup S$. By the continuity of $F$, $F(s) \geq k$ that is, $s \in S$. It follows that $s \neq d$ (because $F(d) < k$), therefore $s < d$. Choose a sequence $x_n \in (s, d)$ such that $x_n \to s$. Since $x_n > s = \sup S$, necessarily $x_n \notin S$, therefore $F(x_n) < k$; passing to the limit, $F(s) \leq k$ and therefore $F(s) = k$. Then

$$F(d) < F(s) < F(c) \quad \text{and} \quad F(x_n) < F(s);$$

in particular $s \neq c$, thus $c < s < d$. Since $F(s) = k$, it follows from the choice of $k$ that $s \notin N$; by (ii), $F'(s) > 0$. But

$$\frac{F(x_n) - F(s)}{x_n - s} < 0 \text{ for all } n,$$

and passage to the limit yields $F'(s) \leq 0$, a contradiction.

We now know that $F$ is increasing. If $F$ is not strictly increasing, then it is constant on an interval $(c, d)$, and $(c, d) \subset N$ because $F' = 0$ on $(c, d)$; this is ruled out by (iii). ◇

**Corollary 1.** *If* $F: [a, b] \to \mathbb{R}$ *is absolutely continuous and if, for almost every* $x \in (a, b)$, $F$ *is differentiable with* $F'(x) > 0$, *then* $F$ *is strictly increasing.*

**Proof.** By hypothesis, there exists a negligible set $N \subset (a, b)$ such that, at every $x \in (a, b) - N$, $F$ is differentiable and $F'(x) > 0$. Since $F$ is AC, $F(N)$ is also negligible (§1, Lemma 2). A negligible set obviously has empty interior, so $F$ is strictly increasing by the theorem. ◇

**Corollary 2.** *Suppose* $F: [a, b] \to \mathbb{R}$ *is absolutely continuous.*

(1) *If, for almost every* $x \in (a, b)$, $F$ *is differentiable with* $F'(x) \geq 0$, *then* $F$ *is increasing.*

(2) *If, for almost every* $x \in (a, b)$, *F is differentiable with* $F'(x) = 0$, *then F is constant.*

**Proof.** (1) Assuming $a \leq c < d \leq b$, we must show that $F(c) \leq F(d)$. For each $r > 0$, let $F_r : [a, b] \to \mathbb{R}$ be the (absolutely continuous) function $F_r(x) = F(x) + rx$. By hypothesis, $(F_r)'(x) = F'(x) + r \geq r > 0$ for almost every $x \in (a, b)$, so $F_r$ is strictly increasing by Corollary 1. In particular, $F_r(c) < F_r(d)$, that is, $F(c) + rc < F(d) + rd$; letting $r \to 0$ we have $F(c) \leq F(d)$.

(2) By (1), $F$ is increasing; but $-F$ also satisfies the hypotheses of (2), so $-F$ is also increasing, whence $F$ is constant.    ◇

In particular, (2) of Corollary 2 proves item (II) of the Introduction. The next two corollaries are not needed for the Lebesgue theory but they push Theorem 2 in an interesting direction [BOU, Ch. I, §2, no. 2 and Ch. II, §1]:

**Corollary 3.** *If* $F : [a, b] \to \mathbb{R}$ *is continuous and if, for all but countably many* $x \in (a, b)$, *F is differentiable with* $F'(x) > 0$, *then F is strictly increasing.*

**Proof.** By assumption, there exists a countable set $N \subset (a, b)$ such that, for every $x \in (a, b) - N$, $F$ is differentiable and $F'(x) > 0$. Since $F(N)$ is also countable, and countable sets have empty interior, $F$ is strictly increasing by the theorem.    ◇

**Corollary 4.** *Suppose* $F : [a, b] \to \mathbb{R}$ *is continuous.*

(1) *If, for all but countably many* $x \in (a, b)$, *F is differentiable with* $F'(x) \geq 0$, *then F is increasing.*

(2) *If, for all but countably many* $x \in (a, b)$, *F is differentiable with* $F'(x) = 0$, *then F is constant.*

**Proof.** The basic format is the same as for Corollary 2, with "absolutely continuous" replaced by "continuous," and "almost every" replaced by "all but countably many."    ◇

Finally, the arguments of Theorem 2 can be refined so as to yield the following generalization of a theorem in [McS, p. 200, 34.1]:

**Theorem.** *Let* $F : [a, b] \to \mathbb{R}$, *let* D *be one of the derivate operators[4]* $D^+$, $D_+$, $D^-$, $D_-$ *and let* $N = \{x \in (a, b) : (DF)(x) \leq 0\}$. *If*

(i) $F(x) \leq \liminf_{t \to x+} F(t)$ *for all* $x \in [a, b)$,

---

[4]For example, $(D^+ F)(x)$ is the upper right-hand derivate of $F$ at $x$ ([McS, p. 188, 31.1], [H&S, p. 257, 17.2]).

(ii) $\limsup_{t \to x^-} F(t) \leq F(x)$ *for all* $x \in (a, b]$, *and*

(iii) $F(N)$ *has empty interior,*

*then* $F$ *is increasing. If, in addition,*

(iv) N *has empty interior,*

*then* $F$ *is strictly increasing.*

**Corollary.** *Suppose* $F: [a, b] \to \mathbb{R}$ *satisfies* (i) *and* (ii) *of the Theorem, and* D *is one of the four derivate operators. If* $(DF)(x) > 0$ ($\geq 0$, $= 0$) *for all but countably many* $x \in (a, b)$, *then* $F$ *is strictly increasing* (*increasing, constant*).

This corollary (the "increasing" part, with empty exceptional set) plays a role in McShane's proof that every function of bounded variation is differentiable almost everywhere [McS, p. 207, 34.2].


# 3. Proof of (III)

The theorem of this section—*an indefinite integral is a primitive of its integrand*—completes the program outlined in the Introduction.

A quick sketch of the strategy: A Lebesgue-integrable function is approximable by simple functions (linear combinations of characteristic functions of measurable sets); a measurable set is approximable by open supersets and closed subsets; the characteristic functions of open sets and closed sets are semicontinuous. The theorem on primitives is elementary for a *continuous* integrand; the basic trick is to make do with semicontinuous approximations of the integrand (at the cost of introducing a negligible set).

In essence, the proof follows that in [McS, p. 198, 33.3]. The main differences are due to the fact that many of the details in McShane's exposition are absorbed in the definition of integral [McS, p. 72, 14.1s and p. 75, 15.1s]; this has the advantage of shortening the proof of the theorem on primitives, at the cost, however, of limiting the definition of integral to functions defined on intervals of the line (or at least on topological spaces). In contrast, the present exposition assumes that the general machinery of integration has already been developed, via measures and simple functions (cf. [H], [B]), but no further than the monotone convergence theorem (and its easy consequence, the Radon–Nikodym theorem for finite measures).

**Lemma 1.** *Let* $f: [a, b] \to \mathbb{R}$ *be Lebesgue-integrable and let* $\varepsilon > 0$. *There exist a Lebesgue-integrable function* $h: [a, b] \to \mathbb{R}$ *and a lower semicontinuous*[5] *function* $k: [a, b] \to \mathbb{R} \cup \{+\infty\}$ *such that*

---

[5][H&S, p. 88, 7.20], [McS, p. 39].

(i) $h = k$ a.e., $f \leq k$ everywhere on $[a, b]$ (therefore $f \leq h$ a.e.), and

(ii) $\int_a^b h\, d\lambda \leq \int_a^b f\, d\lambda + \varepsilon$.

**Proof.** The message of the lemma is that the integral of $f$ can be approximated by the integrals of functions $h$ that are equal a.e. to a lower semicontinuous majorant of $f$. {If one allows integrable functions with infinite values, then $f$ is the limit in mean of its l.s.c. majorants. The functions $k$ are "$U$-functions" in the sense of [McS], where they figure in the definition of integrability.}

The lemma is proved by reduction to special cases: in increasing order of generality, (a) $f$ the characteristic function of a measurable set; (b) $f$ simple and $\geq 0$; (c) $f$ integrable and $\geq 0$; (d) $f$ integrable (the general case).

(a) Say $f = \varphi_E$, $E \subset [a, b]$ Lebesgue-measurable. By the regularity of Lebesgue measure, there exist a closed set K and an open set U of $\mathbb{R}$ with $K \subset E \subset U$ and $\lambda(U - K) < \varepsilon$ (here $\lambda$ is not confined to subsets of $[a, b]$). The characteristic function of $U \cap [a, b]$ (as a function on $[a, b]$) meets the requirements for $h$ and $k$. The characteristic function of K, which is an upper semicontinuous minorant of $f$, plays a role later in the proof; for the moment, we note that

$$\lambda(E) \leq \lambda(U) \leq \lambda(U - K) + \lambda(K) < \varepsilon + \lambda(K),$$

therefore

$$\int \varphi_K d\lambda > \int f\, d\lambda - \varepsilon.$$

(To simplify the notation, we omit the limits of integration when they are $a$ and $b$.)

(b) Assuming $f$ is simple and $\geq 0$, write

$$f = c_1 \varphi_{E_1} + \cdots + c_n \varphi_{E_n},$$

where the $E_i$ are pairwise disjoint measurable sets and the $c_i$ are $> 0$. For each $i$, $K_i \subset E_i \subset U_i$ with $K_i$ closed, $U_i$ open in $\mathbb{R}$, and $\lambda(U_i - K_i) < \varepsilon/nc_i$. Writing $A_i = U_i \cap [a, b]$, the function

$$c_1 \varphi_{A_1} + \cdots + c_n \varphi_{A_n}$$

meets the requirements for $h$ and $k$. For use in part (c), we note that the function

$$g = c_1 \varphi_{K_1} + \cdots + c_n \varphi_{K_n}$$

is an upper semicontinuous minorant of $f$ such that

$$\int g\, d\lambda > \int f\, d\lambda - \varepsilon.$$

(c) Choose an increasing sequence of simple functions $f_n \geq 0$ such that $f_n(x) \uparrow f(x)$ for every $x \in [a, b]$. By (b), there exist simple functions $g_n$ and $h_n$, with $g_n$ u.s.c. and $h_n$ l.s.c., such that $0 \leq g_n \leq f_n \leq h_n$ (everywhere) and

$$\int (h_n - g_n)d\lambda < \varepsilon/2^n \text{ for all } n.$$

Define the pointwise suprema

$$G_n = \sup(g_1, \ldots, g_n), \quad H_n = \sup(h_1, \ldots, h_n);$$

$G_n$ is u.s.c., $H_n$ is l.s.c.,

$$0 \leq g_n \leq G_n \leq \sup(f_1, \ldots, f_n) = f_n \leq h_n \leq H_n$$

and

$$0 \leq \int f_n d\lambda - \int G_n d\lambda \leq \int (h_n - g_n)d\lambda < \varepsilon/2^n,$$

therefore

$$\lim_{n \to \infty} \int G_n d\lambda = \lim_{n \to \infty} \int f_n d\lambda = \int f \, d\lambda$$

(the second equality, by the monotone convergence theorem). Also,

$$0 \leq H_n - G_n \leq \sum_{i=1}^{n}(h_i - g_i),$$

thus

$$0 \leq \int (H_n - G_n)d\lambda \leq \sum_{i=1}^{n} \int (h_i - g_i)d\lambda < \sum_{i=1}^{n} \varepsilon/2^i < \varepsilon,$$

therefore,

$$\int H_n d\lambda < \int G_n d\lambda + \varepsilon \leq \int f \, d\lambda + \varepsilon \tag{1}$$

(the latter inequality because $G_n \leq f_n \leq f$). Since $H_n \uparrow$, it follows from (1) and the monotone convergence theorem that there exists an integrable function $h \geq 0$ such that $H_n \uparrow h$ a.e., and passage to the limit in (1) yields

$$\int h \, d\lambda \leq \int f \, d\lambda + \varepsilon.$$

The function $k = \sup H_n$ is l.s.c. [McS, p. 41, 7.5],

$$f = \sup f_n \leq \sup H_n = k,$$

and $k = h$ a.e. (because $H_n \uparrow h$ a.e.), thus $h$ and $k$ meet the requirements of the lemma.

(d) In general, $f = f_1 - f_2$ with $f_1$, $f_2$ integrable and $\geq 0$. Applying (c) to $f_1$, there exist an integrable function $h_1 \geq 0$ and an l.s.c. function $k_1$ such that $h_1 = k_1$ a.e., $f_1 \leq k_1$ everywhere on $[a, b]$, and

$$\int h_1 d\lambda \leq \int f_1 d\lambda + \varepsilon/2. \tag{2}$$

Applying the proof of (c) to $f_2$, there exists a u.s.c. simple function $g_2$, such that $0 \leq g_2 \leq f_2$ and

$$\int (f_2 - g_2) d\lambda < \varepsilon/2. \tag{3}$$

Thus $-g_2$ is l.s.c., therefore so is $k_1 - g_2$, and

$$f = f_1 - f_2 \leq f_1 - g_2 \leq k_1 - g_2.$$

The functions $h = h_1 - g_2$ and $k = k_1 - g_2$ meet the requirements of the lemma; for, $h = k$ a.e. (because $h_1 = k_1$ a.e.) and

$$h - f = (h_1 - g_2) - (f_1 - f_2) = (h_1 - f_1) + (f_2 - g_2),$$

therefore

$$\int (h - f) d\lambda = \int (h_1 - f_1) d\lambda + \int (f_2 - g_2) d\lambda < \varepsilon/2 + \varepsilon/2$$

by (2) and (3).   ◊

The next lemma gives an estimate of the growth of a continuous increasing function in terms of its derivates:

**Lemma 2** [McS, p. 196, 32.3]. *If* $F\colon [a, b] \to \mathbb{R}$ *is continuous and increasing,* $r > 0$ *and*

$$A = \{x \in (a, b)\colon (\overline{D}F)(x) > r\},$$

*then* $F(b) - F(a) \geq \frac{1}{2} r \lambda^*(A)$.

**Proof.** Here $\overline{D}$ denotes upper derivate,

$$(\overline{D}F)(x) = \limsup_{t \to x, t \neq x} \frac{F(t) - F(x)}{t - x}$$
$$= \max\{(D^- F)(x), (D^+ F)(x)\}$$

for all $x \in (a, b)$, where $D^-$ and $D^+$ indicate upper left-hand and upper right-hand derivate ([McS, p. 189, 31.3], [H&S, p. 257, 17.2]); $\lambda^*(A)$ denotes the outer measure of A (actually A is a Borel set, but we don't have to know this). Let

$$B = \{x \in (a, b)\colon (D^+ F)(x) > r\},$$
$$C = \{x \in (a, b)\colon (D^- F)(x) > r\};$$

since $A = B \cup C$ and $\lambda^*(B \cup C) \leq \lambda^*(B) + \lambda^*(C)$, we need only show that each of B and C has outer measure $\leq [F(b) - F(a)]/r$. We give the proof for B (the proof for C is similar).

Let U be the set of all $x \in (a, b)$ such that

$$\frac{F(t) - F(x)}{t - x} > r \quad \text{for some } t \in [a, b] \text{ with } t > x.$$

Defining $G: [a, b] \to \mathbb{R}$ by $G(x) = F(x) - rx$, we see that

$$U = \{x \in (a, b): G(t) > G(x) \text{ for some } t \in [a, b] \text{ with } t > x\}.$$

If U is empty then $G$ is decreasing, $D^+G \leq 0$ on $[a, b)$, $D^+F \leq r$ on $[a, b)$, $B = \emptyset$ and the desired inequality $r\lambda^*(B) \leq F(b) - F(a)$ is trivial. If U is nonempty then, by F. Riesz's "Rising sun lemma" [SzN, p. 107], $U = \bigcup(a_n, b_n)$ for a countable (possibly finite) family of pairwise disjoint open intervals such that $G(a_n) \leq G(b_n)$ for all $n$, that is, $r(b_n - a_n) \leq F(b_n) - F(a_n)$ for all $n$. For each $n$,

$$r \sum_{i=1}^{n}(b_i - a_i) \leq \sum_{i=1}^{n}[F(b_i) - F(a_i)] \leq F(b) - F(a)$$

(the latter inequality because $F$ is increasing); since this is true for every $n$, $r\lambda(U) \leq F(b) - F(a)$. To complete the proof, we need only show that $B \subset U$. If $x \in B$, that is,

$$\limsup_{t \to x^+} \frac{F(t) - F(x)}{t - x} > r,$$

then there exists a sequence $t_n > x$ such that $t_n \to x$ and

$$\frac{F(t_n) - F(x)}{t_n - x} \to (D^+F)(x) > r;$$

if $n$ is any index such that

$$\frac{F(t_n) - F(x)}{t_n - x} > r,$$

then, since $t_n > x$, $x \in U$ by the definition of U.   $\Diamond$

**Lemma 3.** *If* $h: [a, b] \to \mathbb{R}$ *is Lebesgue-integrable,* $k: [a, b] \to \mathbb{R} \cup \{-\infty, +\infty\}$ *is lower semicontinuous,* $h = k$ *a.e., and* H *is the indefinite integral of* $h$,

$$H(x) = \int_a^x h \, d\lambda \quad (a \leq x \leq b),$$

*then* $\underline{D}H \geq k$ *everywhere on* $[a, b]$.

**Proof.** [McS, p. 197, 33.1]. Here $\underline{D}$ denotes lower derivate,

$$(\underline{D}H)(x) = \liminf_{t \to x, t \neq x} \frac{H(t) - H(x)}{t - x}$$
$$= \min\{(D_-H)(x), (D_+H)(x)\}$$

for all $x \in (a, b)$; at the endpoints,

$$(\underline{D}H)(a) = (D_+H)(a), \quad (\underline{D}H)(b) = (D_-H)(b).$$

Given $x \in [a, b]$, we are to show that $(\underline{D}H)(x) \geq k(x)$; we can suppose that $k(x) > -\infty$.

Suppose first that $a \leq x < b$. Let $r$ be a real number with $k(x) > r$. By the lower semicontinuity of $k$, the set

$$V = \{t \in [a, b] : k(t) > r\}$$

is a neighborhood of $x$ in the metric space $[a, b]$. For almost every $t \in V \cap (x, +\infty)$ we have $h(t) = k(t) > r$, therefore

$$\int_x^t h \, d\lambda \geq r(t - x)$$

for *every* $t \in V \cap (x, +\infty)$, that is,

$$\frac{H(t) - H(x)}{t - x} \geq r \quad \text{for all } t \in V \cap (x, +\infty);$$

thus

$$r \leq \inf_{t \in V \cap (x, +\infty)} \frac{H(t) - H(x)}{t - x} \leq \liminf_{t \to x+} \frac{H(t) - H(x)}{t - x} = (D_+H)(x).$$

Since this is true for every $r < k(x)$, we conclude that

$$k(x) \leq (D_+H)(x) \quad \text{for all } x \in [a, b). \tag{1}$$

A similar argument, with $(x, +\infty)$ replaced by $(-\infty, x)$, shows that

$$k(x) \leq (D_-H)(x) \quad \text{for all } x \in (a, b]. \tag{2}$$

From (1), (2) and the formulas at the beginning of the proof, we see that

$$k(x) \leq (\underline{D}H)(x)$$

for all $x \in [a, b]$.   $\Diamond$

The foregoing proof contains the key to the success of semicontinuity in this circle of ideas: an inequality $h(x) > r$ says nothing about points other than $x$, but lower semicontinuity projects the inequality $k(x) > r$ into an entire neighborhood of $x$ (and carries $h$ along with it almost everywhere). Also, integration shows its aspect of 'smoothing operator,' transforming an 'a.e.' inequality into an 'everywhere' inequality.

**Theorem 3.** *If $f: [a, b] \to \mathbb{R}$ is Lebesgue-integrable then its indefinite integral*

$$F(x) = \int_a^x f \, d\lambda \quad (a \leq x \leq b)$$

*is a primitive for $f$.*

**Proof** [McS, p. 198, 33.3]. We already know that $F$ is absolutely continuous (§1, Theorem 1); the problem is to show that, for almost every $x \in (a, b)$, $F$ is differentiable at $x$ with $F'(x) = f(x)$.

The first step is to prove that $\underline{D}F \geq f$ a.e., in other words, that the set

$$A = \{x \in [a, b]: (\underline{D}F)(x) - f(x) < 0\}$$

is negligible; given any $r > 0$, it suffices to show that the set

$$A_r = \{x \in [a, b]: (\underline{D}F)(x) - f(x) < -r\}$$

is negligible ($A$ is the union of the sequence of sets $A_{1/n}$ for $n = 1, 2, 3, \ldots$). Given any $\varepsilon > 0$, it is enough to show that $\lambda^*(A_r) < \varepsilon$. By Lemma 1, there exist an integrable function $h$ and a lower semicontinuous function $k$ such that $h = k$ a.e., $f \leq k$ everywhere on $[a, b]$, and

$$\int_a^b h \, d\lambda < \int_a^b f \, d\lambda + \frac{1}{2} r \varepsilon. \tag{1}$$

Let $H$ be the indefinite integral of $h$; since $h - f \geq 0$ a.e. and $H - F$ is the indefinite integral of $h - f$, $H - F$ is increasing (therefore all of its derivates are $\geq 0$). Let

$$E_r = \{x \in [a, b]: [\overline{D}(H - F)](x) > r\};$$

applying Lemma 2 to $H - F$, we have

$$(H - F)(b) - (H - F)(a) \geq \frac{1}{2} r \lambda^*(E_r),$$

that is,

$$\frac{1}{2} r \lambda^*(E_r) \leq H(b) - F(b). \tag{2}$$

But (1) says that $H(b) - F(b) < \frac{1}{2} r \varepsilon$, thus, in view of (2),

$$\lambda^*(E_r) < \varepsilon. \tag{3}$$

Note that $E_r$ depends on $H$, therefore on $\varepsilon$, but $A_r$ does not. We are going to show that $A_r \subset E_r$ (whence $\lambda^*(A_r) \leq \lambda^*(E_r) < \varepsilon$, completing the proof that $A_r$—and therefore A—is negligible). From $F = H - (H - F)$, we have

$$\underline{D}F \geq \underline{D}H - \overline{D}(H - F) \tag{4}$$

at every point of $[a, b]$ where the right side is defined [McS, p. 190, 31.4]. Given $x \notin E_r$, we must show that $x \notin A_r$, that is,

$$(\underline{D}F)(x) - f(x) \geq -r.$$

Since $H - F$ is increasing and $x \notin E_r$, we have

$$0 \leq [\overline{D}(H - F)](x) \leq r, \tag{5}$$

therefore the subtraction in (4) is permissible at $x$ and we see from (4) and (5) that

$$(\underline{D}F)(x) \geq (\underline{D}H)(x) - [\overline{D}(H - F)](x) \geq (\underline{D}H)(x) - r;$$

but $(\underline{D}H)(x) \geq k(x)$ by Lemma 3, and $k(x) \geq f(x)$ by the choice of $k$, therefore

$$(\underline{D}F)(x) \geq f(x) - r$$

as we wished to show.

We now know that A is negligible, that is,

$$\underline{D}F \geq f \text{ a.e.} \tag{6}$$

Applying (6) to $-f$ (whose indefinite integral is $-F$), we have $\underline{D}(-F) \geq -f$ a.e., that is, $-(\overline{D}F) \geq -f$ a.e., thus

$$\overline{D}F \leq f \text{ a.e.} \tag{7}$$

From (6) and (7) we see that

$$f \leq \underline{D}F \leq \overline{D}F \leq f \text{ a.e.},$$

(recall that $\liminf \leq \limsup$), thus

$$\underline{D}F = \overline{D}F = f \text{ a.e.};$$

in other words, for almost every $x$, $F'(x)$ exists and is equal to $f(x)$.   $\Diamond$

Theorem 3 is item (III) of the Introduction and completes the proof of (I)–(IV). The proofs of (I) and (II) are relatively straightforward and (IV) is an easy consequence of (I) and (III); thus (III) is the prime target for further simplification (but we may be getting close to the bone).

Finally, here is a stripped-down version of the "Fundamental theorem of calculus" whose proof does not require the Radon–Nikodym theorem (at the cost of losing the characterizations (I) and (IV)):

**Theorem.** *If* $f \colon [a, b] \to \mathbb{R}$ *is Lebesgue-integrable, then* (1) *the indefinite integral*

$$x \mapsto \int_a^x f \, d\lambda \quad (a \le x \le b)$$

*is a primitive for* $f$, *and* (2) *if* $F$ *is any primitive of* $f$, *then*

$$\int_a^b f \, d\lambda = F(b) - F(a).$$

**Proof.** (1) This is (III); in particular, the proof of absolute continuity (the "if" part of the proof of Theorem 1 of §1) does not require the Radon–Nikodym theorem.

(2) Suppose $F$ is a primitive for the Lebesgue-integrable function $f$. If $G$ is the indefinite integral of $f$ then, by (1), the function

$$H(x) = F(x) - F(a) - G(x) \quad (a \le x \le b)$$

is a primitive for the zero function that vanishes at $a$, therefore $H$ is the zero function by (II). In particular, $0 = H(b) = F(b) - F(a) - G(b)$, which is the desired formula of (2).   $\Diamond$

The role of the Radon–Nikodym theorem in this circle of ideas is now sharply in focus: it is the basis for (I) (which characterizes the indefinite integrals that occur in the Theorem as the AC functions that vanish at $a$). In fact, (I) is *equivalent* to the Radon–Nikodym theorem for finite signed measures on the Borel sets of $[a, b]$, with Lebesgue measure (restricted to the Borel sets) as the "base" measure. The crux of the matter is that, for a finite signed measure $\nu$ on the Borel sets of $[a, b]$, the following conditions are equivalent: (a) $\nu \ll \lambda$ (that is, $\lambda(B) = 0 \Rightarrow \nu(B) = 0$); (b) the function $F(x) = \nu([a, x])$ is absolutely continuous and $F(a) = \nu(\{a\}) = 0$ [H, §43]. Thus, the representation via (I) of the AC functions $F$ occurring in (b) amounts to the representation of finite signed measures $\nu$ satisfying (a) as indefinite integrals with respect to $\lambda$ (in the sense of [H, p. 97]. This explains why the Radon–Nikodym theorem is also called the "Lebesgue–Radon–Nikodym theorem."

# 4. The Lebesgue Integral without Measure or Integration

Items (II) and (IV) of the Introduction—the "uniqueness theorem" and the characterization of Lebesgue-integrability via the existence of primitives— open the door to a bizarre approach to the Lebesgue theory on $[a, b]$ that *avoids Lebesgue measure and the Lebesgue integral* (at least at the outset) and virtually reduces the "Fundamental theorem of calculus" to a definition. The program is as follows:

(1) Define negligible sets and absolute continuity in the usual way.

(2) Prove the "uniqueness theorem" (II) as in Section 2.

(3) Turn (IV) upside down and *define* a function $f: [a, b] \rightarrow \mathbb{R}$ to be "Lebesgue-integrable" if there exists an absolutely continuous function $F$ such that $F'(x) = f(x)$ for almost every $x$, then *define* the "integral" of $f$ by the formula

$$\int_a^b f = F(b) - F(a)$$

(well-defined, by the uniqueness theorem).

This is the ultimate calorie-free proof of the "Fundamental theorem": the only thing that has to be proved is the uniqueness theorem (II); the rest is all definitions. It is a way to go. Is it the way to go? The standard convergence and completeness theorems seem far down the road (in fact, it is not obvious from this perspective that if $f$ is "Lebesgue-integrable" then so is $|f|$).

One can algebrize matters even further, by easing absolute continuity out of the picture (more precisely, by replacing it by three weaker properties): by a theorem of S. Banach, a function $F: [a, b] \rightarrow \mathbb{R}$ is AC if and only if it is continuous, of bounded variation, and maps negligible sets to negligible sets [H&S, p. 288, 18.25]. There is an obstacle to taking Banach's criterion as the definition of "absolute continuity": it is not at all obvious (though certainly true) that the sum of two functions satisfying Banach's criterion maps negligible sets to negligible sets, a proposition that is needed for defining "integral" by a formula $F(b) - F(a)$ (via an analog of the uniqueness theorem (II)). The Jordan decomposition theorem suggests a way around this obstruction:

**Definition 1.** Call a function $F: [a, b] \rightarrow \mathbb{R}$ ≪absolutely continuous≫ (briefly, ≪AC≫) if it can be written in the form $F = G - H$, where $G$ and $H$ are strictly increasing continuous functions on $[a, b]$ that map negligible sets to negligible sets. (We leave to the end the proof that this is equivalent to the classical definition of absolute continuity.)

It is obvious that constant functions are ≪AC≫ $(c = (c + x) - x)$ and that if $F$ is ≪AC≫ then $cF$ is ≪AC≫ for every real number $c$. The case of sums will be taken up shortly.

**Definition 2.** Call a function $f: [a, b] \rightarrow \mathbb{R}$ ≪Lebesgue-integrable≫ if there exists an ≪AC≫ function $F$ such that, for almost every $x \in (a, b)$, $F'(x)$ exists and is equal to $f(x)$.

With the preceding notations, to justify defining the ≪integral≫ of $f$ to be $F(b) - F(a)$ we need to check two things: (i) if $F$ and $G$ are ≪AC≫ then so is $F - G$; (ii) if $F$ is ≪AC≫ and $F' = 0$ a.e. then $F$ is constant. The crux of the matter is to give an elementary proof of the following theorem:

**Theorem.** *If $G$ and $H$ are strictly increasing continuous functions on $[a, b]$ that map negligible sets to negligible sets, then $G + H$ and $G - H$ map negligible sets to negligible sets.*

**Proof.** (i) Consider first the function $K = G + H$. At any rate, $K$ is continuous and strictly increasing. If $I = [c, d]$ is a closed subinterval of $[a, b]$, then $K(I) = [K(c), K(d)]$ and $\lambda(K(I)) = K(d) - K(c)$; similarly, $\lambda(G(I)) = G(d) - G(c)$ and $\lambda(H(I)) = H(d) - H(c)$, therefore

$$\lambda(K(I)) = \lambda(G(I)) + \lambda(H(I)). \qquad (*)$$

Let N be a negligible set; we know that $G(\mathrm{N})$ and $H(\mathrm{N})$ are negligible, and the problem is to show that $K(\mathrm{N})$ is negligible. Let $\varepsilon > 0$. Since $G(\mathrm{N})$ is negligible, $G(\mathrm{N}) \subset \bigcup \mathrm{P}_n$ with $(\mathrm{P}_n)$ a sequence of closed intervals such that $\sum \lambda(\mathrm{P}_n) \leq \varepsilon$. Replacing $\mathrm{P}_n$ by $\mathrm{P}_n \cap G([a, b])$, we can suppose that $\mathrm{P}_n \subset [G(a), G(b)]$. As in the proof of §1, Lemma 2, we can further suppose that the $\mathrm{P}_n$ are nonoverlapping. Since $G$ is an order isomorphism of $[a, b]$ onto $[G(a), G(b)]$, $\mathrm{P}_n = G(\mathrm{I}_n)$ with $(\mathrm{I}_n)$ a sequence of nonoverlapping closed subintervals of $[a, b]$ such that $\mathrm{N} \subset \bigcup \mathrm{I}_n$.

To summarize, given a negligible set N and $\varepsilon > 0$, there exists a sequence $(\mathrm{I}_n)$ of nonoverlapping closed subintervals of $[a, b]$ such that $\mathrm{N} \subset \bigcup \mathrm{I}_n$ and $\sum \lambda(G(\mathrm{I}_n)) \leq \varepsilon$. Similarly (applying the foregoing to the function $H$), there exists a sequence $(\mathrm{J}_n)$ of nonoverlapping closed subintervals of $[a, b]$ such that $\mathrm{N} \subset \bigcup \mathrm{J}_n$ and $\sum \lambda(H(\mathrm{J}_n)) \leq \varepsilon$. Then

$$\mathrm{N} \subset \bigcup_{m,n} (\mathrm{I}_m \cap \mathrm{J}_n). \qquad (**)$$

We now have a covering of $G(\mathrm{N})$ by a double sequence $G(\mathrm{I}_m \cap \mathrm{J}_n)$ of nonoverlapping closed intervals, and similarly for $H(\mathrm{N})$ and the $H(\mathrm{I}_m \cap \mathrm{J}_n)$. By $(**)$, $K(\mathrm{N}) \subset \bigcup K(\mathrm{I}_m \cap \mathrm{J}_n)$; in view of $(*)$, it will suffice to show that

$$\sum_{m,n} \lambda(G(\mathrm{I}_m \cap \mathrm{J}_n)) \leq \varepsilon \quad \text{and} \quad \sum_{m,n} \lambda(H(\mathrm{I}_m \cap \mathrm{J}_n)) \leq \varepsilon.$$

Fix $m$ and $n$; we assert that

$$\sum_{k=1}^{n} \lambda(G(\mathrm{I}_m \cap \mathrm{J}_n)) \leq \lambda(G(\mathrm{I}_m)). \qquad (***)$$

At any rate, the intervals $\mathrm{I}_m \cap \mathrm{J}_1, \ldots, \mathrm{I}_m \cap \mathrm{J}_n$ (possibly empty) are nonoverlapping; to simplify the notation, we can suppose that they are nonempty and are arranged from left to right on the number line. Thus, if $I_m = [c_0, d_0]$ and $I_m \cap J_k = [c_k, d_k]$ for $k = 1, \ldots, n$, we are supposing that

$$c_0 \leq c_1 \leq d_1 \leq c_2 \leq d_2 \leq \cdots \leq c_n \leq d_n \leq d_0;$$

then (telescoping sum)

$$G(d_0) - G(c_0) = [G(c_1) - G(c_0)] + [G(d_1) - G(c_1)] + \cdots + [G(d_0) - G(d_n)]$$

$$= \sum_{k=1}^{n} [G(d_k) - G(c_k)] + \sum_{k=1}^{n-1} [G(c_{k+1}) - G(d_k)]$$

$$+ [G(c_1) - G(c_0)] + [G(d_0) - G(d_n)]$$

$$\geq \sum_{k=1}^{n} [G(d_k) - G(c_k)]$$

(the inequality because $G$ is increasing), whence $(\ast\ast\ast)$. It follows that

$$\sum_{n=1}^{\infty} \lambda(G(I_m \cap J_n)) \leq \lambda(G(I_m))$$

for each $m$, therefore

$$\sum_{m,n} \lambda(G(I_m \cap J_n)) \leq \sum_{m} \lambda(G(I_m)) \leq \varepsilon.$$

Similarly,

$$\sum_{m,n} \lambda(H(I_m \cap J_n)) \leq \varepsilon.$$

It then follows from $(\ast)$ that

$$\sum_{m,n} \lambda(K(I_m \cap J_n)) \leq \varepsilon + \varepsilon,$$

which shows that $K(N)$ is negligible.

(ii) Let $F = G - H$. With the preceding notations, we have $F(N) \subset \bigcup F(I_m \cap J_n)$ and

$$\lambda(F(I_m \cap J_n)) \leq \lambda(G(I_m \cap J_n)) + \lambda(H(I_m \cap J_n))$$

by Lemma 1 of §1, therefore $\sum \lambda(F(I_m \cap J_n)) \leq 2\varepsilon$, which shows that $F(N)$ is negligible.    $\Diamond$

**Corollary 1.** (i) *The $\ll$AC$\gg$ functions form a vector space for the pointwise linear operations.*

(ii) *Every $\ll$AC$\gg$ function maps negligible sets to negligible sets.*

**Proof.** (i) Let $F_1$ and $F_2$ be $\ll$AC$\gg$ and, for $i = 1, 2$, write $F_i = G_i - H_i$ as in Definition 1; then $F_1 + F_2 = (G_1 + G_2) - (H_1 + H_2)$, where $G_1 + G_2$ and $H_1 + H_2$ are continuous, strictly increasing and, by (i) of the theorem, map negligible sets to negligible sets, therefore $F_1 + F_2$ is $\ll$AC$\gg$ by definition. The status of scalar multiples was noted following Definition 1.

(ii) If $F$ is $\ll$AC$\gg$ and $F = G - H$ as in Definition 1, then $F$ maps negligible sets to negligible sets by (ii) of the theorem.    $\Diamond$

**Corollary 2.** *If $F: [a, b] \to \mathbb{R}$ is $\ll$AC$\gg$ and if, for almost every $x \in (a, b)$, $F$ is differentiable with $F'(x) = 0$, then $F$ is constant.*

**Proof.** The proof of Corollary 1 in §2 carries over with AC replaced by $\ll$AC$\gg$. So does the proof of Corollary 2 in §2; the key point is that the identity function $x \mapsto x$ trivially maps negligible sets to negligible sets, therefore (by the preceding Corollary 1) so do the functions $F_r(x) = F(x) + rx$ that figure in the proof of Corollary 2 of §2.    $\Diamond$

**Definition 3.** With $f$ and $F$ as in Definition 2, define the $\ll$integral$\gg$ of $f$ to be $F(b) - F(a)$.

If also $H$ is $\ll$AC$\gg$ and $H' = f$ a.e., then $F - H$ is $\ll$AC$\gg$ (Corollary 1) and $(F - H)' = 0$ a.e., therefore $F - H$ is constant (Corollary 2); thus the $\ll$integral$\gg$ of $f$ is well-defined.

Finally, we observe that Definitions 1–3 are equivalent to the classical definitions:

**Proposition.** AC $\Leftrightarrow$ $\ll$AC$\gg$.

**Proof.** $\Rightarrow$: Suppose $F: [a, b] \to \mathbb{R}$ is AC. By the Jordan theorem, $F = G - H$ with $G$ and $H$ increasing and AC. Replacing $G$ and $H$ by the functions $x \mapsto G(x) + x$ and $x \mapsto H(x) + x$, we can suppose they are strictly increasing, and we know from §1, Lemma 2 that they map negligible sets to negligible sets; thus $F$ is $\ll$AC$\gg$ by Definition 1.

$\Leftarrow$: Suppose $F$ is $\ll$AC$\gg$ and write $F = G - H$ as in Definition 1. By the technique in the "only if" part of §1, Theorem 1, the function $x \mapsto G(x) - G(a)$ is an indefinite integral, hence $G$ is AC by the "if" part of Theorem 1; similarly $H$ is AC, therefore so is $F$.    $\Diamond$

Inspecting Definitions 1–3 in the light of the preceding proposition, we see that the definition of the Lebesgue integral over $[a, b]$ has been effectively reduced to the concept of negligible set (and such items from elementary analysis as the intermediate value theorem).

## REFERENCES

[B] BERBERIAN, S.K.: *Measure and integration.* New York: Macmillan 1965. Reprinted New York: Chelsea 1970.

[BOU] BOURBAKI, N.: *Fonctions d'une variable réelle. Théorie élémentaire.* Paris: Hermann 1976.

[H] HALMOS, P.R.: *Measure theory.* New York: Van Nostrand 1950. Reprinted New York: Springer-Verlag 1974.

[H&S] HEWITT, E., and STROMBERG, K.: *Real and abstract analysis.* New York: Springer-Verlag 1965.

[L] LEBESGUE, H.: *Leçons sur l'intégration et la recherche des fonctions primitives.* Second edition. Paris: Gauthier-Villars 1928.

[McS] McSHANE, E.J.: *Integration.* Princeton: Princeton University Press 1944.

[R] ROYDEN, H.L.: *Real analysis.* Third edition. New York: Macmillan 1988.

[SzN] Sz.-NAGY, B.: *Introduction to real functions and orthogonal expansions,* New York: Oxford University Press 1965.

*Department of Mathematics*
*University of Texas*
*Austin, Texas 78712*

# Mathematics from Fun & Fun from Mathematics: An Informal Autobiographical History of Combinatorial Games

## Richard K. Guy

Richard Guy, 1963

Perhaps I've never done any serious mathematics. Perhaps there isn't any serious mathematics. The subject of combinatorics is only slowly acquiring respectability and combinatorial games will clearly take longer than the rest of combinatorics. Although he didn't actually use the word, it must have been the *finiteness* of chess that led Hardy [25] to contrast it with genuine mathematics:

There are masses of chess-players in every civilized country — in Russia, almost the whole educated population; and every chess-player can recognize and appreciate a 'beautiful' game or problem. Yet a chess problem is *simply* an exercise in pure mathematics (a game not entirely, since psychology also plays a part), and everyone who calls a problem 'beautiful' is applauding mathematical beauty, even if it is beauty of a comparatively lowly kind. Chess problems are the hymn-tunes of mathematics (p. 87)

A chess problem is genuine mathematics, but it is in some way 'trivial' mathematics. (p. 88)

Chess problems are *unimportant*. (pp. 88–89)

No chess problem has ever affected the general development of scientific thought. (p. 89)

Much of combinatorics is finite. But not necessarily trivial, as the proofs of the four color theorem and of the nonexistence of a projective plane of order 10 demonstrate. Not that Hardy would have been very excited about either of these proofs. Colleagues in combinatorial set theory assure me that infinite problems are usually easier than finite ones.

My interest in combinatorics might have been foreseen from my interest in chess, especially in the endgame where complete analysis was often possible, even before the advent of computers. While still an undergraduate I began winning prizes in international endgame competitions. After WW2, Thomas Rayner Dawson, the fairy chess king, wanted to hand over the editorship of the Endings section of the *British Chess Magazine,* since it was one of the least of his many interests, and in his retirement from his job as a rubber chemist he didn't enjoy very good health. I had been his most regular British contributor (almost the only one in those days) so I was invited to take it on.

Dawson was also an amateur mathematician [11–17] and interested in a wide variety of games and puzzles. He lived in Thornton Heath, an easy cycle ride from Forest Hill, where I lived, so I occasionally visited him, and he showed me many things. One that intrigued me particularly was the game he had invented [9,10] and which we now call Dawson's Chess.

This is chess played on a $3 \times n$ board with white pawns on the first rank and black pawns on the third. It was posed as a *losing* game (last-player-losing, now called **misère**) so that capturing was obligatory. Fortunately (because we *still* don't know how to play Misère Dawson's Chess) I assumed, as a number of writers of that time and since have done, that the misère analysis required only a trivial adjustment of the normal (last-player-winning) analysis. This arises because Bouton, in his original analysis of Nim [7], had observed that only such a trivial adjustment was necessary to cover both normal and misère play. Before I say what it is, let us remind ourselves how to play Nim (well). It is played with heaps of beans. Two players move alternately. A move is to choose a heap and remove as many beans from it as you wish (perhaps the whole heap, if you like, but at least one bean). The winner is the person to take the last bean in normal play; the last player is the loser in misère play. A good strategy is to play so that the nim-sum of the numbers of beans in the heaps is zero. To find the **nim-sum** of numbers, write them in binary and add them without carrying. For example,

$$3 \overset{*}{+} 5 \overset{*}{+} 6 = \text{(in binary)} \ 11 \overset{*}{+} 101 \overset{*}{+} 110 = 0,$$

and if there are heaps of 3, 5 and 6 beans, you had better hope that it is your opponent's turn to move, whether it's normal play or misère play.

Play the same strategy until there's<br>
just one heap which contains more than one bean.<br>
Then take the whole of that heap, or all but one bean,<br>
leaving an **odd** or an **even** number of heaps

(each containing one bean) according as you are playing
**misère** or **normal** Nim.

But even for **impartial** games, in which the same options are available to both players, regardless of whose turn it is to move, Grundy and Smith [22] showed that the general situation in misère play soon gets very complicated, and Conway [8, p. 140] confirmed that the situation can only be simplified to the microscopically small extent noticed by Grundy and Smith. See also [6, pp. 393–426], [2] and [29]. Even as I write, Thane Plambeck phones that he is completing the misère analysis of several games which had earlier seemed intransigent.

At first sight Dawson's Chess doesn't look like an impartial game, but if you know how pawns move at Chess, it's easy to verify that it's equivalent to the game played with rows of skittles in which, when it's your turn, you knock down any skittle, together with its immediate neighbors, if any.

Dawson, with help from someone else, probably W.E. Lester, back in the thirties, well before [30] and [21] appeared, had analyzed the (misère) game far enough to notice a tendency to exhibit a periodicity, with period 14 (see the table at the foot of p. 418 of [6]). In normal play this periodicity is less noticeable, but the possibility of periodicity suggested putting positions into equivalence classes. Moreover, I'd read more than one edition of Rouse Ball, and knew of Michael Goldberg's very significant diagram — compare [4] with [5] — for a partial analysis of Dudeney's Kayles [19] (see also [28]). Once you realize that Goldberg's diagram is a partial manifestation of the Fano configuration (Fig. 1), it's not too difficult to (re)discover the Sprague–Grundy theory, at least for this specific game. Sprague [30] and Grundy [21] independently published in comparatively obscure journals at a time when the world had other things on its mind. I know that one of them, and probably both, did not live long enough to learn of the other's work.

Even today the Sprague–Grundy theory is insufficiently widely known.

> Every (position in an) impartial game
> is equivalent to a **nim-heap**
> (a heap of beans in the game in Nim).

And we all know how to play one-heap Nim: just grab the whole heap! "All" that we need to know is the **nim-value** of the position, the number of beans in the heap.

> The nim-value of a (position in a)n impartial game
> is the mex of the nim-values
> of the options.

> The **mex** of a set of non-negative integers
> is the least non-negative integer *not* in the set.

Richard K. Guy

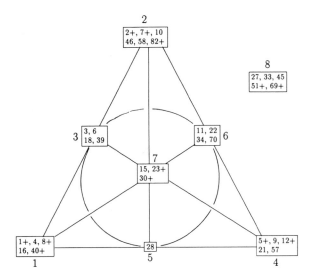

FIGURE 1. The analysis of the game of Kayles. The numbers in the boxes are numbers of skittles in a row: an entry $a+$ denotes the progression $a$, $a + 12$, $a + 24$, $a + 36, \ldots$ . The large numbers near the boxes are the nim-values of the corresponding rows of skittles. The nim-sum of three of these on a line (e.g., $1 \overset{*}{+} 5 \overset{*}{+} 4$) is zero. For example, the position with rows of 16, 28, and 21 skittles is a $\mathcal{P}$ position (previous-player-winning position) — if it's your turn to move you will lose against a good opponent. Note that boxes 3, 5 and 6 are "in line": they comprise the "rare values" in Kayles [6, p. 109]. The nim-sum of the complementary set of four nim-values *not* on a line is also zero. And, of course, two rows of equal length (or two from the same box, e.g., 3 and 6) form a $\mathcal{P}$ position. Michael Goldberg's diagram was a subset of this figure, omitting the 5 and 8 boxes, and numbers greater than 20.

From the practical point of view all we need to know is a move in the impartial game which changes the nim-value of the position to zero (if it's not zero already, in which case you hope that it's your opponent's turn to move). Expressed in this way, the Sprague–Grundy theory seems neither very deep nor of much practical value. But many games, and Kayles and Dawson's Chess are two of them, naturally split up in the course of play into separate components, leading to the idea of the **sum** (or **disjunctive compound**) of games.

> When it's your turn to move in the **sum**
> of a number of component games
> you choose one component
> and make a legal move in it.

Now that you know the definition of sum and how to play Nim (notice that Nim is the sum of one-heap games of Nim) you can see the significance of the Sprague–Grundy theory.

> The nim-value of the (game) sum
> of several impartial games
> is the nim-sum of the nim-values
> of the individual games.

Now the nim-values of rows of skittles of various lengths in Kayles (played with rows of skittles; when it's your turn, choose a row and knock down one skittle or two contiguous ones from it — note that if they are not on the end of the row, then the row splits into two separate component rows) and the nim-values of an opening position in Dawson's Chess on a $3 \times n$ board — each played with normal play, last-player-winning — turn out to exhibit periodicity. The periods are 12 and 34, respectively, but they are each beset by early exceptions, so that, even when you are equipped with a theorem of Guy and Smith [24], you need to analyze the games to rows of lengths 166 and 173, respectively, before you are sure that the periodicity, if you suspected it, really persists.

I took these calculations to Theodor Estermann, with whom I was studying number theory at University College, London, and asked him if they were significant. "I don't know," he said, "but I can tell you someone who does," and he sent me across to the Galton Laboratory. In 1949 this was headed by the famous geneticist Lionel Penrose, whom I already knew because he had composed end-game studies and because I'd met him at British Chess Federation tourneys where sons Oliver and Jonathan were playing. Jonathan has since won the British Chess Championship more times than any other person. I also discussed self-reproducing machines, impossible objects, and a great variety of mathematical puzzles with Lionel and son Roger, who is now more famous than any of them. But I digress: the person I now sought was not a member of the Penrose family, but Cedric Austen Bardell Smith, whom I soon came to realize I had often seen a dozen years before, seated keenly in the company of Leonard Brooks and Arthur Stone in the front row of lectures in Cambridge. The fourth member of the Blanche Descartes quartet (Blanche Descartes neé Filet de Carte-Blanche [18]) was William Thomas Tutte, F.R.S. (*alias* Mr. Graph-Theory), but he was a chemist in those days.

Cedric was delighted that I had rediscovered the theory of his friend Patrick Michael Grundy (we didn't learn of Roland Percival Sprague's anticipation until after Grundy was dead). Moreover, I had an uncountable infinity of games to which the theory could be applied. Grundy had only a few rather trivial examples, together with Grundy's Game (split a heap into two unequal non-empty heaps) for which we still do not have a complete analysis, in spite of the calculation of the first ten million nim-values by my son Mike Guy.

I went to Singapore in 1951 and our joint paper [24] wasn't published until 1956, Cedric having first made the mistake of trying to sell it to the London Mathematical Society, which in those days seemed only to publish papers on groups, forms, Banach spaces, and summability. Meanwhile Adams and Benson [1] also analyzed Kayles, as well as some artificial variants formed by changing the first few nim-values, which sometimes had the effect of changing the length of the period. Which games are periodic, and what is the relationship between the rules of the game and the length and structure of the period, remain as considerable mysteries.

In 1960 Mike Guy went to Cambridge and immediately interested John Horton Conway in combinatorial games. From then on I visited Cambridge most years and discussed games and other things with Conway. In 1965 I went to Calgary, where Kenyon [26] and Austin [3] each found games with simple rules, but long periods. For some recent, even more surprising results, see [20].

In 1967 I met Elwyn Ralph Berlekamp at a Combinatorics Conference in Chapel Hill. He had recently made his remarkable analysis of the game of Dots-and-Boxes [6, pp. 507–550] in which he used, among many other things, my analyses of both Kayles and Dawson's Chess. He suggested that we write a book on combinatorial games, so I said that I knew just the person to help us. Conway and I now started to produce informal chapters of such a book, often using material supplied by Berlekamp. Up till then I had mainly been interested in impartial games, but Conway vigorously attacked partizan games, in which the options for the two players were not in general the same. As the theory developed, he was delighted to find imbedded in it a Dedekind-like approach to the binary rationals and the reals and a von Neumann-type definition of the Cantor transfinite ordinals. This led to [8] and [27].

Conway's definition of a (partizan combinatorial) game is deceptively simple, and leads to incredible ramifications. A **game** is simply an ordered pair of sets of games:

$$G = \{G^L \mid G^R\},$$

where $G^L$ is shorthand (analogous to the usual function notation) for the set of Left options (which may be empty or infinite) and $G^R$ is the set of Right options, Left and Right being the two players. This definition is inductive, the basis being the empty set, so the zeroth game is the **Endgame**

$$\{\emptyset \mid \emptyset\} = \{ \mid \},$$

in which neither player has an option, which we will call **Zero** and denote by 0. Now we have Zero and the empty set, so $2^2$ games are born on day 1: Zero again, and

$$\{0 \mid \} = 1, \qquad \{ \mid 0\} = -1, \qquad \{0 \mid 0\} = *.$$

With 4 games we have $2^4$ subsets and $2^8$ candidates for day 2 games. In the event, only 22 of these turn out to be distinct, but games proliferate super-exponentially. The formal definition of the sum of two games $G$ and $H$ is

$$G + H = \{G^L + H, G + H^L \mid G^R + H, G + H^R\},$$

which is just the symbolic form of our earlier verbal definition. If we also define the **negative** of $G$ by

$$-G = \{-G^R \mid -G^L\}$$

then we have a very large partially ordered commutative Group of games. The partial order is defined as follows:

$G \geq H$ just if $G - H \geq 0$, i.e., $G + (-H) \geq 0$,
i.e., just if Left wins the game $G - H$
(i.e., just if Left can arrange to be the last player),
*provided Right starts.*

An important subgroup is the group of (surreal) numbers [27].

A game is a **number**
just if all the options are numbers
and each Left option
is less than every Right option.

With the proper definition of multiplication and of reciprocal [8, pp. 19–22], Conway establishes that

The Class **No** of all numbers forms a totally ordered Field.

After an incubation period of 15 years, [6] appeared. It was an interesting collaboration between three very different personalities. Anyone who knows us only slightly would find it hard to guess our several contributions. The extrovert John Conway might be expected to make brilliant intuitive leaps, while the much quieter Elwyn Berlekamp clearly proceeds in a steady logical manner. Just the opposite is the case! John works through thousands of examples, before finally distilling the exact pattern, arriving at the formulation which is not only clear to him, but which he is able to make clear to the man-in-the-street, while Elwyn makes conjectures whose proofs are sometimes completed by others after considerable discussion and after some small but essential modifications have been made.

The subject continues to develop; it recently acquired some respectability in that a Short Course was given at the American Mathematical Society Summer Meeting in Columbus. The proceedings [23] will contain amplified notes for each of the 8 lectures, a list of 30-odd unsolved problems, and a bibliography prepared by Aviezri Fraenkel and containing nearly 400 items.

## REFERENCES

1.  E.W. Adams and E.C. Benson, Nim-type games, Carnegie Inst. Tech., Report 13, 1956.

2.  Dean Thomas Allemang, *Machine Computation with Finite Games*, M.Sc. thesis, Cambridge, England, 1984.

3.  Richard Bruce Austin, *Impartial and Partisan Games*, M.Sc. thesis, University of Calgary, 1976.

4.  W.W. Rouse Ball (revised H.S.M. Coxeter), *Mathematical Recreations and Essays*, 11th edition, Macmillan, New York, 1939, p. 40.

5.  W.W. Rouse Ball and H.S.M. Coxeter, *Mathematical Recreations and Essays*, 12th edition, University of Toronto Press, Toronto, 1974; reprinted Dover, New York, 1987; pp. 36–40.

6.  E.R. Berlekamp, J.H. Conway, and R.K. Guy, *Winning Ways for your Mathematical Plays*, Academic Press, New York, 1982.

7.  Charles L. Bouton, Nim, a game with a complete mathematical theory, *Ann. Math. Princeton* **3** (1901–02), 35–39.

8.  J.H. Conway, *On Numbers and Games*, Academic Press, New York, 1976.

9.  T.R. Dawson, Problem 1603, *Fairy Chess Review = The Problemist, Fairy Chess Supplement*, Vol. 2, No. 9 (Dec. 1934), 94; solution, Vol. 2, **2** No. 10 (Feb. 1935), 105.

10. T.R. Dawson, *Caissa's Wild Roses*, 1935, 13.

11. T.R. Dawson, "Match-stick" geometry, *Math. Gaz.* **23** (1939), 161–168.

12. T.R. Dawson, Note 1481. Some algebraic computation, *Math. Gaz.* **24** (1940), 291–294.

13. T.R. Dawson, Note 1484. Unequal lines are sometimes equal, *Math. Gaz.* **24** (1940), 342–343.

14. T.R. Dawson, Note 1493. An unexpected converse, *Math. Gaz.* **24** (1940), 356–357.

15. T.R. Dawson, Note 1580. On Note 1522, *Math. Gaz.* **26** (1942), 61.

16. T.R. Dawson, Ornamental squares and triangles, *Math. Gaz.* **30** (1946), 19–21.

17. T.R. Dawson, Note 1892. How not to state a theorem, *Math. Gaz.* **30** (1946), 92–93.

18. Blanche Descartes, Why are series musical? *Eureka* **16** (1953), 18–20; reprinted, **27** (1964), 29–31.

19. H.E. Dudeney, *Canterbury Puzzles,* London, 1910, pp. 118, 220.

20. Anil Gangolli and Thane Plambeck, A note on periodicity on some octal games, *Int. J. Game Theory* **18** (1989), 311–320.

21. P.M. Grundy, Mathematics and games, *Eureka* **2** (1939), 6–8; reprinted, **27** (1964), 9–11.

22. P.M. Grundy and Cedric A.B. Smith, Disjunctive games with the last player losing, *Proc. Cambridge Philos. Soc.* **52** (1956), 527–533; MR **18**, 546b.

23. Richard K. Guy (editor), Combinatorial games, *Proc. Symp. Appl. Math.* **43**, American Mathematical Society, Providence, RI, 1991.

24. Richard K. Guy and Cedric A.B. Smith, The *G*-values of various games, *Proc. Cambridge Philos. Soc.* **52** (1956), 514–526; MR **18**, 546a.

25. Godfrey Harold Hardy, *A Mathematician's Apology,* reprinted with a Foreword by C.P. Snow, Cambridge University Press, Cambridge, 1967.

26. John Charles Kenyon, *Nim-like Games and the Sprague–Grundy Theory,* M.Sc. thesis, University of Calgary, 1967.

27. Donald Ervin Knuth, *Surreal Numbers,* Addison-Wesley, Reading, MA, 1974.

28. Sam Loyd, *Cyclopedia of Tricks and Puzzles,* New York, 1914, p. 232.

29. William L. Sibert and John H. Conway, Mathematical Kayles, *Int. J. Game Theory* (submitted).

30. R.P. Sprague, Über mathematische Kampfspiele, *Tôhoku Math. J.* **41** (1935–36), 438–444; Zbl. **13**, 290.

*Mathematics and Statistics Department*
*University of Calgary*
*2500 University Drive NW*
*Calgary, Alberta*
*CANADA T2N 1N4*

# Quantification Theory in a Nutshell

## Raymond M. Smullyan

Raymond Smullyan, 1968

I first knew Professor Halmos when I was a student at the University of Chicago. I recall sitting in on a course he gave on axiomatic set theory and that towards the beginning he announced that he was about to prove the existence of the empty set. I, brash student that I was, interrupted him and said: "But that's obvious! If there wasn't an empty set, then the set of all empty sets would be empty and we would have a contradiction!" I was very proud of myself, until Halmos wisely answered: "That's a bad pun!" I then realized that my fallacy consisted in the unwarranted assumption that there was such a thing as the set of all empty sets. [Incidentally, if one took this assumption as an axiom, instead of the weaker axiom that there exists a set, then the existence of the empty set would follow (by the very argument I gave) without use of the separation axiom (aussunderungsaxiom) necessary for the usual proof. But let that pass!]

It was in this course that I first learned the Curry paradox — a method of proving any proposition whatsoever. For those who don't know the method, let $p$ be any proposition whatsoever — say, that Santa Claus exists. Now let $S$ be the following sentence:

If this sentence is true then Santa Claus exists.

Thus $S$ says that $S$ implies $p$, where $p$ is the proposition that Santa Claus exists. If $S$ is true, then $S$ really does imply $p$, hence $p$ is then implied by a true proposition, hence is also true. This proves that if $S$ is true, so is $p$. But that's just what $S$ says, hence $S$ is true! And since $S$ does imply $p$ (as we have shown), $p$ must also be true — i.e., Santa Claus exists.

Someone (Hegel?) once defined a paradox as a truth standing on its head. Well, this paradox of Curry has indeed led me to realize a truth that has surprised many a logician to whom I have told it — namely, that conjunction is definable from implication and equivalence. How? Well, $p \wedge q$ is logically equivalent to the formula $p \equiv (p \supset q)$ (as the reader can verify by, say, a truth table). [For Curry's paradox, since $S$ says that $S$ implies $p$, we have $S \equiv (S \supset p)$, hence we get both $S$ and $p$.]

And so, we have seen Curry's method of proving any proposition whatsoever. I have invented another method, but unfortunately, only a magician can use it (I was a magician at the time that I was a student at Chicago). To prove to you that Santa Claus exists, I take a deck of cards, show you one — say the queen of hearts — and lay it face down on your palm. I then say: "You certainly grant that *either* this card is the queen of hearts *or* Santa Claus exists — at least *one* of those two propositions is true?" Of course you assent. I then ask you to turn over the card, and you find that it is *not* the queen of hearts, and so I triumphantly say: "So therefore Santa Claus exists!"

Of course I pulled this trick on Halmos, and I also used the same method to prove to the philosopher Rudolph Carnap (who was at Chicago at the same time) the existence of God. Carnap's reaction was delightful; he said: "Ah, yes! Proof by legerdemain. Same as the theologians use."

Saunders Maclane was also at the University of Chicago at the time, and of course he also knew I was a magician. After not having seen me for some time, he once met me in the hall of the math building and said: "Well Smullyan, how's tricks?" And after a brief pause, added: "Or is that an impertinent question?"

One of my fondest memories of Halmos is his definition of an $\omega$-inconsistent mother. As he first explained to the class, a mathematical system is called $\omega$-inconsistent if there is some property $P$ of natural numbers such that the system can prove that there exists at least one natural number that has the property, yet for each natural number $n$, the system can prove that $n$ doesn't have the property (thus all members of the infinite set $\{\exists n P(n),$ $\sim P(0), \sim P(1), \ldots, \sim P(n), \ldots\}$ are provable). Then he gave the example of the $\omega$-inconsistent mother who forbids her child to do this thing, or to do that thing, or the other thing, — indeed, whatever the child asks to do, the mother forbids. The child then asks: "Isn't there *anything* I can do?" The mother replies: "Oh, yes, there is *something* you can do, but it is not this, nor that, nor ..."

I recall that Halmos was quite fond of logic puzzles, so here is one that I would like to dedicate to him. It might appropriately be called: "Are all saints distinguished?" The puzzle concerns a well-orderable universe $U$ with infinitely many inhabitants (perhaps nondenumerably many) but we are given that each inhabitant has at most denumerably many children. [That's a reasonable requirement, isn't it?] We are also given that the child–parent relation is well-founded, in the sense that given any set $A$ of inhabitants, a sufficient condition that all inhabitants belong to $A$ is that for any inhabitant $x$, if all children of $x$ belongs to $A$, so does $x$. [This is equivalent to saying that each inhabitant can be assigned an ordinal in such a manner that each parent has a higher ordinal than any of his or her children.] A set of inhabitants is called *agreeable* if no member dislikes any other member.

The curious thing about this universe is that on any day each inhabitant either lies during the entire day or tells the truth during the entire day, but not both. His behavior may change from day to day, but is constant during any one day. Now, here are the "truth-telling" laws of the universe:

$T_1$: A father tells the truth on a given day if all his children do.

$T_2$: A mother tells the truth on a given day if at least one of her children does.

$T_3$: Given any agreeable set of childless inhabitants, there is at least one day on which they all tell the truth.

Now, the inhabitants of this universe have formed various clubs — each club being countable. We will make the convenient but inessential assumption that every subset of a club is a club. A person $x$ is said to be *compatible* with a club $C$ if $C \cup \{x\}$ (the result of adjoining $x$ to the set $C$) is a club. The clubs obey the following conditions:

$C_1$: If a father is compatible with a club $C$ then every one of his children is compatible with $C$.

$C_2$: If a mother is compatible with a club $C$, then at least one of her children is compatible with $C$.

$C_3$: For any club $C$, the set of all childless people in $C$ is agreeable.

The problem is to prove that for any club $C$, there is at least one day on which all its members tell the truth.

Does this problem remind you of any famous theorem you know?

There is another part of the problem: There is another universe $U^0$ (called the *dual* of $U$) in which all the given laws of $U$ hold, interchanging "lying" with "truth-telling." In this dual universe, a set of inhabitants is called a *society* if it is not a club. A person is called *distinguished* if he belongs to a society that contains him as the only member. Finally, a person is called a *saint* if he tells the truth on all days. The problem now is to prove that all saints are distinguished.

By now, many of you recognize this as Gödel's completeness theorem in disguise (saints correspond to formulas that are true in all interpretations and distinguished inhabitants correspond to formulas that are provable). Of course more can be said: In this dual universe, given any club, there must be at least one day on which all the members lie.

The first part of the problem corresponds to a powerful result of modern logic known as the *model existence theorem* (the "clubs" are, roughly speaking, the consistent sets). Let me now restate the problem in purely set-theoretic terms, leaving out all anthropomorphisms.

Instead of a set $U$ of *people* we consider an arbitrary infinite set $U$. Instead of the child–parent relation we consider an arbitrary binary relation $C(x, y)$, which we read: $x$ is a *component* of $y$. Again the relation is given to be well founded, in the sense already described, and each element $y$ of $U$ has at most denumerably many components. An element having no components will be called *simple*. All nonsimple elements are classified under two heads — *conjunctive* elements and *disjunctive* elements (these are respectively the fathers, mothers, for the club problem). Instead of the particular relation $x$ dislikes $y$ we consider an arbitrary binary relation $N(x, y)$. [In one application to mathematical logic, $N(x, y)$ is the relation that the formula $x$ is the *negation* of the formula $y$.] We call a set *agreeable* if $N(x, y)$ holds for no $x, y$ in the set.

Next, we are given that certain sets of elements are called *truth sets*. [For the puzzle, the truth set of a day is the set of all the inhabitants who tell the truth on that day. For mathematical logic, instead of *days*, we have *interpretations*, and the truth set of an interpretation is the set of all formulas true under the interpretation.] A subset $A$ of $U$ is called *satisfiable* if it is a subset of at least one truth set. We are given the following facts about the truth sets:

$T_1$: For any truth set $T$, a conjunctive element is in $T$ if all its components are.

$T_2$: For any truth set $T$, a disjunctive element is in $T$ if at least one of its components is.

$T_3$: Every agreeable set of simple elements is satisfiable.

**Remark.** In all applications that I know, the "if" in $T_1$ and $T_2$ is actually "if and only if." But these stronger conditions are not necessary for the theorem that follows.

Lastly we have a collection $\mathcal{C}$ of countable subsets of $U$ that we will call *good* sets (these are the clubs, for the puzzle, and are roughly the countable consistent sets in one application to mathematical logic) and we are given conditions $C_1$, $C_2$, $C_3$ of the puzzle reading "good set" for "club" (and likewise in the definition of *compatible*). Again we make the convenient but inessential assumption that every subset of a good set is good.

**Theorem G** [The Good Set Theorem]. *Under these given conditions, every good set is satisfiable.*

This theorem (whose proof I will discuss later) is a purely set theoretic generalization of the result known as the *model existence theorem*, which in turn not only generalizes Gödel's completeness theorem, but has a host of other applications as well.

To begin with, let us consider first-order logic with $\sim$, $\wedge$, $\vee$, $\forall$, $\exists$, as primitives. We add denumerably many constants (parameters) $a_0, a_1, a_2, \ldots, a_n,$

... to the language and our "formulas" may contain constants as well. By a *simple* formula we shall mean an atomic formula or its negation. By a *conjunctive* formula we shall mean one of any of the five forms $\sim\sim X$, $X \wedge Y$, $\sim (X \vee Y)$, $\forall x F(x)$, $\sim \exists x F(x)$, and by a *disjunctive* formula we shall mean one of any of the four forms $X \vee Y$, $\sim (X \wedge Y)$, $\exists x F(x)$, $\sim \forall x F(x)$. The *components* of a formula are given as follows.

| Formula | Components |
|---------|-----------|
| $\sim\sim X$ | $X$ |
| $X \wedge Y$ | $X, Y$ |
| $X \vee Y$ | $X, Y$ |
| $\sim(X \wedge Y)$ | $\sim X, \sim Y$ |
| $\sim(X \vee Y)$ | $\sim X, \sim Y$ |
| $\forall x F(x)$ | $F(a_0), F(a_1), \ldots, F(a_n), \ldots$ |
| $\exists x F(x)$ | $F(a_0), F(a_1), \ldots, F(a_n), \ldots$ |
| $\sim\forall x F(x)$ | $\sim F(a_0), \sim F(a_1), \ldots, \sim F(a_n), \ldots$ |
| $\sim\exists F(x)$ | $\sim F(a_0), \sim F(a_1), \ldots, F(a_n), \ldots$ |

As can be easily seen, under any interpretation (of all predicates and constants) a conjunctive formula is true if and only if all its components are true, and a disjunctive formula is true if and only if at least one of its components is true. Also, given any set $S$ of atomic formulas and negations of atomic formulas, such that no atomic formula and its negation are both in $S$, there is an interpretation in which all elements of $S$ are true (and also one in which they are also false), thus the truth sets do satisfy the laws $T_1$, $T_2$, $T_3$.

Next, a collection $\mathcal{C}$ of countable sets of formulas is called an *abstract consistency property* — more briefly, a *consistency property* — if conditions $C_1, C_2, C_3$ hold, reading "element of $\mathcal{C}$" for "good set." The model existence theorem for first-order logic (which is a special case of Theorem G) is that if $\mathcal{C}$ is any consistency property, every element of $\mathcal{C}$ is satisfiable. Out of this has come a host of applications: Let us first consider ordinary consistency: A set of $S$ of formulas is said to be *consistent* in an axiom system if no negation of any finite conjunction of members of $S$ is provable in the system. Now, given any standard axiom system for first-order logic, the collection $\mathcal{C}$ of all consistent sets $S$ that have the additional property that there are infinitely many parameters that do not occur in any element of $S$ — this collection $\mathcal{C}$ is a consistency property (indeed, it is fairly easy to design an axiom system that ensures this) and so by the model existence theorem, every member of $\mathcal{C}$ is satisfiable (from which, incidentally, it easily follows that every consistent set — even one without the above mentioned additional property — is satisfiable). From this, in turn, follows Gödel's completeness theorem that every valid formula (formula that is true under all interpretations) is provable, because if $X$ is valid, its negation $\sim X$ is not satisfiable, hence $\{\sim X\}$ is not consistent, hence $\sim\sim X$ is provable, hence $X$ is provable.

There are many other abstract consistency properties than the one we have just considered, and the model existence theorem applied to different ones yields all sorts of interesting results other than completeness. For example, consider the collection $\mathcal{C}$ of all sets $S$ such that every finite subset of $S$ is satisfiable and such that infinitely many parameters occur in no element of $S$. This $\mathcal{C}$ is also a consistency property, and so the model existence theorem guarantees that every member of this collection $\mathcal{C}$ is satisfiable, from which easily follows the fact (known as the *compactness* theorem for first-order logic) that a set is satisfiable if and only if all its finite subsets are satisfiable.

Another application of the model existence theorem yields Craig's interpolation lemma: If $X \supset Y$ is valid, and if neither $Y$ nor $\sim X$ is valid, then there is a formula $Z$ such that $X \supset Z$ and $Z \supset Y$ are both valid and all predicates and parameters of $Z$ are in both $X$ and $Y$. To prove this, call a formula $Z$ an *interpolation* formula for an ordered pair $(A_1, A_2)$ of sets of formulas if every predicate and parameter of $Z$ occurs in at least one element of $A_1$ and at least one element of $A_2$ and if $A_1 \cup \{Z\}$ and $A_2 \cup \{\sim Z\}$ are both unsatisfiable. Now call a set $A$ *Craig consistent* if it is finite and can be partitioned into two satisfiable nonempty sets $A_1, A_2$ such that there is no interpolation formula for $(A_1, A_2)$. The collection of all Craig consistent sets is another consistency property, hence by the model existence theorem, every Craig consistent set is satisfiable, from which easily follows Craig's interpolation lemma. Many refinements of this lemma can be proved, using other consistency properties — indeed, Kiesler's book *Model Theory for Infinitary Logics* [1] is replete with applications of the model existence theorem for infinitary logics in which we allow denumerable conjunctions and disjunctions of formulas. In this setup, a conjunction, disjunction (whether finite or denumerable) is, respectively, of conjunctive, disjunctive type, and the components are again, respectively, the conjuncts, disjuncts. And negations of conjunctions, disjunctions are, respectively, of disjunctive, conjunctive type and the components are, respectively, the negations of the conjuncts, disjuncts. Under this meaning of conjunctive and disjunctive elements and their components, Theorem G is the model existence theorem for language with denumerable conjunctions and disjunctions (but without identity or function symbols) and the model existence theorem in this form is due to M. Makkai and was published in his 1969 paper: *An application of a method of Smullyan to logics on admissable sets* [2]. My own role in the history of this theorem is briefly as follows:

In looking for a consistency proof for arithmetic, Gentzen gave an axiom system for first-order logic and showed that any formula provable in his system had a special proof (called a "cut-free proof") having the property that it used only subformulas of the formula to be proved. This result was known as Gentzen's *Haupsatz*. Now, in the early sixties it occurred to me that one should be able to get Gentzen's Haupsatz, Gödel's completeness theorem, the compactness theorem, the Skolem–Löwenheim theorem (every satisfiable set of formulas of first-order logic is satisfiable in a denumerable domain), a variant of Herbrand's theorem, and Craig's interpolation lemma all from one common construction. This I did in my 1963 paper: "A Unifying Principle in Quantification Theory" [3], in which I introduced the notion of a *consistency property* and showed that all the above results are easy corollaries of the fact that for any such property, any set having that property is satisfiable in a denumerable domain. However, my definition of a consistency property differed somewhat from that of Makkai: For one thing, my definition had a totally unnecessary feature that made it not generalizable to infinitary logics — namely that the property be compact (a set has the property if and only if all its finite subsets have the property). My proof, however, made absolutely no use of this excess baggage in the definition, and when this was realized, the generalization to infinitary logics (in which compactness fails) was perfectly straightforward.

I have always shared Halmos's feelings that the mathematically interesting aspects of first-order logic should be presented in a form that does not appeal to the peculiarities of the formulas of the language (logical connectives, quantifiers, scope, free and bound occurrences of variables, substitution for free occurrences, and other linguistic monstrosities). Theorem G of this article (which I have not published before) is my attempt at a purely set-theoretic formulation. Perhaps someone working in algebraic logic might find a direct application (at least, it would be nice!).

Of course Theorem G, as it stands, is nonconstructive, but a constructive version of it is attainable as follows: Assume that the universe $U$ is denumerable and that the following items are recursive: the set of conjunctive elements; the set of disjunctive elements; the relations $C(x,y)$, $N(x,y)$; and the set of all $x$ having infinitely many components. Finally, assume that there is a recursive function $\varphi$ that maps every finite set $S$ and every element $x$ to some element $\varphi(S,x)$ such that if $x$ is disjunctive and has infinitely many components and if $x$ is compatible with $S$, then $\varphi(S,x)$ is a component of $x$ that is compatible with $S$. {For first-order logic, I am thinking of $\varphi[S, \exists x F(x)]$ as $F(a)$, and $\varphi[S, \sim\forall x F(x)]$ as $\sim F(a)$, where $a$ is a parameter (say the first in some preassigned enumeration) that does not occur in $F(x)$ or in any element of $S$.} With these additional restriction (which do indeed hold for first-order logic) the set of all finite sets that are *not* good (which for first-order logic are the finite *inconsistent* sets) is recursively enumerable. From this it easily follows that the set of all valid formulas of first-order logic is recursively enumerable, which is really the essence of Gödel's completeness theorem and (as many logicians believe) is of greater significance than the completeness of this particular or that particular axiom system for first-order logic.

**Postscript.** Concerning the proof of Theorem G, the key lemma is that any good set $G$ is a subset of a set $S$ (a countable one, in fact) having the following three properties: $(P_1)$ $S$ contains with each conjunctive element all of its components; $(P_2)$ $S$ contains with each disjunctive element at least one of its components; $(P_3)$ the set of simple elements of $S$ is agreeable. The proof of this (at least one proof of this) uses only conditions $C_1$, $C_2$, and $C_3$ and is briefly as follows: I shall sketch the proof only for the case that $G$ is denumerable (the modifications for a finite $G$ are obvious). Well, let $G$ be enumerated in some order $a_1, a_2, \ldots, a_n, \ldots$. We then construct an infinite sequence $b_1, b_2, \ldots, b_n, \ldots$ by the following inductive scheme. We take $b_1 = a_1$. That concludes the first stage of the construction. Now suppose the $n$th stage has been completed and we have a finite sequence $b_1, \ldots, b_k$ — call it $\theta_n$ — such that $n \leq k$ and that the set $G \cup \{b_1, \ldots, b_k\}$ is good. We then "attack" the $n$th term $b_n$ and do the following: If it is atomic, we let $\theta_{n+1}$ be the sequence $b_1, \ldots, b_k, a_{n+1}$. If $b_n$ is disjunctive, we take some component $x$ of $b_n$ (the first, say, in the well ordering of $U$) such that $G \cup \{b_1, \ldots, b_k, x\}$ is good and we let $\theta_{n+1} = b_1, \ldots, b_k, x, a_{n+1}$. If $b_n$ is conjunctive and has only finitely many components $s_1, \ldots, s_t$, we let $\theta_{n+1} = b_1, \ldots, b_k, s_1, \ldots, s_t, a_{n+1}$. If $b_n$ is conjunctive and has infinitely many components (and this is the subtle case!) we let $y$ be the first component of $b_n$ (in the well ordering of $U$) that is not in the set $\{b_1, \ldots, b_k\}$ and we let $\theta_{n+1} = b_1, \ldots, b_k, y, b_n, a_{n+1}$. [The purpose of "repeating" $b_n$ is to guarantee that we will attack it again — in fact again and again infinitely many times, and so all its components will be eventually thrown in.] We then let $S$ be the infinite set $\{b_1, \ldots, b_n, \ldots\}$ and it is easy to verify that $S$ has the desired properties $P_1$, $P_2$, $P_3$ (and, of course, $G$ is a subset of $S$).

Having proved the lemma, the rest is easy: By $P_3$ and $T_3$, the set of simple elements of $S$ is a subset of some truth-set $T$. For any nonsimple element $x$ of $S$, if all components of $x$ that are in $S$ are also in $T$, then $x$ must be in $T$ (this follows by $P_1$ and $T_1$ if $x$ is conjunctive, and by $P_2$ and $T_2$ if $x$ is disjunctive), hence by well-foundedness, all elements of $S$ are in $T$, and thus $G$ (which is a subset of $S$) is a subset of $T$.

## REFERENCES

1. H.J. Keisler, *Model Theory for Infinitary Logics*, North-Holland, Amsterdam, 1971.

2. M. Makkai, An application of a method of Smullyan to logics on admissible sets, *Bull. l'Acad. Polon. Sci. Ser. Math.* **17** (1969), 341–346.

3. R. Smullyan, A unifying principle in quantification theory, *Proc. Natl. Acad. Sci.* **49** (1963), 828–832.

*Department of Philosophy*
*Indiana University*
*Bloomington, IN 47405*

# The Legendre Relation for Elliptic Integrals

## Peter Duren

Peter Duren, 1989

In standard notation, the complete elliptic integrals of the first and second kinds are

$$K = K(k) = \int_0^{\pi/2} \frac{d\theta}{\sqrt{1 - k^2 \sin^2 \theta}}$$

$$= \int_0^1 \frac{dt}{\sqrt{1 - t^2}\sqrt{1 - k^2 t^2}}$$

and

$$E = E(k) = \int_0^{\pi/2} \sqrt{1 - k^2 \sin^2 \theta}\, d\theta$$

$$= \int_0^1 \frac{\sqrt{1 - k^2 t^2}}{\sqrt{1 - t^2}}\, dt.$$

The parameter $k$ ($0 < k < 1$) is called the *modulus,* and $k' = \sqrt{1 - k^2}$ is the *complementary modulus.* Symmetrically related to $K$ and $E$ are $K' = K(k')$ and $E' = E(k')$. The *Legendre relation* asserts that for all $k$,

$$K'E + E'K - K'K = \frac{\pi}{2}. \tag{1}$$

This beautiful identity, discovered by Adrien Marie Legendre (1752–1833) as early as 1811, is not very well known today. It recently attracted the writer's interest when, quite unexpectedly, it played a crucial part in the solution of an extremal problem for univalent functions [9]. The present paper is an attempt to uncover the mathematical significance of the Legendre relation, to trace its historical roots, and to give three very different proofs.

## Legendre's Derivation

Any effort to understand the Legendre relation must begin at the source, the classical works of Legendre. How was the master led to discover his remarkable relation? The answer is disappointing: he seems to have come across it by accident, while searching for applications of his systematic method for the evaluation of elliptic integrals.

The Legendre relation appeared in Legendre's integral calculus text [13] and again in his monumental treatise on elliptic integrals [14, p. 61]. In the latter source his discussion begins with two formulas cited from Euler's integral calculus text. With the notation

$$I(p) = \int_0^1 t^p (1-t^2)^{-1/2} dt, \qquad -1 < p < \infty,$$

the two formulas are

$$I\left(-\frac{2}{3}\right) I\left(\frac{1}{3}\right) = \frac{3\pi}{2}; \qquad I\left(-\frac{1}{3}\right) I\left(\frac{2}{3}\right) = \frac{3\pi}{4}. \tag{2}$$

Actually, these are special cases of the formula

$$I(p)I(p+1) = \frac{1}{p+1} \frac{\pi}{2}, \tag{3}$$

which is easily derived from Euler's famous relation

$$B(x,y) = \frac{\Gamma(x)\Gamma(y)}{\Gamma(x+y)} \tag{4}$$

between the beta function

$$B(x,y) = \int_0^1 t^{x-1} (1-t)^{y-1} dt, \qquad x > 0,\ y > 0,$$

and the gamma function

$$\Gamma(x) = \int_0^\infty t^{x-1} e^{-t} dt, \qquad x > 0.$$

Simply let $u = t^2$ to see that

$$I(p) = \frac{1}{2} B\left(\frac{p}{2} + \frac{1}{2}, \frac{1}{2}\right) = \frac{\Gamma(\frac{p}{2} + \frac{1}{2})\Gamma(\frac{1}{2})}{2\Gamma(\frac{p}{2} + 1)}. \tag{5}$$

Then

$$I(p+1) = \frac{\Gamma(\frac{p}{2} + 1)\Gamma(\frac{1}{2})}{(p+1)\Gamma(\frac{p}{2} + \frac{1}{2})}, \tag{6}$$

by the identity $\Gamma(x+1) = x\Gamma(x)$. Multiplying Eq. (5) by Eq. (6) and recalling that $\Gamma(\frac{1}{2}) = \sqrt{\pi}$, one obtains Eq. (3).

Legendre now uses one of his formulas for the reduction of elliptic integrals to normal form to conclude that

$$I\left(-\frac{2}{3}\right) = 3^{3/4} K\left(\sin\frac{\pi}{12}\right) \tag{7}$$

and

$$I\left(\frac{1}{3}\right) = 3^{-1/4}\left[(1-\sqrt{3})K\left(\cos\frac{\pi}{12}\right) + 2\sqrt{3}\,E\left(\cos\frac{\pi}{12}\right)\right]. \tag{8}$$

[One sees by the half-angle formula that $\sin\pi/12 = (\sqrt{3}-1)/2\sqrt{2}$ and $\cos\pi/12 = (\sqrt{3}+1)/2\sqrt{2}$.] Multiplying Eq. (7) by Eq. (8) and using the first of Euler's formulas (2), he obtains the equation

$$\sqrt{3}K\left[(1-\sqrt{3})K' + 2\sqrt{3}E'\right] = \frac{3\pi}{2} \tag{9}$$

for the modulus $k = \sin\pi/12$. The same procedure leads to similar formulas for $I\left(-\frac{1}{3}\right)$ and $I\left(\frac{2}{3}\right)$, whereupon the second part of Eq. (2) gives

$$\sqrt{3}K'\left[2\sqrt{3}E - (1+\sqrt{3})K\right] = \frac{3\pi}{2} \tag{10}$$

for $k = \sin\pi/12$. Legendre now adds Eqs. (9) and (10) to arrive at his relation (1) for this special value of $k$.

Remarking that $P = K'E + E'K - K'K$ also has the value $\pi/2$ for $k = 1/\sqrt{2}$ and as $k \to 0$, Legendre proceeds to compute the derivative $dP/dk$. With the formulas

$$kk'^2\frac{dK}{dk} = E - k'^2K; \qquad k\frac{dE}{dk} = E - K, \tag{11}$$

a straightforward calculation shows that $dP/dk = 0$, so $P$ is constant. This is Legendre's proof of the Legendre relation.

## Differential Equations

Legendre's proof is straightforward but totally unilluminating. It *verifies* the Legendre relation but does nothing to *explain* it. There is, however, a different approach which exploits the fact that $K$ and $K'$ are solutions to the same hypergeometric differential equation and so satisfy a Wronskian relation. It turns out that the Wronskian relation is none other than the Legendre relation.

The *hypergeometric function* of Gauss is

$$F(a,b;c;x) = 1 + \frac{ab}{c}x + \frac{a(a+1)b(b+1)}{2!c(c+1)}x^2 + \cdots, \qquad |x| < 1.$$

It satisfies the linear differential equation

$$x(1-x)y'' + [c - (a+b+1)x]\,y' - aby = 0, \tag{12}$$

known as the *hypergeometric equation*. Here $a$, $b$, and $c$ are arbitrary parameters, with $c$ not a negative integer. It is easily seen that if $a+b+1 = 2c$, any solution $y(x)$ of Eq. (12) also provides the solution $y(1-x)$.

Expanding the integrands in binomial series and integrating term by term, one can verify that the elliptic integrals $K$ and $E$ are hypergeometric functions of $x = k^2$:

$$K(k) = \frac{\pi}{2} F\left(\frac{1}{2}, \frac{1}{2}; 1; k^2\right), \qquad E(k) = \frac{\pi}{2} F\left(-\frac{1}{2}, \frac{1}{2}; 1; k^2\right).$$

A neater proof of a more general result goes by expressing the series for $F(a, b; c; x)$ in terms of the gamma function, using Euler's formula (4) to substitute beta integrals, then interchanging the order of summation and integration to arrive at the formula (*cf.* [7, p. 214] or [18, p. 293])

$$B(b, c - b) F(a, b; c; x) = \int_0^1 t^{b-1} (1 - t)^{c-b-1} (1 - xt)^{-a} dt, \qquad 0 < b < c.$$

We note in passing that the differentiation formulas (11) are now a direct consequence of the *contiguous relation*

$$\left(x\frac{d}{dx} + a\right) F(a, b; c; x) = aF(a + 1, b; c; x)$$

and the differential equation (12).

It is well known that any two solutions $y_1$ and $y_2$ to a differential equation $y'' + py' + qy = 0$ have a *Wronskian* $W = y_1'y_2 - y_1y_2'$ of the form

$$W(x) = W(x_0) \exp\left\{-\int_{x_0}^x p(t)dt\right\}, \tag{13}$$

where $x_0$ is a convenient base-point. For the hypergeometric equation (12) with $a = \frac{1}{2}$, $b = \frac{1}{2}$, and $c = 1$, the Wronskian relation (13) takes the form

$$W(x) = W\left(\frac{1}{2}\right) \exp\left\{-\int_{1/2}^x \frac{1 - 2t}{t(1 - t)} dt\right\},$$

and an easy calculation gives

$$W(x) = \frac{1}{4} W\left(\frac{1}{2}\right) \frac{1}{x(1 - x)}, \qquad 0 < x < 1. \tag{14}$$

Now choose the particular solutions

$$y_1(x) = K(k) = \frac{\pi}{2} F\left(\frac{1}{2}, \frac{1}{2}; 1; x\right);$$

$$y_2(x) = K'(k) = \frac{\pi}{2} F\left(\frac{1}{2}, \frac{1}{2}; 1; 1 - x\right),$$

where $x = k^2$. (Note that $K'$ is also a solution because $a + b + 1 = 2c$ in this case.) The first differentiation formula (11) yields

$$\frac{dy_1}{dx} = \frac{E - (1 - x)K}{2x(1 - x)}; \qquad \frac{dy_2}{dx} = -\frac{E' - xK'}{2x(1 - x)},$$

which imply that the Wronskian $W$ of $K$ and $K'$ is given by

$$2x(1 - x)W(x) = K'E + E'K - K'K. \tag{15}$$

A comparison of Eqs. (14) and (15) shows that the expression in the Legendre relation is constant. In fact, this interprets the Legendre relation as a Wronskian relation.

The constant value is most easily determined by computing the limit of $K'E + E'K - K'K$ as $k \to 0$. However, it is more interesting to make a direct calculation for $k = k' = 1/\sqrt{2}$. The expression then reduces to

$$2K(1/\sqrt{2})E(1/\sqrt{2}) - K(1/\sqrt{2})^2.$$

That this has the value $\pi/2$ is easily concluded from the formulas

$$K(1/\sqrt{2}) = \frac{1}{4\sqrt{\pi}}\Gamma\left(\frac{1}{4}\right)^2; \tag{16}$$

$$E(1/\sqrt{2}) = \frac{1}{8\sqrt{\pi}}\left[\Gamma\left(\frac{1}{4}\right)^2 + 4\Gamma\left(\frac{3}{4}\right)^2\right], \tag{17}$$

since the familiar identity

$$\Gamma(x)\Gamma(1 - x) = \frac{\pi}{\sin \pi x}$$

gives in particular $\Gamma\left(\frac{1}{4}\right)\Gamma\left(\frac{3}{4}\right) = \sqrt{2}\pi$.

To obtain Eq. (16), write

$$K(1/\sqrt{2}) = \sqrt{2}\int_0^1 \frac{dt}{\sqrt{1 - t^2}\sqrt{2 - t^2}} = \sqrt{2}\int_0^1 \frac{du}{\sqrt{1 - u^4}},$$

by the substitution $u = t/\sqrt{2 - t^2}$. Now let $v = u^4$ to get

$$K(1/\sqrt{2}) = \frac{\sqrt{2}}{4}\int_0^1 v^{-3/4}(1 - v)^{-1/2}dv = \frac{\sqrt{2}}{4}B\left(\frac{1}{4}, \frac{1}{2}\right)$$

$$= \frac{\sqrt{2}}{4}\frac{\Gamma\left(\frac{1}{4}\right)\Gamma\left(\frac{1}{2}\right)}{\Gamma\left(\frac{3}{4}\right)} = \frac{1}{4\sqrt{\pi}}\Gamma\left(\frac{1}{4}\right)^2,$$

because $\Gamma\left(\frac{1}{2}\right) = \sqrt{\pi}$ and $\Gamma\left(\frac{1}{4}\right)\Gamma\left(\frac{3}{4}\right) = \sqrt{2}\pi$.

A similar calculation leads to the formula (17) for $E(1/\sqrt{2})$. Write

$$\sqrt{2}\,E(1/\sqrt{2}) = \int_0^1 \frac{\sqrt{2 - t^2}}{\sqrt{1 - t^2}}dt = \int_0^1 \frac{1 + u^2}{\sqrt{1 - u^4}}du,$$

by the substitution $u = \sqrt{1 - t^2}$, again let $v = u^4$ to express the last integral in terms of the beta function, and use Eq. (4) to obtain Eq. (17). Thus the Legendre relation is completely identified as a Wronskian relation.

This interpretation seems to have escaped Legendre, although he was aware [14, p. 63] that $K$ and $E$ are hypergeometric functions. The proof given above was suggested by an exercise in the text of Chaundy [7, p. 277]. A similar proof via differential equations appeared in the book of Durège [8, pp. 269–275] as early as 1861.

# Elliptic Functions

The most illuminating proof of the Legendre relation comes through the theory of elliptic functions. Shortly after the publication of Legendre's famous treatise on elliptic integrals, Abel and Jacobi started independently to consider the *inverses* of the incomplete elliptic integrals, the doubly periodic meromorphic functions now called *elliptic functions*. Weierstrass later constructed another version of the theory. It turned out that most of the known results on elliptic integrals had simpler and more natural formulations within the framework of elliptic functions.

The Legendre relation is no exception. An equivalent version arises naturally out of the Weierstrass theory of elliptic functions, where it is proved by a simple application of the residue theorem. However, it is not so easy to see that the new version is actually equivalent to the Legendre relation in original form. Before we can establish the connection, we must digress to recall some basic aspects of the Weierstrass and Jacobi theories.

Weierstrass based his theory of elliptic functions on the special example

$$\wp(z) = \frac{1}{z^2} + \sum_{\omega \neq 0} \left( \frac{1}{(z-\omega)^2} - \frac{1}{\omega^2} \right),$$

where $\omega_1$ and $\omega_2$ are prescribed periods with $\text{Im}\{\omega_2/\omega_1\} > 0$ and the sum extends over all periods $\omega = n\omega_1 + m\omega_2$ except 0. It is well known (see, for instance, Ahlfors [1, Chap. 7]) that $\wp$ satisfies the differential equation

$$\wp'(z)^2 = 4[\wp(z) - e_1][\wp(z) - e_2][\wp(z) - e_3], \tag{18}$$

where $e_1$, $e_2$, $e_3$ are distinct complex numbers with $e_1 + e_2 + e_3 = 0$. In fact,

$$e_1 = \wp(\omega_1/2), \quad e_2 = \wp(\omega_2/2), \quad e_3 = \wp(\omega_3/2), \tag{19}$$

where $\omega_3 = \omega_1 + \omega_2$. Conversely, it can be shown either by theta functions or with the help of the modular function $J(\tau)$ (called the *absolute invariant*), that to any prescribed system of distinct complex numbers $e_1$, $e_2$, $e_3$ with sum 0, there correspond periods $\omega_1$, $\omega_2$ whose associated $\wp$-function satisfies Eqs. (18) and (19). This is sometimes called the *Weierstrass inversion problem*. (See [12, p. 227] or [6, p. 91].)

The addition formula for the $\wp$-function is

$$\wp(z + u) = \frac{1}{4} \left( \frac{\wp'(z) - \wp'(u)}{\wp(z) - \wp(u)} \right)^2 - \wp(z) - \wp(u).$$

Important special cases are

$$\wp\left(z + \frac{\omega_1}{2}\right) = \frac{(e_1 - e_2)(e_1 - e_3)}{\wp(z) - e_1} + e_1; \tag{20}$$

$$\wp\left(z + \frac{\omega_2}{2}\right) = \frac{(e_2 - e_1)(e_2 - e_3)}{\wp(z) - e_2} + e_2, \tag{21}$$

deduced by Eqs. (18) and (19), and the fact that $e_1 + e_2 + e_3 = 0$.
Related to the $\wp$-function is the *Weierstrass zeta function*

$$\zeta(z) = \frac{1}{z} + \sum_{\omega \neq 0}\left(\frac{1}{z - \omega} + \frac{1}{\omega} + \frac{z}{\omega^2}\right),$$

the unique odd function for which $\wp(z) = -\zeta'(z)$. The values

$$\eta_1 = \zeta(\omega_1/2), \qquad \eta_2 = \zeta(\omega_2/2) \tag{22}$$

of $\zeta$ at the half-periods are easily seen to serve as correction terms in a kind of quasiperiodic property:

$$\zeta(z + \omega_1) = \zeta(z) + 2\eta_1, \qquad \zeta(z + \omega_2) = \zeta(z) + 2\eta_2. \tag{23}$$

Integrating $\zeta(z)$ around the boundary of any period-parallelogram which encloses a point of the period module, and applying both Eq. (23) and the residue theorem, one obtains the simple identity

$$\omega_2\eta_1 - \omega_1\eta_2 = \pi i, \tag{24}$$

also known in the literature as the *Legendre relation*. This terminology goes back to Weierstrass, who showed in his lectures that Eq. (24) corresponds to the Legendre relation (*cf.* [16, p. 34] or [17, p. 131]).

We now find ourselves in a rather curious situation. It is fairly easy to verify the Legendre relation in either of the forms (1) or (24), but much more difficult to see that they are expressions of the same mathematical phenomenon. Authors of texts on elliptic functions often follow Weierstrass in using theta functions to establish the correspondence (*cf.* Akhiezer [2]). As we shall see, however, it is possible to give a direct proof, without appeal to theta functions, that the quasiperiod form (24) of the Legendre relation implies the original form (1) and is essentially equivalent to it. In the course of the proof we shall encounter some surprising identities.

The original form of the Legendre relation involves the complete elliptic integrals $K$ and $K'$, which are closely allied with Jacobi's *sine amplitude function* $\mathrm{sn}(z)$. This suggests the need for a connection between the Weierstrass and Jacobi theories. Jacobi's function $w = \mathrm{sn}(z)$ depends on a parameter $k$ ($0 < k < 1$) and may be defined initially as the conformal mapping of the rectangle with vertices $\pm K$ and $(\pm K + iK')$ onto the upper half-plane

$\operatorname{Im}\{w\} > 0$, with $\operatorname{sn}(0) = 0$, $\operatorname{sn}(\pm K) = \pm 1$, $\operatorname{sn}(\pm K + iK') = \pm 1/k$, and $\operatorname{sn}(iK') = \infty$. It is the inverse of the Schwarz–Christoffel mapping

$$z = \int_0^w \frac{dw}{\sqrt{1 - w^2}\,\sqrt{1 - k^2 w^2}}$$

(*cf.* Nehari [15], p. 280). Thus sn satisfies the differential equation

$$\operatorname{sn}'(z)^2 = (1 - \operatorname{sn}^2 z)(1 - k^2 \operatorname{sn}^2 z). \tag{25}$$

It is understood that $K$ and $K'$ are the standard elliptic integrals with modulus $k$. The function sn can be extended by Schwarz reflection to a meromorphic function in the whole plane with primitive periods $4K$ and $2iK'$.

Suppose now that distinct *real* numbers $e_1$, $e_2$, and $e_3$ are given with $e_2 < e_3 < e_1$ and $e_1 + e_2 + e_3 = 0$. Define $k$ by

$$k^2 = \frac{e_3 - e_2}{e_1 - e_2}, \qquad 0 < k < 1. \tag{26}$$

Then we claim that

$$\wp(z) = e_2 + \frac{e_1 - e_2}{\operatorname{sn}^2(\sqrt{e_1 - e_2}\,z)} \tag{27}$$

is the Weierstrass $\wp$ function with invariants $e_1$, $e_2$, $e_3$ and with primitive periods

$$\omega_1 = \frac{2K}{\sqrt{e_1 - e_2}}, \qquad \omega_2 = \frac{2iK'}{\sqrt{e_1 - e_2}}. \tag{28}$$

This is the basic link between the Weierstrass and Jacobi theories (*cf.* Enneper [10, p. 38]; Weierstrass [16, p. 32], [17, p. 100]; Whittaker and Watson [18, p. 505]; Hille [11, p. 156]).

To establish the link, let $f(z)$ denote the right-hand side of Eq. (27), and observe that $f$ is a doubly periodic function with primitive periods $\omega_1$ and $\omega_2$ as given by Eq. (28), since $\operatorname{sn}(z + 2K) = -\operatorname{sn}(z)$. Furthermore, $f(z) - e_2$ has a double pole at 0 and a double zero at $\omega_2/2$, and it has no other zeros or poles up to translation by periods. Let $\wp(z)$ be the Weierstrass $\wp$-function with the same primitive periods $\omega_1$ and $\omega_2$; and let $\tilde{e}_1 = \wp(\omega_1/2)$, $\tilde{e}_2 = \wp(\omega_2/2)$, $\tilde{e}_3 = \wp(\omega_3/2)$ be its invariants. Then it follows from Liouville's theorem that

$$f(z) - e_2 = c\,[\wp(z) - \tilde{e}_2],$$

where $c$ is a constant. But $f$ has the Laurent expansion $f(z) = 1/z^2 + \cdots$ at the origin, so $c = 1$. Thus $f(z) = \wp(z) + b$, where $b$ is a constant. In particular, one finds from the definition of $f$ that $e_1 = \tilde{e}_1 + b$, $e_2 = \tilde{e}_2 + b$, $e_3 = \tilde{e}_3 + b$. However, the invariants of $\wp$ satisfy $\tilde{e}_1 + \tilde{e}_2 + \tilde{e}_3 = 0$, and $e_1 + e_2 + e_3 = 0$ by construction, so $b = 0$ and $f(z) = \wp(z)$. Incidentally, it is entertaining to verify directly from Eqs. (27) and (25) that $\wp$ satisfies its differential equation (18).

On the basis of the connecting formula (27), one can now calculate expressions for the elliptic integrals $E$ and $E'$ which involve the values $\eta_1$ and $\eta_2$ of the Weierstrass zeta function at the half-periods. Indeed, the substitution $t = \mathrm{sn}(\sqrt{e_1 - e_2}\, z)$ gives by Eqs. (25), (26), (27), and (28)

$$E = \int_0^1 \frac{\sqrt{1 - k^2 t^2}}{\sqrt{1 - t^2}}\, dt = \sqrt{e_1 - e_2} \int_0^{\omega_1/2} \left[1 - k^2 \mathrm{sn}^2(\sqrt{e_1 - e_2}\, z)\right] dz$$

$$= \sqrt{e_1 - e_2} \left\{ \frac{\omega_1}{2} + (e_2 - e_3) \int_0^{\omega_1/2} \frac{dz}{\wp(z) - e_2} \right\}.$$

Introducing the addition formula (21) and recalling the definition (22) of $\eta_1$ and $\eta_2$, one finds after further manipulation that

$$\sqrt{e_1 - e_2}\, E = \frac{e_1 \omega_1}{2} + \eta_1, \tag{29}$$

using the fact easily deduced from Eq. (23) that $\zeta(\omega_3/2) = \eta_1 + \eta_2$.

The calculation of $E'$ is similar. First let $u = (1 - k'^2 t^2)^{-1/2}$ to write

$$E' = \int_0^1 \frac{\sqrt{1 - k'^2 t^2}}{\sqrt{1 - t^2}}\, dt = \int_1^{1/k} \frac{du}{u^2 \sqrt{u^2 - 1}\sqrt{1 - k^2 u^2}}.$$

Then let $u = \mathrm{sn}(\sqrt{e_1 - e_2}\, z)$ and use Eqs. (25) and (27) to obtain

$$i\sqrt{e_1 - e_2}\, E' = (e_1 - e_2) \int_{\omega_1/2}^{\omega_3/2} \frac{dz}{\mathrm{sn}^2(\sqrt{e_1 - e_2}\, z)} = \int_{\omega_1/2}^{\omega_3/2} [\wp(z) - e_2]\, dz.$$

Since $\wp(z) = -\zeta'(z)$, this gives

$$\sqrt{e_1 - e_2}\, E' = i\left( \frac{e_2 \omega_2}{2} + \eta_2 \right). \tag{30}$$

Finally, a straightforward calculation with the formulas (28), (29), and (30) leads to the identity

$$2i(K'E + E'K - K'K) = \omega_2 \eta_1 - \omega_1 \eta_2. \tag{31}$$

This establishes the equivalence of the two forms of the Legendre relation: Legendre's original statement (1) and the quasiperiod version (24).

Actually, the quasiperiod version is more general because it does not require the period lattice to be rectangular. The identity (31) was derived under an arbitrary choice of Weierstrass invariants $e_1$, $e_2$, $e_3$ *on the real line*, satisfying $e_2 < e_3 < e_1$ and $e_1 + e_2 + e_3 = 0$, so that the modulus $k$ defined by Eq. (26) is an arbitrary real number in the interval $0 < k < 1$. It makes no essential difference if the invariants are prescribed to lie on another line through the origin, but more general choices cannot be allowed.

One remarkable byproduct of the connecting formulas (26), (27), and (28) is the explicit solution of the Weierstrass inversion problem when the invariants $e_1$, $e_2$, $e_3$ are prescribed on a line through the origin.

# Epilogue: The Calculation of $\pi$

The famous *arithmetic-geometric mean algorithm* of Gauss can be used to calculate complete elliptic integrals rapidly and precisely. In this way the Legendre relation becomes an effective tool for the calculation of $\pi$. For details the reader is referred to the fascinating book of Borwein and Borwein [4] and to the recent articles ([3] and [5]).

*Acknowledgments.* This work was supported in part by the National Science Foundation under Grant DMS-8701751. The author thanks Richard Askey, Stephan Huckemann, Max Schiffer, and James Wendel for helpful discussions.

## REFERENCES

1. L.V. Ahlfors, *Complex Analysis,* Third Ed., McGraw-Hill, New York, 1979.

2. N.I. Akhiezer, *Elements of the Theory of Elliptic Functions,* Izdat. "Nauka," Moscow, 1970 [in Russian]; English transl., American Mathematical Society, Providence, RI, 1990.

3. G. Almkvist and B. Berndt, Gauss, Landen, Ramanujan, the arithmetic-geometric mean, ellipses, $\pi$, and the *Ladies Dairy, Am. Math. Monthly* **95** (1988), 585–608.

4. J.M. Borwein and P.B. Borwein, *Pi and the AGM,* John Wiley and Sons, New York, 1987.

5. J.M. Borwein, P.B. Borwein, and D.H. Bailey, Ramanujan, modular equations, and approximations to pi or how to compute one billion digits of pi, *Am. Math. Monthly* **96** (1989), 201–219.

6. K. Chandrasekharan, *Elliptic Functions,* Springer-Verlag, Berlin, 1985.

7. T.W. Chaundy, *Elementary Differential Equations,* Oxford University Press, Oxford, 1969.

8. H. Durège, *Theorie der Elliptischen Functionen,* B.G. Teubner, Leipzig, 1861.

9. P.L. Duren and M.M. Schiffer, Univalent functions which map onto regions of given transfinite diameter, *Trans. Am. Math. Soc.* **323** (1991), 413–428.

10. A. Enneper, *Elliptische Functionen, Theorie und Geschichte,* Zweite Auflage, neu bearbeitet und herausgegeben von F. Müller, Louis Nebert, Halle, 1890.

11. E. Hille, *Analytic Function Theory,* Vol. II, Ginn and Co., Boston, 1962; reprinted by Chelsea Publishing Co., New York, 1987.

12. A. Hurwitz and R. Courant, *Vorlesungen über Allgemeine Funktionentheorie und Elliptische Funktionen,* Vierte Auflage, Springer-Verlag, Berlin, 1964.

13. A.M. Legendre, *Exercises de Calcul Intégral,* Tome I, Paris, 1811.

14. A.M. Legendre, *Traité des Fonctions Elliptiques,* Tome I, Paris, 1825.

15. Z. Nehari, *Conformal Mapping,* McGraw-Hill, New York, 1952; Dover reprint, 1975.

16. K. Weierstrass, *Formeln und Lehrsätze zum Gebrauche der Elliptischen Functionen,* bearbeitet und herausgegeben von H.A. Schwarz, Julius Springer, Berlin, 1893.

17. K. Weierstrass, *Mathematische Werke von Karl Weierstrass, Fünfter Band. Vorlesungen über die Theorie der Elliptischen Functionen,* bearbeitet von J. Knoblauch, Mayer und Müller, Berlin, 1915.

18. E.T. Whittaker and G.N. Watson, *A Course of Modern Analysis,* Fourth Ed., Cambridge University Press, Cambridge, 1958.

*Department of Mathematics*
*University of Michigan*
*Ann Arbor, MI 48109*

# Index of Names